T0220517

Applications of Synchrotron Radiation Techniques to Materials Science V

MATERIALS RESEARCH SOCIETY
SYMPOSIUM PROCEEDINGS VOLUME 590

Applications of Synchrotron Radiation Techniques to Materials Science V

Symposium held November 29–December 3, 1999, Boston, Massachusetts, U.S.A.

EDITORS:

Stuart R. Stock
Georgia Institute of Technology
Atlanta, Georgia, U.S.A.

Susan M. Mini
Northern Illinois University
DeKalb, Illinois, U.S.A.
and
Argonne National Laboratory
Argonne, Illinois, U.S.A.

Dale L. Perry
Lawrence Berkeley National Laboratory
Berkeley, California, U.S.A.

Materials Research Society
Warrendale, Pennsylvania

CAMBRIDGE UNIVERSITY PRESS
Cambridge, New York, Melbourne, Madrid, Cape Town,
Singapore, São Paulo, Delhi, Mexico City

Cambridge University Press
32 Avenue of the Americas, New York NY 10013-2473, USA

Published in the United States of America by Cambridge University Press, New York

www.cambridge.org
Information on this title: www.cambridge.org/9781107413344

Materials Research Society
506 Keystone Drive, Warrendale, PA 15086
http://www.mrs.org

First published 2000
First paperback edition 2013

Single article reprints from this publication are available through
University Microfilms Inc., 300 North Zeeb Road, Ann Arbor, MI 48106

CODEN: MRSPDH

ISBN 978-1-107-41334-4 Paperback

CONTENTS

*Invited Paper

SCATTERING AND DIFFRACTION

*Invited Paper

SURFACES

*Invited Paper

MICROBEAM DIFFRACTION, MICROTOMOGRAPHY, TOPOGRAPHY

*Invited Paper

PREFACE

This volume is a collection of papers presented at Symposium R, "Applications of Synchrotron Radiation Techniques to Materials Science V," held November 29–December 3 at the 1999 MRS Fall Meeting in Boston, Massachusetts. The symposium brought together the materials science community and the characterization techniques that use synchrotron radiation, much like the earlier MRS symposia on this topic (MRS Symposium Proceedings Volume 143, 1988 Fall Meeting, edited by R. Clarke, J. Gland, and J.H. Weaver; Volume 307, 1993 MRS Spring Meeting, edited by D.L. Perry, N.D. Shinn, R.L. Stockbauer, K.L. D'Amico, and L.J. Terminello; Volume 375, 1994 MRS Fall Meeting, edited by L.J. Terminello, N.D. Shinn, G.E. Ice, K.L. D'Amico, and D.L. Perry; Volume 437, 1996 MRS Spring Meeting, edited by L.J. Terminello, S.M. Mini, H. Ade, and D.L. Perry; Volume 524, 1998 MRS Spring Meeting, edited by S.M. Mini, S.R. Stock, D.L. Perry, and L.J. Terminello).

Each year, synchrotron facilities, both in the United States and in other countries, are utilized for more applications of synchrotron radiation as they pertain to materials science. Both basic and applied research possibilities are manyfold, including studies of materials mentioned below and those that are yet to be discovered. The combination of synchrotron-based spectroscopic techniques with ever-increasing high-resolution microscopy allows researchers to study very small domains of materials in an attempt to understand their chemical and electronic properties. This is especially important for composites and related materials involving material bonding interfaces.

Topics covered in this proceedings include surfaces, interfaces, electronic materials, metal oxides, metal sulfides, radiation detector materials, thin films, carbides, polymers, alloys, nanoparticles, and metal composites. Results reported at this symposium relate recent advances in x-ray absorption and scattering, imaging, tomography, microscopy, and diffraction methods.

Recently, Professor J.B. Cohen, a strong advocate for the materials synchrotron radiation community, died suddenly. He planned a full-time return to teaching and research after over a decade of carrying heavy administrative burdens. He will be missed by the entire materials community, not just those of us working with synchrotron radiation. As an acknowledgment of his leadership, the editors of this volume dedicate it to his memory.

Stuart R. Stock
Susan M. Mini
Dale L. Perry

January 2000

ACKNOWLEDGMENTS

The symposium organizers wish to thank the following for funding used to support this endeavor:

Blake Industries

Lawrence Livermore National Laboratory
Chemistry and Materials Science Department

Northern Illinois University
The Graduate School and the College of Liberal
Arts and Sciences

The organizers also wish to thank the session chairs (Dale E. Alexander, Paul Fenter, Tony W.H. van Buuren, Uta Ruett, Mark R. Antonio, Stephen R. Wasserman, Debra R. Rolison, and Paul J. Schilling), and the symposium assistant, Cora Lind, who generously gave of their time to make the event so successful, as well as Marlene White for her help in preparing the manuscript for publication.

MATERIALS RESEARCH SOCIETY SYMPOSIUM PROCEEDINGS

MATERIALS RESEARCH SOCIETY SYMPOSIUM PROCEEDINGS

Prior Materials Research Society Symposium Proceedings available by contacting Materials Research Society

Spectroscopy

X-RAY SPECTRO-MICROSCOPY AND MICRO-SPECTROSCOPY IN THE 2100 eV TO 12000 eV REGION (INVITED)

N. MÖLDERS*, P.J. SCHILLING** and J.M. SCHOONMAKER*
* Center for Advanced Microstructures and Devices, Louisiana State University, Baton Rouge, LA 70806, nmolder@lsu.edu
** Mechanical Engineering Department, University of New Orleans, New Orleans, LA 70148, USA

ABSTRACT

An x-ray microprobe beamline was recently developed and commissioned at the Center for Advanced Microstructures and Devices (CAMD), Louisiana State University. It achieves a moderate horizontal and vertical focal spot size of 18.8 µm x 7.0 µm (σ), respectively. The beamline and end-station are designed and optimized to perform (*i*) spatially-resolved x-ray fluorescence spectroscopy (spectro-microscopy) using the broad intense spectrum of the white synchrotron radiation, and (*ii*) spatially-resolved x-ray absorption spectroscopy (micro-spectroscopy) in the energy region of 2100 eV to 12000 eV. These dual capabilities enable K-edge measurements and mapping, in non-vacuum conditions, of low-Z elements down to Cl, S, and P that are of both environmental interest and technological importance. In this paper, an application of this novel synchrotron tool to elucidate the elemental distribution (microstructure) and chemical state (speciation) of Mn, Cl, S, and P-containing particulates emitted from automobile engines burning methylcyclopentadienyl manganese tricarbonyl- (MMT-) added fuel will be discussed in detail. Future opportunities of this microbeam technique in materials science and materials characterization will also be outlined.

INTRODUCTION

The storage ring of the Center for Advanced Microstructures and Devices (CAMD) at the Louisiana State University was designed and optimized for soft x-ray lithography [1]. The ring usually operates at 1.3 GeV with an average injection current of 200 mA, having a critical photon energy of ~ 1660 eV. In order to take full advantage of the high photon flux in the low-energy region of the CAMD storage ring, we designed, installed and commissioned a x-ray microprobe beamline and end-station using an achromatic mirror focusing system in Kirkpatrick-Baez configuration [2-4]. The large emittance of the CAMD storage ring combined with the desire to accept as much incident radiation as possible in the energy range between 2000 eV to 6000 eV, triggered the choice of a Kirkpatrick-Baez (KB) mirror focusing system [5]. Due to this set-up we are able to perform (*i*) spatially-resolved elemental determination via x-ray fluorescence spectroscopy (micro-XRF) using high intensity polychromatic (white) radiation and (*ii*) spatially-resolved x-ray absorption near-edge structure spectroscopy (micro-XANES) using monochromatic radiation of a double crystal monochromator. Resolution

3

measurements show that we obtain a horizontal and vertical focal spot size of 18.8 μm x 7.0 μm (σ), respectively, and a monochromatic photon flux of ~ 10^8 photons per second per 100 mA in the low-energy energy region.

In this paper we show part of a study investigating the elemental distribution and the chemical state of particulate matter with a spatial resolution of ~ 26 μm (σ). The particulate matter was produced by an automobile engine burning regular gasoline fuel with the additive methylcyclopentadienyl manganese tricarbonyl- (MMT-). Micro-XRF results were used to locate concentrations of constituent elements. Micro-XANES measurements at the appropriate elemental K-edges were then performed at selected points-of-interest (POI's) corresponding to areas of high concentration. The K-edges scanned include sulfur and phosphorus. This thus serves as an excellent example of the capabilities of the system, including elemental mapping using polychromatic light, micro-XANES using monochromatic light, and covering the K-edges of lighter elements (S and P).

EXPERIMENTAL SET-UP OF THE MICROPROBE END-STATION

A detailed view of the microprobe end-station is presented in Figure 1. A double crystal monochromator (DCM) in Golovchenko configuration (not indicated in Figure 1) is located upstream of the end-station and connected through a vacuum pipe, a thin Kapton window and a bellows to the end-station. The set-up allows us to either use monochromatic or direct white radiation by simply translating the DCM crystals in/out of the white beam and translating the kinematic table vertically. The entire end-station is housed in a Plexiglas enclosure (HE), which is purged with helium during operation. A motorized 4-jaw entrance slit (HS) defines the horizontal and vertical acceptance of the KB-system. The sample (SH) with a maximal size of 45 mm x 15 mm is mounted under 45° to the incident focused beam. To collect data in transmission a photo-diode (PD) is positioned behind the sample in the direct beam. Fluorescence data, on the other hand, is acquired with an energy-dispersive germanium detector (GE), which is also coupled in to the helium atmosphere. To obtain an optical microscope image of the sample, a long-working distance microscope (MI) is pointed through a view port under 90° to the sample. To collect the primary intensity of the focused x-ray beam, an ion-chamber (IC) is located between the KB-system and the sample [5].

SAMPLE PREPARATION AND DATA ACQUISITION

The particulate samples were collected on Teflon-coated fiberglass filters in such a way that an approximately 36 mm diameter area was exposed to the exhaust particulates. The filters were attached to the exhaust of a 1997 Ford Taurus with a 3.0 L V6 engine operating on gasoline containing MMT at 0.3125 g-Mn/gal. Further details of the sample preparation appear in the bulk XANES investigation of these samples by Ressler et al. [6]. The particulate sample investigated here corresponds to Sample #12 in that study.

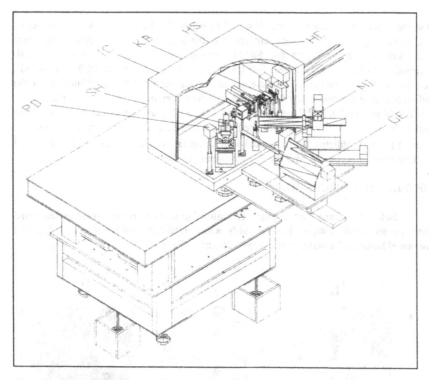

Figure 1. The experimental setup of the microprobe end-station.

The sample was mounted on the x-z sample stage, positioned under 45° to the incident focused beam and the energy-dispersive germanium detector (Ge-detector). The distance between the sample and the Ge-detector varied depending upon data collection (micro-spectroscopy or spectro-microscopy) between 5 mm and 50 mm. The experimental chamber was purged with helium to suppress air absorption.

The elemental maps depicting the elemental distributions in the particulate samples were collected with focused, white/polychromatic x-ray beam with a horizontal and vertical focal spot size of 26 μm x 17 μm (σ). The grazing angles of the focusing mirrors in the Kirkpatrick-Baez focusing system were set to 8 mrad, resulting in a high-energy cut off at ~ 8000 eV. To elucidate the elemental distribution of P, S, and Mn in the particulate samples, regions-of-interest (ROI's) were set around the characteristic x-ray fluorescence lines in the multi-channel analyzer software and the integrated counts of the 3 ROI's were simultaneously acquired while scanning the sample through the incident focused beam in the x and z directions.

For the micro-XANES experiments at the K-edges of Mn, S, and P, the Si(111) crystals of the LNLS double crystal monochromator were translated into the white beam and the experimental chamber was lowered to adjust for the resulting vertical offset of ~ 20.4 mm [6]. Prior to the data acquisition at the S and P K-edges, harmonic rejection was performed by detuning the Si(111) crystals approximately 50% of the peak intensities of the rocking curves. For the energy calibration of the DCM, at the Mn K-edge the first inflection point of elemental Mn was set to 6539 eV. Similarly for the calibration at the S and P K-edges the sharp white lines of the zinc sulfate ($ZnSO_4$) at 2481.4 eV and $ZnPO_4$ at 2150 eV were used. Micro x-ray absorption near-edge structure (micro-XANES) spectra were collected in fluorescence mode at selected points-of-interest (POI's), utilizing the integrated ROI count rate of the appropriate K_α elemental fluorescence line collected with the energy-dispersive Ge-detector.

RESULTS AND DISCUSSION

Before beginning data acquisition, knife-edge scans were performed to determine the beam size on the sample. For this study, a moderately focused beam was used, with a measured horizontal x vertical spot size of 26 μm x 17 μm (σ).

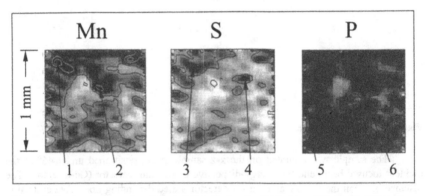

Figure 2. Fluorescence maps from Sample #12, indicating the distributions of Mn, and S. Points-of-interest (POI's) from which micro-XANES spectra were collected are labeled.

Fluorescence maps were collected from 3 areas located along the radius of the filter, with each area having dimensions of 1 mm vert. x 1 mm hor. (at 45° to the beam direction). Each area was scanned using steps of 20 μm vertical and 30 μm horizontal (at 45° to beam direction). Fluorescence data were simultaneously collected using regions-of-interest (ROI's) corresponding to the K_α lines of P, S, and Mn, (and others). These maps reflected a trend with a higher concentration of Mn-particulate matter near the center of the filter than at the edge. Therefore, the points-of-interest (POI's) for this sample were all taken from this central region. The Mn, S, and P fluorescence maps for the 1 mm x 1 mm area near the center of Sample #12 appear in Figure 2. Distinct particles or agglomerations of manganese, sulfur, and phosphorus can be observed in the appropriate maps. From visual comparison of these elemental distributions, specific

POI's were selected for micro-XANES measurements. These spots are labeled in Figure 2.

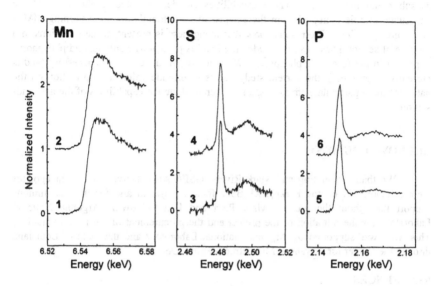

Figure 3. Micro-XANES spectra collected at the K-edges of Mn, S, and P from Sample #12 (Group II). The labels correspond to POI's indicated in Figure 2.

Micro-XANES measurements were performed at the Mn, S, and P K-edges at the appropriate POI's labeled in Figure 2. The resulting spectra are shown in Figure 3. The Mn spectra look similar to the bulk spectrum for this sample reported by Ressler et al. [6]. In that study, this spectrum was described in terms of a mixture of three phases: Mn_3O_4, $MnSO_4 \cdot H_2O$, and $Mn_5(PO_4)[PO_3(OH)]_2 \cdot 4H_2O$. The similarity between the micro-XANES and the bulk XANES spectra suggests a relatively homogeneous distribution of the constituent phases down to a scale of ~ 26 μm, the σ-value of the focused x-ray beam at the sample. The S spectra both exhibit a prominent sulfate peak at about 2481 eV, and an indication of a feature at a lower energy that could indicate the presence of a more reduced species (most likely a sulfide). In the P XANES, the sample spectra reveal the presence of phosphate, represented by a strong white line at about 2150 eV.

CONCLUSIONS

The design and implementation of the CAMD microprobe beamline have been successfully completed, with the instrument demonstrating the desired capabilities (x-ray micro-fluorescence mapping with polychromatic radiation, and micro-XANES with monochromatic radiation), including the ability to measure in the lower energy region (down to about 2000 eV) to assess the distribution and speciation of lighter elements such as sulfur and phosphorus. These capabilities are demonstrated in the study of the particulate sample collected from the exhaust of an automobile engine burning MMT-containing gasoline. The results show that manganese is present in the emissions in a mixture of several phases, sulfur is identified and is present as a sulfate, and phosphorus is present in the form of a phosphate. While a more detailed study is necessary (and is currently in progress), the current study serves to provide valuable information on the nature of these particulate emissions, and to demonstrate the capabilities of the new end-station.

ACKNOWLEDGMENT

We thank Peter Eng and Mark Rivers, GSE/CARS, University of Chicago, for their collaboration on the Kirkpatrick-Baez focusing system and for their continuous support throughout the project; Mark Petri and Len Leibowitz, Argonne National Laboratory for the initiation of the project and the procurement of the instrumentation; This work was supported by Argonne National Laboratory and the State of Louisiana through the Center for Advanced Microstructures and Devices.

REFERENCES

1. R. L. Stockbauer, P. Ajmera, E. D. Poliakoff, B. C. Craft, and V. Saile, Nuclear Instruments and Methods in Physics Research A **291**, p. 505-510 (1990).
2. P. Kirkpatrick and A. Baez, Journal of the Optical Society of America **38**, p. 766-774 (1948).
3. P. Kirkpatrick, A. Baez, and A. Newell, Physical Review **73**, 535-536 (1948).
4. P. J. Eng, M. Newville, M. L. Rivers, and S. R. Sutton, in *X-Ray Microfocusing: Applications and Technique*, edited by I. McNulty (SPIE Proceeding **3449**, Bellingham, WA 1998), p. 145-156.
5. N. Mölders, H. O. Moser, V. Saile, and P. J. Schilling, "Spatially-Resolved X-Ray Spectroscopy at CAMD," Report No. FZKA **6314** (1999).
6. T. Ressler, J. Wong, and J. Roos, accepted for publication in Environmental Science and Technology (1999).

SURFACE INTERACTIONS OF ACTINIDE IONS WITH GEOLOGIC MATERIALS STUDIED BY XAFS

E. R. SYLWESTER, E. A. HUDSON, P. G. ALLEN
Glenn T. Seaborg Institute for Transactinium Science, Lawrence Livermore National Laboratory, L-231, P.O. Box 808, Livermore, CA 94551, USA

ABSTRACT

We have investigated the interaction of the actinyl ion, UO_2^{2+}, with silica, alumina, and montmorillonite surfaces under ambient atmosphere and aqueous conditions using X-ray Absorption Fine Structure (XAFS) Spectroscopy. In acid solution (pH ~ 3.5), the uranyl ion shows a strong interaction with the silica and alumina surfaces, and a relatively weak association with the montmorillonite surface. The extent of direct surface interaction is determined by comparing structural distortions in the equatorial bonding environment of the uranyl ion relative to the structure of a "free" uranyl aquo complex. Based on this formalism, surface complexation on silica and alumina occurs through an inner-sphere mechanism with surface oxygen atoms binding directly to the equatorial region of the uranyl ion. In contrast, sorption on montmorillonite occurs by an outer sphere mechanism in which the uranyl ion retains the simple aquo complex structure and binds to the surface via ion-exchange. In near-neutral solutions (pH ~ 6), sorption on all of the materials is dominated by an inner-sphere mechanism. The formation of surface oligomeric species is also observed on silica and alumina.

INTRODUCTION

Naturally occurring uranium is mined for fuel for nuclear reactors and so is present in the ecosphere as a mineral as well as a waste product of nuclear industry. It has a long half-life and is observed to have coordination chemistry consisting of multiple stable oxidation states and stable solid and aqueous forms within the ecosphere [1-8]. Under standard environmental conditions, uranium typically occurs in the hexavalent form as the mobile, aqueous uranyl ion (UO_2^{2+}). Adsorption of this ion onto geologic materials has been extensively studied since this process can have a significant effect on radionuclide transport properties in the environment [3-4, 6, 8-26].

Traditional investigations of geochemical adsorption concentrate on the macroscopic aspects of the interaction of a particular ionic or molecular species with a mineral surface (i.e, uptake, K_d), but give little direct structural information on the chemical environment around a single atom. X-ray absorption fine structure (XAFS) spectroscopy which includes the extended x-ray absorption fine structure (EXAFS) and the x-ray absorption near-edge structure (XANES), is a technique which can be used to determine local structure and oxidation state of an atom in chemical environments where long range order does not exist, such as liquids, amorphous solids, and surface complexes. Thus it is an ideal tool for the direct determination of the structure of surface-adsorbed complexes and can be used to study the mechanism of adsorption on the molecular scale [1, 4, 6, 8-12, 14, 16-17].

EXPERIMENT

Sample Preparation

The montmorillonite material used in this study was prepared from a calcium-montmorillonite (SAz-1; source locality: Cheto, Arizona) obtained from the Source Clay Minerals Repository (University of Missouri, Columbia, Missouri). Prior to loading with uranium, the <2 μm fraction was selectively obtained via centrifugation and was converted to the sodium-form by contacting with 2 M NaCl solutions several times. The clay was rinsed free of chloride ions and subsequently freeze-dried. The same material was used in previous sorption experiments [21]. Its external surface area, calculated from a multipoint N_2-BET isotherm determined using a Coulter SA3100 surface area analyzer, is 97 m^2/g. The reported cation exchange capacity for SAz-1 montmorillonite is 1.2 meq/g [27].

Γ-Al_2O_3 and amorphous SiO_2, with reported purities of 99.8 % and 99.7 %, respectively, were obtained from Alfa Aesar. The surface areas of the dry, as-received γ-Al_2O_3 and SiO_2 material measured with the Coulter SA3100 surface area analyzer are 87 and 370 m^2/g, respectively. These materials were loaded with uranium without any pretreatment step.

One-liter uranium(VI) solutions in polycarbonate bottles were made from reagent grade $UO_2(NO_3)_2 \cdot 6H_2O$ (Mallinkrodt) and deionized water with a resistivity greater than 17 MΩ·cm. Of the one-liter solutions, 30-mL aliquots were taken and acidified with HCl for later analysis of uranium concentration. Uranium loaded sorbents were prepared by mixing 0.97 gram of the solid with the remaining 970-mL solutions.

The pH values were adjusted by addition of small amounts of 0.5 M NaOH or 0.1 M HNO_3. For each solid, samples were prepared at different pH values so that the effect of pH upon adsorption mechanism(s) could be evaluated. NaCl was added to montmorillonite mixtures G, H, and I to determine if a change in the structure of adsorbed uranium species can be discerned from EXAFS data as the ion-exchange mechanism is suppressed.

The mixtures were kept open to atmospheric CO_2(g) and were agitated at ambient temperature conditions using a gyratory shaker. After about five days of reaction time, the solid phase was separated by filtration and the aqueous phase was sampled for measurement of the pH and final uranium concentration. After filtration, the wet paste samples were kept in sealed plastic vials prior to EXAFS analysis. Chemical data for all prepared samples is summarized in Table 1.

Table 1. Summary of uranium sorption data

Sample	Sorbent	[NaCl] (molal)	Final pH	Initial $[UO_2^{2+}]$ (molal)	Final $[UO_2^{2+}]$ (molal)	Uranium Uptake (%)	Uranium Loading (moles of U per g of solid)
A	silica	0	6.46	4.15E-5	1.02E-6	97.5	4.05E-05
B	silica	0	3.14	0.979E-2	0.970E-2	0.92	9.00E-05
C	alumina	0	6.50	4.18E-5	4.62E-8	99.9	4.18E-05
D	alumina	0	3.48	0.991E-2	0.958E-2	3.33	3.30E-04
E	montmorillonite	0	3.24	0.987E-2	0.949E-2	3.85	3.80E-04
F	montmorillonite	0	4.11	1.00E-4	1.72E-7	99.8	9.98E-05
G	montmorillonite	0.01	4.11	0.979E-4	0.196E-4	80.0	7.83E-05
H	montmorillonite	0.10	4.06	1.00E-4	0.891E-4	10.9	1.09E-05
I	montmorillonite	0.10	6.41	1.02E-4	0.276E-4	72.9	7.44E-05

XAFS Data Acquisition and Analysis

Uranium L_{III}-edge X-ray absorption spectra for samples A and C were collected on beamline X23A2 at the National Synchrotron Light Source (NSLS) using a Si(311) double-crystal monochromator. Spectra for all other samples were collected at the Stanford Synchrotron Radiation Laboratory (SSRL) on beamline II-3 using a Si (220) double-crystal monochromator. Beam size was typically cropped to 9 mm (horizontal) by 1 mm (vertical). All spectra were collected in transmission mode at room temperature using argon-filled ionization chambers, and energy calibrated by simultaneous measurement of the transmission spectrum of a solid uranyl nitrate hexahydrate standard that was placed in the x-ray beam downstream from the sample. The white line maximum in the spectrum of the standard was assumed to be at 17175 eV [28].

The XANES and EXAFS data were extracted from the raw absorption spectra by standard methods described elsewhere [29] using the suite of programs, EXAFSPAK developed by Graham George of SSRL. Non-linear least squares curve fitting analysis was performed using EXAFSPAK to fit the raw k^3-weighted EXAFS data.

The theoretical modeling code FEFF7.2 [30] was used to calculate the backscattering phases and amplitudes of the individual neighboring atoms for the purposes of curve-fitting the raw data. Input files for FEFF7.2 were prepared from the structural modeling code Atoms 2.46b [31]. All of the interactions modeled in the fits for uranyl sorbed onto silica, alumina, or montmorillonite were derived from FEFF7.2 single-scattering (SS) or multiple-scattering (MS) paths calculated from the model compound uranyl orthosilicate, $(UO_2)_2SiO_4 \cdot 2H_2O$, which has previously been shown to produce accurate backscattering distances and coordination numbers for uranyl-silica interactions [12]. In the case of the alumina samples the equivalent SS (U-Al) path was used. The coordination number (N) for each O_{ax} MS path was directly linked to the N of the SS axial oxygen path, while the Debye-Waller factor (σ^2) and path length (R) for each MS path was linked at twice the σ^2 and R of the SS axial oxygen path [32]. In analysis of samples that appeared to have two separate equatorial oxygen shells, the values of σ^2 for the two shells were linked together in the fit to avoid correlation problems between N and σ^2 for each equatorial shell. The amplitude reduction factor, S_0^2, was held fixed at 1.0 for all fits. The shift in the threshold energy, ΔE_0, was allowed to vary as a global parameter for all atoms in all of the fits.

RESULTS

XANES

The Uranium L_{III} XANES spectra (not shown) exhibit features which have been identified as characteristic of the uranyl structure [33-37], and which agree with the features in the reference spectrum of a "free" uranyl ion. These features include a white line peak centered at 17175 eV and a shoulder centered at approximately 17190 eV. The feature at 17190 eV has previously been shown to result from multiple scattering resonances of the linear uranyl ion structure, specifically due to the short U-O_{ax} bonds [4, 28, 32, 36]. In this respect, the XANES data serve to confirm that in each sample the uranyl ion structure is retained, thereby providing a reference point for interpreting the EXAFS spectra.

EXAFS

The k^3-weighted experimental EXAFS data and fits for all of the samples are shown in Figure 1. The spectrum for the UO_2^{2+} aquo reference is also included for comparison purposes. All but two of the samples (H and I) show good signal-to-noise ratios out to a k value of ~13 or further. All of the spectra are dominated by a low frequency oscillation attributable to the backscattering from the 2 (axial) oxygens of the uranyl structure. The montmorillonite samples E-H are dominated in the higher k region by frequency patterns which match those of the aqueous uranyl spectrum included as a reference; the single exception being sample I, which does not seem to resemble the aqueous uranyl reference. In comparison to the montmorillonite samples, the two alumina samples show a broad feature at k=6.5-7 significantly different from the two-feature aqueous uranyl signature seen in the same region, and a slight shift of the phase for higher k features. Sample A shows a complex high frequency pattern dissimilar in both phase and amplitude to any of the other samples.

The Fourier transforms (FTs) of the EXAFS spectra and fits for all samples are also be shown in Figure 1. Fourier transforms represent a pseudo-radial distribution function of the uranium near-neighbor region, and the peaks appear at lower R values relative to the true near-neighbor distances as a result of the electron scattering phase shifts which are different for each neighboring atom ($\alpha \leq 0.5$ Å). The most prominent peak in all samples is at 1.3-1.4 Å and arises from the backscattering caused by oxygen atoms nearest to the uranium, identified as the axial oxygens of the linear uranyl group [9, 14-15, 17, 32, 38]. Backscattering from oxygen atoms lying in the equatorial plane of the UO_2^{2+} ion appears as a single broad peak at 1.9 Å in the montmorillonite samples, and as two separate peaks at approximately 1.8Å and 2.2 Å in the silica and alumina samples. A peak at 2.6 Å attributable to the backscattering from the nearest Si neighbor appears only in sample A. A small, higher-R feature is also visible most prominently in sample A but also in sample C and occurs at ca. 4.0 Å. This peak may tentatively be assigned to backscattering from a uranium neighbor atom [32].

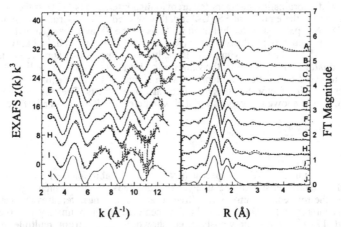

Figure 1. Experimental (dashed) and fitted (solid line) U L_{III} EXAFS spectra for samples A-I, and the aqueous uranyl reference sample (J) taken from Allen et al. [1997]. LEFT: raw k^3-weighted EXAFS data. RIGHT: Fourier Transforms of the raw data.

Non-linear least squares curve fitting was performed on the raw EXAFS data for all samples in an attempt to more thoroughly identify individual atomic components contributing to the EXAFS spectra. Best fits to the data are shown in Table 2, along with the k range used in the fit for each sample. In addition to the limits placed on the parameters detailed in section 2.2, the total coordination number for the equatorial shells N_{eq1} and N_{eq2} of sample B was fixed at 5 in order to avoid correlation problems between N and σ^2 for the two equatorial shells.

All samples show clear evidence for the persistence of uranium in the uranyl form, with the large peak around 1.80 Å seen in many previous studies and identified here as originating from backscattering from the two (axial) oxygens bonded linearly with the central uranium atom. Curve fits confirm this assignment, giving 2 oxygen atoms at ca. 1.80 Å for each sample. Radial distances (i.e. bond lengths) of 1.76 Å to 1.81 Å were fitted, with average bond lengths of 1.77 Å for the silica samples, 1.80 Å for the alumina, and 1.78 Å for the montmorillonite samples. Of the seven samples where the total equatorial oxygen coordination number ($N_{eq,tot}$) was allowed to vary, these coordination numbers lie between 4.8 and 6.6, in good qualitative agreement with previous studies which have shown $N_{eq,tot}$ for uranyl to be 5-6 [4, 9, 12-14, 17, 32-34, 38].

Montmorillonite

Based on the spectra in Figure 1 and the curve-fitting results in Table 2, the montmorillonite samples E-H all show a structure very similar to that found for the aqueous uranyl ion (ca. 5 O at 2.40 Å), with no splitting in the equatorial oxygen shell and no detectable near-neighbor uranium, silicon, or aluminum shells. Strong inner-shell

Table 2. EXAFS best least-squares fitting results for all samples.

Sample	k	Shell	N[a]	R (Å)[a]	σ^2 (Å^2)[b]	ΔE_0
A	3 to 14.3	U-O$_{ax}$	2.00	1.76	0.00075	-10.9
		U-O$_{eq1}$	3.49	2.26	0.0034	
		U-O$_{eq2}$	1.68	2.48	0.0034	
		U-Si	0.97	3.08	0.0040	
		U-U	1.79	3.97	0.0050	
B	3 to 13	U-O$_{ax}$	2.00	1.78	0.0022	-5.8
		U-O$_{eq1}$	2.05	2.28	0.0081	
		U-O$_{eq2}$	2.95	2.46	0.0081	
C	3 to 14	U-O$_{ax}$	2.00	1.81	0.0018	-4.6
		U-O$_{eq1}$	2.64	2.32	0.0074	
		U-O$_{eq2}$	3.13	2.47	0.0074	
		U-U	0.43	4.01	0.0050	
D	3 to 13.5	U-O$_{ax}$	2.00	1.80	0.0026	-2.3
		U-O$_{eq1}$	2.81	2.37	0.0033	
		U-O$_{eq2}$	2.00	2.53	0.0033	
E	3 to 13	U-O$_{ax}$	2.00	1.79	0.0027	-6.3
		U-O$_{eq}$	5.42	2.43	0.0102	
F	3 to 12.5	U-O$_{ax}$	2.00	1.78	0.0014	-4.1
		U-O$_{eq}$	6.59	2.42	0.0096	
G	3 to 13	U-O$_{ax}$	2.00	1.77	0.0020	-4.8
		U-O$_{eq}$	5.99	2.41	0.0090	
H	3 to 12	U-O$_{ax}$	2.00	1.77	0.0020	-12.0
		U-O$_{eq}$	6.35	2.41	0.0130	
I	3 to 12	U-O$_{ax}$	2.00	1.77	0.00154	-11.1
		U-O$_{eq}$	2.97	2.30	0.00254	
		U-O$_{eq}$	2.72	2.48	0.00254	

[a]The 95% confidence limits for the bond lengths (R) and coordination numbers (N) for each shell are: U-O$_{ax}$: ±0.004 Å; U-O$_{eq1}$: ±0.02 Å and ±34%; U-O$_{eq2}$: ±0.03 Å and ±32%; U-O$_{eq(mont)}$: ±0.01 Å and ±16%; U-Si: ±0.03 Å and ±49%; and U-U: ±0.02 Å and ±32%, respectively.
[b]σ is the EXAFS Debye-Waller term which accounts for the effects of thermal and static disorder through damping of the EXAFS oscillations by the factor $\exp(-2k^2\sigma^2)$.

interactions between the surface and the uranyl are expected to result in changes in the geometrical configuration of the equatorial oxygen atoms and possibly the presence of either aluminum or silicon shells in the EXAFS and FT spectra. In the model compound uranyl orthosilicate, the nearest silicon atoms occur at 3.16 Å and 3.80 Å [12], and correspond to inner-sphere bidentate and monodentate binding of the SiO$_x$ surface groups to the equatorial region of the uranyl group, respectively.

Our observation of a single equatorial shell and no other high-R features is consistent with previous experiments for uranyl sorption on montmorillonite at pH 2 [17] and 3.3 [9]. The EXAFS structural results observed can be wholly explained by uranyl forming mononuclear aquo-complexes binding to the surface via outer-sphere interactions.

The uranium uptake and NaCl concentration for montmorillonite samples F, G, and H given in Table 1 indicate that a cation exchange mechanism is important. Decreasing uranium uptake corresponds with increasing NaCl concentration, evidence that Na^+ competes with UO_2^{+2} for the ion exchange sites. For samples E, F, and G, ion exchange should be the predominant mechanism, and a single equatorial shell with no near neighbor U, Si, or Al shells would be consistent with this mechanism. Previous studies have indicated that the contribution of a surface hydrolysis mechanism (i.e. inner-sphere complexation) to uranium adsorption may be important for the conditions of sample H [21], though we find no evidence for a different adsorption mechanism in this case.

Dent et al. [9] also find no evidence for splitting of the equatorial oxygen shell in the uranyl complex for samples at pH 3.3 and pH 5.5, and through comparison with uranyl solution EXAFS at different pH's find that uranyl sorption on montmorillonite clay is best modeled by outer-sphere complexation.

The lack of any detectable U-Si, U-Al, U-U, or split $U-O_{eq}$ interactions in the montmorillonite samples E-H supports the conclusion that adsorption of the uranyl ion onto montmorillonite at low pH occurs predominantly through an outer-sphere mechansim. For samples E, F, G, and H uranyl sorbed on the montmorillonite most likely exists as a monomeric aquo-complex.

For sample I, the best fit occurs with the inclusion of a second equatorial oxygen shell. The two O_{eq} shells are observed to have radial distances of 2.30 Å and 2.48 Å, in good agreement with the results for samples A and C (discussed below), and with previous studies on kaolinite by Thompson et al. [39] under similar ionic and pH conditions. There is no evidence for either Si or Al atoms in the range of 3-4 Å. There is also no evidence for a near-neighbor U atom at ca. 4.0 Å, which would indicate a precipitate or oligomeric surface complex. The presence of a split equatorial shell and no more distant interactions supports the conclusion that in sample I we are observing uranyl that is adsorbing onto the surface via inner-sphere, mononuclear complexation mechanism, distinct from the outer sphere mechanism observed at lower pH's.

Silica

Both silica samples show a split in the equatorial shell, with resulting average backscatter distances of 2.27 Å and 2.47 Å for the two separate equatorial shells. At pH 6.24, ca.1 Si atom at 3.1 Å and ca. 2 U atoms at 4.0 Å are also detected. This indicates the formation of an oligomeric surface complex or precipitate at the surface, with each uranyl group binding not only to a single silica site but to adjoining uranyl groups as well. The atomic distance of 3.1 Å for U-Si and the coordination number for Si of ca. 1 implies bidentate complexation to a single silica center. For complexation with oxygens bound to separate silicon atoms, a coordination number of ca. 2 for the Si shell, and a longer U-Si backscatter distance would be expected instead. Monodentate complexation with the surface would be expected to give a coordination number for Si of ca. 1 at a longer distance of ~3.80 Å.

At pH 3.14 the equatorial oxygens remain split into two distinct shells, but no near-neighbor silicon or uranium is detected. Similar results have been reported for uranyl samples prepared on silica gel at pH 3.5-4.0 [12], where the splitting of the equatorial oxygen shell was attributed to the formation of an inner-sphere, mononuclear, bidentate surface complex.

The clear contrast between the two equatorial shells observed in the silica sample at pH= 3.14, and the single equatorial shell observed in the montmorillonite samples at pH's ≥3.24 is evidence that a stronger UO_2^{2+}-surface interaction is occurring in the silica sample. The splitting of the equatorial shell in this sample cannot be attributed to the formation of split-shell containing aqueous complexes that might form independently of the surface (i.e. carbonate or nitrate complexes), because no equatorial splitting is detected in the montmorillonite samples at similar pH's.

14

Alumina

Both the low and high pH alumina samples show a split in the equatorial shell, with O backscatterer distances of 2.34 Å and 2.50 Å. At high pH, a near-neighbor U atom is detected with a coordination number ca. 0.4. As with the silica samples, the splitting of the equatorial shell into two distinct shells is attributed to the formation of an inner-sphere surface complex. The observation of near-neighbor uranium at high pH is most likely the result of the formation of polynuclear (surface) complexes, with approximately half of the uranyl groups having a single near-neighbor uranyl.

Some consideration must also be given to the possibility that a fraction of the uranyl in our samples has precipitated out of solution as a uranyl hydroxide such as schoepite or coprecipitated as a hydrated uranyl silicate (soddyite) or aluminate. Thermodynamic modeling of uranyl speciation under the conditions of samples A-I was performed using the Geochemist's Workbench® 3.0, taking into account the possible precipitation of rutherfordine, schoepite, and related phases (i.e, $UO_3 \cdot xH_2O$, x≤2). Results of the modeling indicate that using the final UO_2^{2+} concentrations listed in Table 1 samples A and I could be supersaturated with respect to schoepite. The EXAFS spectra shown in Figure 1 for samples A and I do not agree with the published spectrum for schoepite [32]. This result suggests that schoepite did not form in these samples. Although the EXAFS results taken alone do not exclude the formation of other uranyl-containing phases, our present knowledge of the geochemistry in these systems suggests that this is unlikely.

CONCLUSIONS

We have investigated the adsorption of the uranyl ion (UO_2^{2+}) in contact with silica (SiO_2), alumina (Al_2O_3), and montmorillonite surfaces in the pH range of 3.1-6.5, using X-ray absorption fine structure (XAFS) spectroscopy. In all samples the uranyl ion structure is preserved. For the montmorillonite samples at low pH a single equatorial oxygen shell is observed at ca. 2.4 Å. At high pH (6.41) and high ion concentration (0.1 M NaCl), a split equatorial shell is observed, indicating inner-sphere bonding with the surface. The samples of uranyl on silica and alumina are all observed to have a similar split equatorial shell. A uranium shell is observed in the high pH (~6.5) samples of uranyl on silica and on alumina. A silicon shell is observed in the sample of uranyl on silica at pH 6.5. These results suggest that adsorption of the uranyl ion onto montmorillonite at low pH is occurring via ion exchange, while at high pH and in the presence of a competing cation, inner-sphere complexation with the surface predominates. Adsorption of the uranyl onto the silica and alumina surfaces appears to occur via an inner-sphere, bidentate complexation with the surface, with the formation of polynuclear complexes occurring at higher pH.

ACKNOWLEDGEMENTS

The authors would like to thank R. T. Pabalan (Southwest Research Institute) for his work in preparation of the samples. This research was carried out in part at NSLS, Brookhaven National Laboratory, and SSRL, which are both supported by the U. S. Dept. of Energy, Divisions of Materials Sciences and Chemical Sciences, respectively. This work was performed under the auspices of the U.S. Department of Energy by Lawrence Livermore National Laboratory under Contract W-7405-Eng-48.

REFERENCES

[1] Duff, M. C., Amrheim, C., Soil Sci. Soc. Am. J. 60, 1393 (1996).
[2] Duff M. C., Amrhein C., Bertsch P. M., Hunter D. B., Geochim. Cosmochim. Acta 61(1), 73 (1997).
[3] Giaquinta D. M., Soderholm L., Yuchs S. E., Wasserman S. R., Radiochim. Acta 76, 113 (1997).

[4]   Hudson E. A., Terminello L. J., Viani B. E., Denecke M., Reich T., Allen P. G., Bucher J. J., Shuh D. K., Edelstein N. M. Clays & Clay Min (accepted December 1998).
[5]   Jones D. J., Roziere J., Allen G. C., Tempest P. A., J. Chem. Phys. 84(11), 6075 (1986).
[6]   Payne T. E., Davis J. A., Waite T. D., Radiochim. Acta 74, 239 (1996).
[7]   Sturchio N. C., Antonio M. R., Soderholm L., Sutton S. R., Brannon J. C., Science 281, 971 (1998).
[8]   Ticknor K. V., Radiochim. Acta 64, 22 (1994).
[9]   Dent A. J., Ramsay J. D., Swanton S. W., J. Coll. Int. Sci. 150, 45 (1992)
[10] Piron E., Accominotti M., Domard A., Langmuir 13, 1653 (1997).
[11] McKinley J. P., Zachara J. M., Smith S. C., Turner G. D., Clays & Clay Min., 43, 586 (1995).
[12] Reich T., Moll H., Denecke A., Geipel G., Bernhard G., Mitche H, Allen P. G., Bucher J. J., Kaltsoyannis N., Edelstein N. M., Shuh D. K., Radiochim. Acta 74, 219 (1996).
[13] Thompson H. S., Brown G. E. Jr., Parks G. A., Am. Min. 82, 483 (1997).
[14] Waite T. D., Davis J. A., Payne T. E., Waychunas G. A., Xu N., Geochim. Cosmochim. Acta 58, 5465 (1994).
[15] Wasserman S. R., Giaquinta D. M., Yuchs S.E., Soderholm L., Mat. Res. Soc. Symp. Proc. 465, 473 (1997).
[16] Reich T., Moll H., Arnold T., Denecke M. A., Hennig C., Geipel G., Bernhard G., Nitsche H., Allen P. G., Edelstein N. M., Shuh D. K., J. Elec. Spec. & Rel. Phen. 96, 237 (1998).
[17] Chisholm-Brause C., Conradson S. D., Buscher C. T., Eller P. G., Morris D.E., Geochim. Cosmochim. Acta., 58, 3625 (1994)
[18] Prikryl J. D., Pabalan R. T., Turner D. R., Leslie B. W., Radiochim. Acta 66/67, 291 (1994).
[19] Michard P., Guibal E., Vincent T., Le Cloirec P., Microp. Mat. 5, 309 (1996).
[20] Arnold T., Zort T., Bernhard G., Nitsche H., Chem. Geo. 151, 129 (1998).
[21] Pabalan R. T., Turner D. R., Aq. Geochem. 2, 203 (1997).
[22] Pabalan R. T., Turner D. R., Bertetti F. P., Prikryl J. D., in: Adsorption of Metals by Geomedia, edityed by E. Jenne, Academic Press (1998).
[23] McKinley, J. P., Smith, S. C., Zachara, J. M., Berg, J. M., Chisholm-Brause, C. J., Morris, D. C. Env. Sci. & Tech, accepted 1998.
[24] Tsunashima A., Brindley G. W., Bastovano M., Clays & Clay Min., 29, 10 (1981).
[25] Zachara J. M., McKinley J. P. Aq. Sci. 55, 250 (1993).
[26] Keller-Besrest F., Bénazeth S., Souleau Ch., Mat. Let. 24, 17 (1995).
[27] Van Olphen H., Fripiat J. J., Data handbook for Clay Minerals and Other Non-Metallic Materials, (Pergamon, 1979).
[28] Petiau J., Calas G., Petitmaire D., Bianconi A., Benfatto M., Marcelli A., Phys. Rev. B, 34, 7350 (1986).
[29] Prins. R., Koningsberger D. E., eds., X-ray Absorption: Principles, Applications, Techniques for EXAFS, SEXAFS, and XANES. (Wiley-Interscience, 1988).
[30] Mustre de Leon J., Rehr J. J., Zabinsky S., Albers R. C., Phys. Rev. B 44, 4146 (1991).
[31] Ravel B. (1996): "ATOMS, a program to generate atom lists for XAFS analysis from crystallographic data." University of Washington, Seattle, WA.
[32] Allen P. G., Shuh D. K., Bucher J. J., Edelstein N. M., Palmer C. E. A., Silva R. J., Nguyen S. N., Marquez L. N., Hudson E. A., Radiochim. Acta 75, 47 (1996).
[33] Allen P. G., Bucher J. J., Shuh D. K., Edelstein N. M., Reich T., Inorg. Chem. 36, 4676 (1997).
[34] Allen P. G., Shuh D. K., Bucher J. J., Edelstein N. M., Reich T., Denecke A., Nitsche H., Inorg. Chem. 35, 784 (1996).
[35] Bertram S., Kaindl G., Jove J., Pages M., Gal J., Phys. Rev. Lett. 63, 2680 (1989).
[36] Farges F., Ponader C. W., Calas G., Brown G. E., Jr., Geochim. Cosmochim. Acta 56, 4205 (1992).
[37] Kalwolski G., Kaindl G., Brewer W. D., Krone W., Phys. Rev. B 35, 2667 (1987).
[38] Allen P. G., Bucher J. J., Clark D. L., Edelstein N. M., Ekberg S. A., Gohdes J. W., Hudson E. A., Kaltsoyannis N., Lukens W. W., Neu M. P., Palmer P. D., Reich T., Shuh D. K., Tait C. D., Zwick B. D., Inorg. Chem., 34, 4797 (1995).
[39] Thompson H. S., Parks G. A., Brown G. E. Jr., in Adsorption of Metals by Geomedia, edited by E. A. Jenne, Chap. 16, pp 349-370 (Academic Press, 1998).

IN SITU XAFS OF THE Li$_x$Ni$_{0.8}$Co$_{0.2}$O$_2$ CATHODE FOR LITHIUM-ION BATTERIES

A. J. Kropf and C. S. Johnson
Argonne National Laboratory, Chemical Technology Division, 9700 South Cass Avenue, Argonne, IL, 60439

ABSTRACT

The layered LiNi$_{0.8}$Co$_{0.2}$O$_2$ system is being considered as a new cathode material for the lithium-ion battery. Compared with LiCoO$_2$, the standard cathode formulation, it possesses improved electrochemical performance at a projected lower cost. *In situ* x-ray absorption fine-structure spectroscopy (XAFS) measurements were conducted on a cell cycled at a moderate rate and normal Li-ion operating voltages (3.0-4.1 V). The XAFS data collected at the Ni and Co edges approximately every 30 min. revealed details about the response of the cathode to Li insertion and extraction. These measurements on the Li$_x$Ni$_{0.8}$Co$_{0.2}$O$_2$ cathode (0.29<x<0.78) demonstrated the excellent reversibility of the cathode's short-range structure. However, the Co and Ni atoms behaved differently in response to Li insertion. This study corroborates previous work that explains the XAFS of the Ni atoms in terms of a Ni^{3+} Jahn-Teller ion. An analysis of the metal-metal distances suggests, contrary to a qualitative analysis of the x-ray absorption near-edge structure (XANES), that the Co^{3+} is oxidized to the maximum extent possible (within the Li content range of this experiment) at x = 0.47 ± 0.04, and further oxidation occurs at the Ni site.

INTRODUCTION

The emergence of portable telecommunications, computer equipment, and ultimately electric vehicles has created a substantial need for improvements in energy storage devices, namely, less expensive batteries that operate for a longer time, are smaller in size, and weigh less. The Li-ion battery possesses the highest energy density of all rechargeable batteries available in the marketplace today, yet there is still an array of new materials that are being developed for the Li-ion battery. These new compounds require further investigation and physical/chemical characterization.

The Li-ion cathode material of choice is Li$_{1-x}$CoO$_2$, a hexagonally (space group R$\bar{3}$m) layered oxide material that, when coupled with a graphite or coke anode (Li$_x$C$_6$) and a 1 *M* LiPF$_6$ electrolyte salt in an organic solvent, makes up a Li-ion cell. This battery operates at 4.0 V and has excellent cyclability and long life, but suffers from too low specific capacity [1]. The similar Li$_{1-x}$Ni$_x$O$_2$ suffers from a lower rate capability. The rate problem has been associated with Ni migration into sites within the depleted Li layer for x≤0.5, which causes a deleterious change in the cathode structure [2]. Cobalt has been substituted into the Ni structure to improve the stability of the cathode. Reasons for the improved electrochemical performance accompanying this substitution are not well understood from a structural standpoint.

Significant work has been performed in the area of in situ analysis using x-ray absorption fine-structure spectroscopy (XAFS) and x-ray diffraction (XRD) on materials in the NiO$_2$ and CoO$_2$ families [3-9]. These two methods are complementary. *In situ* XRD gives one a good picture of the long-range structural changes in the cathode, but XAFS can provide an understanding of the oxidation state as well as short-range ordering, which may not be accessible from XRD. A more detailed examination of the structure of the Li$_x$Ni$_{0.8}$Co$_{0.2}$O$_2$ cathode with XAFS during the battery charge cycle can be illuminating in its own right, but also provides an

17

excellent baseline for further studies in which the cathode material is modified by the introduction of additional elements.

EXPERIMENT

Battery (pouch cell) Construction

The pouch-type battery cell used in this work was assembled and sealed in a helium-atmosphere glovebox. Cathode laminates of nominal composition $LiNi_{0.8}Co_{0.2}O_2$ (Sumitomo Metal and Mining Co., Ltd.) were dried at 80°C in a vacuum oven inside the glovebox prior to assembly. All cell leads and separator materials were rigorously dried in a like manner. The pouch material was a heat-sealable (0.0045-in. thick), transparent, polyester film laminate. The pouch cells are limited by the cathode capacity and contain metallic lithium foil (0.008 in.) as the anode. The electrode configuration featured in these cells uses metallic lithium as the anode on either side of the cathode. This design ensures an excellent current distribution throughout the cell and provides low resistance for improved electrochemistry. Lithium disks (5/8-in. diameter) were punched from the foil, then cold pressed onto copper mesh (5/8-in. diameter) with an arm as the electrode lead. The cathode composite was painted onto a Ni mesh (5/8-in. diameter) from an N-methyl pyrrolidone slurry of 74% active oxide material, 18% graphite, and 8% polyvinyl difluoride. The mesh laminate electrode was first dried in the glovebox at room temperature, then further vacuum dried at 80°C before final assembly. The separator material was Celgard 2400, and it was placed on either side of the cathode mesh. Electrolyte [10] was added to the pouch cell, then a slight vacuum was applied to the cell to pull all gas bubbles out of the pouch prior to final sealing. During the XAFS analysis, pressure was applied to the exterior of the pouch to maintain the stability of the electrode assembly.

The cycling of the pouch cell mimicked that of a coin cell. The cathode delivered 195 mAh/g capacity during its first charge, and after the first discharge, the cell was subjected to x-ray absorption analysis during its next full cycle, which was carried out at a C/5 rate (I=0.875 mA/cm²) between 4.1 and 3.0 V vs. Li (Figure 1). The capacity in the second cycle was 134 mAh/g and 120 mAh/g, respectively, in C2 and D2. This corresponds to the cathode cycling approximately 0.5 Li into and out of its structure. After the run, the cycling curve was matched with the corresponding XAFS spectra.

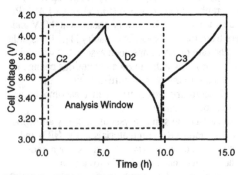

Figure 1. Galvanostatic cycling curve of Li/electrolyte/$Li_xNi_{0.8}Co_{0.2}O_2$ cell during XAFS analysis.

XAFS Measurements

The XAFS measurements were performed on the insertion device (undulator A) beamline of the Materials Research Collaborative Access Team (MRCAT) at the Advanced Photon Source (APS). Using a cryogenically cooled Si(111) double-crystal monochromator, we determined the energy resolution of the monochromatic beam to be ~1.0 eV, as measured by the rocking curve width of the second crystal. An active feedback loop maintained the peak reflectivity of the monochromator, and the resulting large harmonic content in the beam was rejected by means of a platinum-coated mirror. Transmission ion chambers were used to measure the incident (I_0), transmitted (I_t), and reference (I_{ref}) signals. The I_0 chamber was filled with a helium/nitrogen mixture (50%/50%), while the I_t and I_{ref} chambers were filled with a nitrogen/argon mixture (80%/20%). A beam size of 0.5 x 0.5 mm was chosen in order for the beam to pass easily through the Ni mesh on which the cathode was deposited, resulting in an incident beam flux of $\sim 10^{11}$ photons/sec. Nickel and Co metal foils were used to calibrate the energy at each edge.

The undulator gap was tapered and the energy range was set to optimize the flux for the Co and Ni edges individually. The monochromator was step-scanned with an integration time of 0.5-2 sec per data point. Including the time spent moving between data points, each scan required about 12 min., corresponding to $\Delta x = 0.021$ ($Li_x Ni_{0.8}Co_{0.2}O_2$) from the start to the finish of each scan. The data range at the Co K edge was limited to about 12.5 $Å^{-1}$ by the onset of the Ni K edge. The scan range at the Ni edge was chosen to match the usable range at the Co edge. Spectra were collected alternately at each absorption edge.

DATA ANALYSIS

For the x-ray absorption near-edge spectroscopy (XANES), we calibrated each scan using the appropriate metal foil reference scan taken simultaneously with the battery cell data. The calibration procedure was a qualitative best matching of the entire near-edge region of the reference foil scan to one scan in the battery cell series chosen as the baseline. Even though the energy resolution of the x-ray beam was about 1.0 eV and the step size in the edge region was only 0.5 eV, shifts as small as 0.1 eV were reliably detected by this calibration method. Absolute energy calibration was made to the Co foil reference scan, assigning an energy of 7709 eV to the maximum derivative. The near-edge regions of selected spectra are shown in Figure 2.

In the XAFS analysis, the FEFF program (version 7.0) [11] was employed for theoretical calculation of the scattering paths, and the FEFFIT program [12] was employed for non-linear least-squares fitting the Fourier Transform (FT) of the XAFS data. In all cases, the Li contribution to the XAFS was ignored.

As mentioned above, the data range limit for both Ni and Co is about 12.5 $Å^{-1}$. The XAFS measurements of the powdered cathode material (*ex situ*, over a longer range), as well as literature references [5,7], have shown that the XAFS signal is still strong as far out as 18 $Å^{-1}$. Nonetheless, most analyses of similar materials in the past have limited the upper end of the fitting range to 13-15 $Å^{-1}$. The upper limit of the data range used in this discussion is slightly lower; however, it is adequate for the conclusions presented.

It is not possible to resolve two similar paths with distances separated by less than $\Delta R = \pi/(2\Delta k)$. For the data range in this study, $\Delta R > 0.16$ Å could theoretically be resolved. Previous measurements of the Jahn-Teller distortion of the NiO_6 octahedron place the effect at about 0.17 ± 0.03 Å [3,4,6,7,9]. This difference is at the limit of the experimental resolution. As a consequence, we used only one Ni-O path to model the NiO_6 octahedron and relied on secondary indicators (i.e., changes in R and σ^2) to reveal the effects of a Jahn-Teller distortion. Trial model optimizations, including both a long and a short Ni-O bond, demonstrated that adding the long-length Ni-O path improved the data fit, but the improvement was not statistically significant.

A third path has been included in the model: the second nearest metal-oxygen (M-O) path at about 3.4 Å. This path overlaps the M-M path due to the phase-shift difference. Including this path in the fit significantly improves the model. The R-factor, a measure of the difference between the data and the model, is between 0.002 and 0.007 [12]. Values for the R-factor that are less than 0.02 generally indicate a good fit.

RESULTS

Unlike pure $LiNiO_2$ or $LiCoO_2$, $Li_xNi_{0.8}Co_{0.2}O_2$ has a hexagonally layered structure that may not undergo phase changes to inactive secondary monoclinic or hexagonal phases when Li is cycled into and out of its lattice. This effect was shown with previous XRD data collected from an *in situ* coin cell [1,8]. In $Li_xNi_{0.8}Co_{0.2}O_2$, it has been postulated that the Ni and Co interlayer slab has a higher degree of covalency; consequently, the Ni-O and Co-O bond strengths are greater than that for the undoped parent materials [2]. The $LiNi_{0.8}Co_{0.2}O_2$ structure thus does not undergo the phase change to monoclinic symmetry, and its ability to cycle Li into and out of the Li layer within the hexagonal ($R\bar{3}m$) structure is unimpeded.

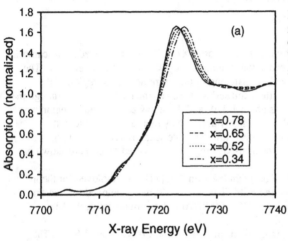

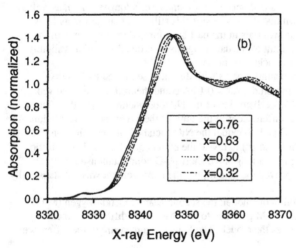

Figure 2. Calibrated and normalized XANES data at the (a) Co and (b) Ni K edges. Reproducibility of the calibration is ±0.1 eV.

The Ni-O and Co-O bond lengths both contract as the Li is removed from the structure during charging (scan numbers 0-23). Both the Co and the Ni become oxidized during this process, each going from a nominal oxidation state of 3^+ to 4^+. During discharge (scan numbers 24-44), this process is reversed electrochemically: Li is reinserted in the lattice structure, the Ni^{4+} and Co^{4+} are reduced to their trivalent states, and their bonds with oxygen return to their original lengths. To a large degree, all of the structural changes visible in the XAFS during the charge are reversed during the discharge. Therefore, in several cases, only data taken during the charge are presented.

XANES

Figures 2(a) and 2(b) show the near-edge regions for selected scans at the Co and Ni edges, respectively. The qualitative difference is immediately apparent. The Ni scans show a steady progression of the entire pattern from lower to higher energy as a function of decreased Li content, indicating an increased average Ni oxidation state, whereas the Co spectra do not follow this progression. Indeed, the shape of the Co absorption spectrum changes markedly from the lower to the higher oxidation state. In both cases, a detailed look at the derivative of the edge spectra (not shown) reveals that more than a simple edge shift is occurring. Figure 3 is a plot of the white line peak position as a function of the scan number. The cell current changed from lithium extraction (oxidizing) to lithium insertion (reducing) during scan number 23, slightly earlier than the white-line position reverses direction for both Co and Ni. This could be evidence for diffusion of the Li ions into the electrode due to the high charge rate. For both metal centers, the change in the peak position is about +1.7 eV as the lithium content changes from x = 0.78 to x = 0.29.

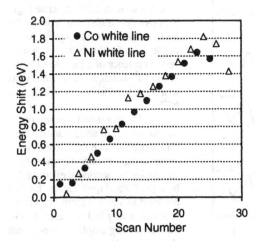

Figure 3. Plot of the white line energy shift versus the scan number. Errors are ±0.1 eV.

For the Co edge, two other common measures of the edge position, the energy at normalized $\mu x = 0.5$ and the inflection point (maximum of the first derivative), do not follow the change in Li content. However, for the Ni edge, the energy at $\mu x = 0.5$ closely follows the trend of Li insertion. On the other hand, the position of the inflection point can be ambiguous. Therefore, for the Ni edge it is possible to say with a high degree of confidence that the oxidation state is changing smoothly and nearly proportionally to the change in Li content. For the Co edge, the situation is not as clear since the common qualitative indicators contradict each other. To determine the behavior of the Co oxidation, it is useful to turn to the XAFS data.

Figure 4 shows normalized $k^2\chi(k)$ data for selected scans throughout the range of x values for both Ni and Co. Only subtle changes take place between x = 0.32 and x = 0.79. Figure 5 shows the FT of the endpoint data. As has been observed previously by Nakai and Nakagome [5], the CoO_6 coordination changes very little with Li content (from an inspection of the FT data). On the other hand, the NiO_6 octahedron changes significantly from one end of the cycle to the other. This has been cited as evidence that Ni^{3+} in these systems is behaving as a Jahn-Teller ion (octahedral, low-spin d^7 configuration). Longer-range XAFS data, for which the radial distance resolution is high enough to distinguish two oxygen distances, have been used by Pickering, et. al. [4] and Rougier, et. al. [9] to demonstrate the existence of a short and long oxygen bond in the $Li_xNi_{1-x}O_2$ material without Co. In our work, the data have been fit quite well with one Ni-O distance (Figure 6). Qualitatively, it is obvious that the Ni-O shell has a reduced amplitude in the high-x region of the cycle, especially compared with the Co-O shell, which exhibits almost no change. The results of the fits are shown in Table 1.

Two supplementary pieces of evidence suggest the existence of two Ni-O bonds in the Ni^{3+} state. First, when one Ni-O path is used to model the data, the Debye-Waller factor for the oxygen shell decreases and approaches the value for the Co-O path as the Li content decreases. This finding indicates smaller disorder in the bond distance for the Ni^{4+} ion. Second, the Co-O distance changes less than the Ni-O distance, although both exhibit a change essentially proportional to the change in Li content. Figure 7(a) plots the metal-oxygen distance as a function of scan number, while Figure 7(b) plots the difference between the Ni-O and Co-O bond lengths. An explanation for this behavior would be the presence of a long Ni-O bond in the Ni^{3+} octahedron, which reverts to a short bond as Ni^{3+} is oxidized to Ni^{4+}.

The metal-metal neighbor results are key to understanding the difference between the Co and Ni atoms in the cathode. For the sake of simplicity, the second shell around both the Co and the Ni atoms has been modeled as a nickel atom. Practically, this makes very little difference as the back-scattering amplitudes and phase shifts of neighboring elements are virtually indistinguishable. In addition, since the ratio of Ni to Co atoms is 4:1, if the Co and Ni are randomly distributed, there will be 4 to 5 Ni atoms and only 1 to 2 Co atoms in the second shell around each metal atom. As was mentioned

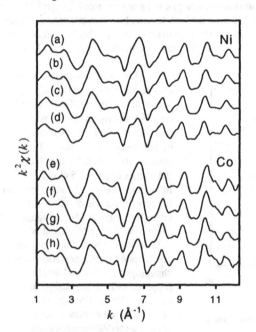

Figure 4. Selected $k^2\chi(k)$ data for Ni and Co; (a) Ni edge, x = 0.32, (b) Ni edge, x = 0.50, (c) Ni edge, x = 0.63, (d) Ni edge, x = 0.76, (e) Co edge, x = 0.34, (f) Co edge, x = 0.52, (g) Co edge, x = 0.65, (h) Co edge, x = 0.78.

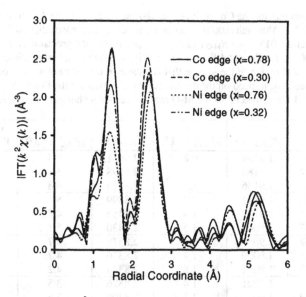

Figure 5. |FT($k^2\chi(k)$)| of the Co and Ni XAFS at either end of charge cycle. (k-range = 2.0-12.25 Å, modified Hanning window function with dk = 0.5 Å^{-1})

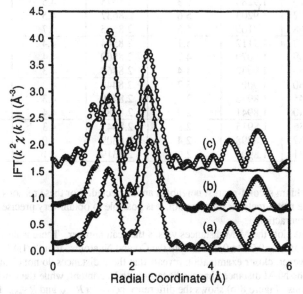

Figure 6. Fit to (a) Co edge, x = 0.78, (b) Ni edge, x = 0.32, and (c) Ni edge, x = 0.76 for scans 1, 2, and 22 (k-range = 2.0-12.25 Å, dk = 0.5 Å^{-1}, R-range = 1.1-3.1 Å). The symbols represent the data and the solid lines represent the fit.

Table 1. Fit parameters for the Co and Ni edges (k-range $= 2.5 - 11.75$ Å^{-1}, $dk = 0.5$ Å^{-1}, R-range $= 1.1 - 3.1$ Å). Although 0.0001 Å accuracy is clearly not reasonable for EXAFS (actually 0.02 Å and 0.015 Å for R_{M-O} and R_{M-M}, respectively), the precision is much greater (0.003 Å and 0.001 Å for R_{M-O} and R_{M-M}, respectively) given that all systematic errors cancel since it is an identical region of the sample that is measured during each scan. Rounding to 0.001 Å would easily be noticed on the difference plots. Errors for the Debye-Waller factor are 0.4×10^{-3} Å^2. Fit parameters for the M-O next-nearest-neighbor path are not shown.

Scan # Co Edge	x	R_{M-O} (Å)	σ^2 (x10^{-3} Å^2)	R_{M-M} (Å)	σ^2 (x10^{-3} Å^2)	White Line Position (eV)
1	0.78	1.9184	0.8	2.8651	5.4	7728.65
3	0.74	1.9180	0.6	2.8598	5.1	7728.66
5	0.70	1.9156	1.0	2.8555	4.8	7728.83
7	0.65	1.9141	1.2	2.8499	5.1	7729.00
9	0.61	1.9109	1.4	2.8455	4.8	7729.16
11	0.57	1.9063	1.5	2.8430	5.1	7729.33
13	0.52	1.9059	1.3	2.8379	4.7	7729.47
15	0.47	1.9060	1.4	2.8376	4.7	7729.60
17	0.42	1.9051	1.8	2.8344	4.2	7729.76
19	0.38	1.9021	1.4	2.8326	3.5	7729.87
21	0.34	1.9013	1.1	2.8309	3.6	7730.02
23	0.30	1.8982	1.4	2.8287	3.9	7730.14
Ni Edge						
2	0.76	1.9280	6.3	2.8675	4.5	8351.54
4	0.72	1.9203	5.6	2.8637	4.5	8351.77
6	0.68	1.9169	5.5	2.8585	4.7	8351.97
8	0.63	1.9112	5.1	2.8544	4.4	8352.27
10	0.59	1.9079	4.7	2.8509	4.4	8352.28
12	0.55	1.9032	4.4	2.8467	4.3	8352.63
14	0.50	1.9001	3.7	2.8449	4.3	8352.68
16	0.45	1.8956	3.3	2.8410	4.3	8352.76
18	0.40	1.8941	2.9	2.8381	4.4	8352.88
20	0.36	1.8915	2.8	2.8343	4.3	8353.04
22	0.32	1.8898	2.4	2.8309	4.0	8353.18
24	0.31	1.8900	2.6	2.8319	4.1	8353.32

previously, the lithium second-nearest neighbors are neglected. For these reasons and because of the strength of the metal-metal scattering path, it is possible to obtain very precise distances for the metal-metal interaction.

Figure 8(a) plots the M-M distances versus the scan number. There is nearly a 0.04 Å change in R_{M-M} from x = 0.78 to 0.29 for both the Co and Ni atoms: about a 1.4% linear contraction. However, closer examination reveals that these distances do not change in precisely the same way. The Ni-M distance changes linearly with Li content, while the Co-M bond exhibits two regions. Figure 8(b) shows the difference between R_{Ni-M} and R_{Co-M}. Between scans 14 and 16 (x = 0.47 ± 0.04) R_{Co-M} begins to change at a lower rate than R_{Ni-M}. We attribute this difference to the Co atoms being essentially oxidized at this point. The remaining decrease in R_{Co-M} is due to the secondary effect of the shrinking of the surrounding NiO_6 octahedrons.

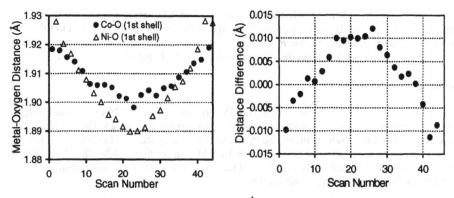

Figure 7. (a) First-shell bond length changes during Li/Li$_x$Ni$_{0.8}$Co$_{0.2}$O$_2$ cell cycling (accuracy, ± 0.02 Å, precision, ± 0.003 Å). (b) Difference between R_{Co-O} and R_{Ni-O} (± 0.005 Å).

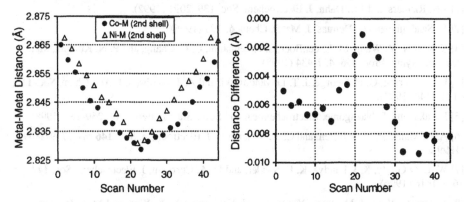

Figure 8. (a) Second-shell metal-metal bond length vs. scan number (accuracy, ± 0.015 Å, precision, ±0.001 Å). (b) Difference between R_{Co-M} and R_{Ni-M} (± 0.0015 Å).

CONCLUSIONS

XAFS has provided us with an excellent tool for analyzing the changes that occur when Li is cycled into and out of the layered lattice of Li$_x$Ni$_{0.8}$Co$_{0.2}$O$_2$ in a Li-ion battery. This material possesses very desirable characteristics for an electrode, such as small volumetric changes and retention of its structure (on charge and discharge), which makes it an excellent choice for high power Li-ion battery applications.

The short-range order in the mixed Ni/Co oxide layered system exhibits good reversibility during Li insertion/removal. The results presented here confirm that only minor changes are occurring in the layered structure of the cathode during Li insertion and removal (0.29≤x≤0.78). The bond lengths vary systematically in response to the change in the metal center's oxidation state. However, the charge compensation mechanism is different for the cobalt atoms compared with the nickel atoms near the top of charge. From the second-shell results, the Co appears to be oxidized by x = 0.47 ± 0.04.

Fitting the Ni-O bond to only one path results in a linear change in distance with the cathode state of charge. However, the overall change in R_{Ni-O} from fully discharged to fully charged is greater than the change in R_{Co-O}. This observation, taken together with changes in the Debye-Waller factor, suggests that there is a second, longer Ni-O path in the Ni^{3+} state.

ACKNOWLEDGMENTS

This research was sponsored by the U.S. Department of Energy, Office of Basic Energy Science/Division of Chemical Science's Electrochemical Energy Storage and Conversion Program. Use of the Advanced Photon Source was supported by the U.S. Department of Energy, Basic Energy Sciences, Office of Science (DOE-BES-SC), under Contract No. W-31-109-Eng-38. The MRCAT is funded by the member institutions and DOE-BES-SC under contracts DE-FG02-94ER45525 and DE-FG02-96ER45589.

REFERENCES

[1] J.N. Riemers and J.R. Dahn, J. Electrochem. Soc. **139**, 2091 (1992).

[2] Y. Saadoune and C. Delmas, J. Mater. Chem. **6**, 193 (1996).

[3] S. Kostov, Y. Wang, M. L. denBoer, S. Greenbaum, C. C. Chang, and P. N. Kumpta, Mat. Res. Soc. Symp. Proc. **496**, 427-434 (1998).

[4] I. J. Pickering, G. N. George, J. T. Lewandowski, and A. J. Jacobson, J. Am. Chem. Soc. **115**, 4137-4144 (1993).

[5] I. Nakai and T. Nakagome, Electrochemical and Solid-State Letters **1** (6), 259-261 (1998).

[6] A. N. Mansour, J. McBreen, and C. A. Melendres, J. Electrochem. Soc. **146** (8), 2799-2809 (1999).

[7] W. E. O'Grady, K. I. Pandya, K. E. Swider, and D. A. Corrigan, J. Electrochem. Soc. **142**, 1613-1616 (1996).

[8] X. Sun, Q. Yang, J. McBreen, Y. Gao, M. V. Yakovleva, X. K. Xing, and M. L. Daroux, submitted to 1999 Joint International Meeting of the Electrochemical Society and the Electrochemical Society of Japan (1999).

[9] A. Rougier, C. Delmas, and A. V. Chadwick, Solid State Comm. **94** (2), 123-127 (1995).

[10] The electrolyte was 1 M LiPF$_6$ dissolved in 50 vol. % ethylene carbonate (EC) and 50 vol. % dimethyl carbonate (DMC) obtained as a solution from Merck Company.

[11] S. I. Zabinsky, J. J. Rehr, A. Ankudinov, R. C. Albers, and M. J. Eller, Phys. Rev. B **52**, 2995 (1995).

[12] E. A. Stern, M. Newville, B. Ravel, Y. Yacoby, and D. Haskel, Physica B **208&209**, 117 (1995).

THE INFLUENCE OF DESULFOVIBRIO DESULFURICANS ON NEPTUNIUM CHEMISTRY

L. SODERHOLM*, C.W. WILLIAMS*, MARK R. ANTONIO*, MONICA LEE TISCHLER**
AND MICHAEL MARKOS**
*Chemistry Division, Argonne National Laboratory, Argonne, IL, 60439.   ls@anl.gov,
mantonio@anl.gov
**Department of Biological Sciences, Benedictine University, Lisle IL

ABSTRACT

Biotic Np(V) reduction is studied in light of its potential role for the environmental immobilization of this hazardous radionuclide. The speciation of Np in *Desulfovibrio desulfuricans* cultures is compared with Np speciation in the spent medium and in the uninoculated medium. Precipitates formed in all three samples. Optical spectroscopy, x-ray diffraction, and x-ray absorption near edge structure (XANES) were used to determine the Np speciation. After 5 days of incubation, there was very little Np left in solution, which was present as Np(V). The precipitate that formed in all samples is an amorphous Np(IV) species, establishing that Np(V) is almost quantitatively reduced. These results demonstrate that the reduction of Np is independent of *Desulfovibrio desulfuricans*. The underlying chemistry associated with these results is discussed.

INTRODUCTION

Neptunium (Np; atomic number 93) is a highly toxic, long-lived radionuclide that is abundant in nuclear waste. Like other light actinide elements, the chemistry of Np is complex and dependent upon speciation[1]. Under environmentally relevant conditions, Np can be found in the (III), (IV), (V), or (VI) oxidation states. In the laboratory, acidic Np solutions at low pH contain primarily the neptunyl $[O=Np(V)=O]^+$ ion. Whereas Np(V) and (VI) are soluble in near neutral pH, Np(IV) forms hydrous oxides or the dioxide, both of which are very insoluble. Therefore, in order to effectively model the fate and transport of Np in the environment, it is important to understand the factors governing Np speciation, and their relative importance as a function of groundwater chemistry. Specifically, it is important to identify the pathways by which neptunyl can be reduced to the insoluble Np(IV) species and to assess the importance of such reduction pathways under environmentally germane conditions.

Although the redox chemistry of Np is predictable under controlled, laboratory conditions, the complexity of natural groundwaters, in terms of dissolved organics and inorganics, as well as the presence of bacteria and catalytic mineral surfaces, prohibits a predictive understanding of Np chemistry and redox speciation in all but the simplest systems. An indication that bacteria may play a role in actinyl redox chemistry comes from the reported reduction of uranyl(VI) to UO_2. This biotic reduction is understood in terms of an enzymatic reaction that is coupled to the electron-transport chains of metal-reducing bacteria[2, 3]. The bacteria involved in the reduction are well known as anaerobic iron, Fe(III), reducers. Thermodynamically, the uranyl(VI) ion is harder to reduce than either the Np(V) ion or the Fe(III) ion, as evidenced by the standard reduction potentials listed in Table I.

A variety of microorganisms have been demonstrated to reduce uranium(VI)[4] and Fe(III)[5] under anaerobic conditions. The microbial influences on uranium and iron chemistries suggest that Np speciation may be significantly altered by microbial growth. Significant microbial reduction of Np(V) would have a marked influence on the fate and transport of this radionuclide in the environment. Based on the literature reports cited above, and our previous experience with the growth of *Desulfovibrio desulfuricans*[6], we have chosen to investigate the influence of this sulfate reducing anaerobic bacterium on Np chemistry.

Table I. The standard reduction potentials of some common metal ions[7]. The more positive the value, the more favored the reaction product. Note that the UO_2^{2+} and Hg^{2+} reactions are two electron reductions as written.

Reduction Half-Cell	Potential (V)
$UO_2^{2+} + 4H^+ + 2e^- \longrightarrow U^{4+} + 2H_2O$	+0.27
$NpO_2^+ + 4H^+ + e^- \longrightarrow Np^{4+} + 2H_2O$	+0.66
$Fe^{3+} + e^- \longrightarrow Fe^{2+}$	+0.771
$Hg^{2+} + 2e^- \longrightarrow Hg^0$	+0.796

EXPERIMENTS

D. desulfuricans (ATCC 29577) were grown anaerobically at 30°C in Modified Starkey's Medium (ATCC 207; http:/www.atcc.org) with lactate. Strict precautions were taken to ensure that Fe was excluded from the growth medium because Fe is known to facilitate the reduction of neptunyl(V). The solution pH was adjusted to 6.2 with hydroxide to stabilize pentavalent Np and to limit hydroxide formation while maintaining an environment suitable for bacterial growth. Neptunyl was added at the time of culture inoculation. Two control experiments were performed: (1) Np was added to uninoculated medium and (2) Np was added to spent medium. After incubation for 5 days in the presence of Np, motile cells were observed at 400x using an Olympus phase contrast microscope.

Optical data were obtained on an Olis-converted Cary-14 spectrophotometer. X-ray diffraction data were obtained on the filtered precipitates using a Scintag theta-theta diffractometer operating with a copper tube and a Peltier detector. The Np L_3-edge (17610 eV) x-ray absorption near edge structure (XANES) data were collected on 12BM-B, the BESSRC bending magnet beamline at the Advanced Photon Source (APS). The beam line is equipped with Si<111> crystals in a double-crystal configuration. Harmonic rejection was accomplished using a Pt mirror, set to reject energies higher than about 25 KeV. Harmonic rejection at these energies is necessary at the APS because of the relatively high flux of high-energy photons. The energy was calibrated by setting the inflection point of the first derivative from the Zr K edge to 17.998 KeV. All data were taken in the fluorescence mode, using a flow-type ion chamber detector (The EXAFS Co.), which was purged with xenon and used without slits or a scattered-radiation filter. No time-dependent spectral changes were observed over multiple sample scans.

RESULTS AND DISCUSSION

The sample of interest and the two controls all showed a white precipitate that formed within the first day of incubation at 30°C. The precipitate made it difficult to quantify microbial growth, however observable motile cells after 5 days of incubation suggests some growth under our experimental conditions. The solutions were all filtered, after which the remaining solute and the precipitate were treated separately. An x-ray powder pattern of the precipitate revealed no diffraction lines. The precipitate appears amorphous at copper-radiation wavelengths.

Optical spectra of the solutions were obtained before and after incubation and filtration. The spectra obtained are compared with standard Np spectra in Figure 1. Optical spectroscopy is often used as a characterization tool for the Np oxidation state in solution because of the well separated signature spectrum available for each oxidation state[8], as demonstrated in the Figure. The spectrum of the Np in the growth medium immediately after inoculation with bacteria is consistent with that expected for $Np(V)O_2^+$. Whereas the spectrum taken on the same sample after incubation and filtration also shows the presence of $Np(V)O_2^+$, the intensity of the signal is significantly reduced. There is no evidence of Np(IV) in solution, even though the molar absorptivity of tetravalent Np is rather high. This result is expected because the solubility of Np(IV) is low at the pH of the solution. In contrast, because of the broad spectral feature associated with Np(VI), the presence of moderate amounts of the neptunyl(VI) moiety cannot be ruled out based on these results alone. Assuming a linear relation between absorbance and concentration (Beer's Law), these data provide evidence that much of the Np(V) has been removed from solution. The Np remaining in solution is indistinguishable by optical spectroscopy from that of the original solution, appearing as Np(V).

XANES spectroscopy supports this finding. The spectra from the Np solutions obtained from both the active bacteria and the spent medium after filtration indicate the presence of very little Np. In contrast, the precipitates obtained from these samples, as well as the control to which no bacteria were added, all show evidence of significant amounts of Np. The XANES spectra obtained from the solid precipitates are compared with signature XANES spectra of Np in the tri-, tetra-, penta-, and hexavalent oxidation states. XANES is a single ion probe that is very sensitive to oxidation state and has successfully been used to characterize Np speciation[9], both in solution and in the solid phase. In most cases, the edge energy, defined as the maximum in the first derivative, is seen to shift to higher energy with increasing valence. The situation is somewhat more complex for U[10], Np[9] and Pu[11], all of which form the dioxo cations $[O=An=O]^{n+}$. In these cases, the edge energy does not increase with increasing formal charge on the actinide ion so that any analysis of valence from XANES data must include the edge features to higher energy. This is demonstrated by the Np data from the standard solutions shown in Figure 2. The Np(IV) and $Np(V)O_2^+$ spectra are similar, and are differentiated in part by the broad peak to higher energy observed for the latter species.

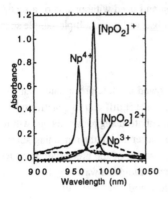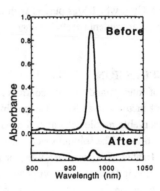

Figure 1. Left: The optical spectra of Np solutions containing Np^{3+}, Np^{4+}, Np(V) and Np(VI). These spectra serve as standards for comparison with the data from the cultures. Right: Optical spectra of the growth medium immediately after inoculation (top) shows the presence of Np(V). All solutions after incubation (bottom) continue to show evidence of only Np(V), however the intensity of the signal is much reduced, indicating a loss of solutions Np.

The data obtained from the three solutions are statistically indistinguishable and are represented in Figure 2 by the data from the Np in the spent medium. The edge energies and the peak shapes from these data are consistent with $Np(V)O_2^+$ in solution. These results directly support the interpretation of the optical data from the same solutions. The data obtained from the solid precipitates are also indistinguishable among themselves. However, they are markedly different from the data obtained from the solutions. The data from all the solid precipitates are consistent with Np(IV).

The XANES data clearly demonstrate the presence of Np(V) in solution and Np(IV) in the solid precipitates. The data from the sample in which Np was added to the culture medium with viable *D. desulfuricans* are indistinguishable from those in which Np was added to spent medium and from those in which Np was added to fresh medium without any *D. desulfuricans*. The precipitates formed in all samples at approximately the same rate during incubation. These results show that the reduction of Np(V) to Np(IV) was independent of the presence of *D. desulfuricans* on the timescale of days. Np(V) is rapidly and almost quantitatively reduced in the presence of the medium alone at 30°C. It has been previously demonstrated that uranyl, exchanged into an organically-coated smectite clay, is reduced to U^{4+} oxides or hydroxides under hydrothermal treatment[15, 16]. Iron(III) has been shown to reduce under similar conditions[17].

The underlying assumption for the experimental protocol chosen herein was that NpO_2^+ would be stable in the culture medium and that the reduction of Np(V) would arise primarily from a biotic process, or as a secondary result of a biotic process. This assumption was based in part on the

30

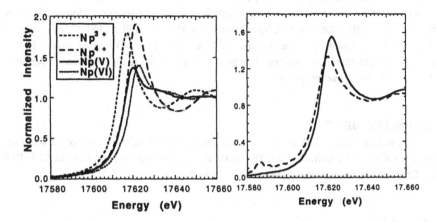

Figure 2. Left: A comparison of the XANES spectra obtained from standard Np solutions in four different oxidation states[9]. It can be seen from the figure that Np^{4+} and $Np(V)$ have similar edge positions. Right: A comparison of the XANES spectra obtained from solution after incubation (dotted line) with that of the solid precipitate (solid line). These XANES data show that Np has been reduced, and in present in the solid phase as Np^{4+}.

reduction potential listed in Table I for $Np(V)$. This standard reduction potential represents the thermodynamic value obtained with Np in an acidic solution (pH=0) of perchloric acid. The culture medium required to grow *D. desulfuricans* differs markedly from this standard solution, and as a result the redox couple is expected to shift[12]. Specifically, published studies on the reduction of $Np(V)$ in 1 M H_2SO_4 report that the formal reduction potential has shifted to 0.91 V[18], indicating that the presence of sulfate significantly stabilizes Np^{4+}. Also important to aqueous Np chemistry are disproportionation equilibrium in which NpO_2^+ disproportionates into Np^{4+} and NpO_2^{2+} . This is an equilibrium reaction that is dependent on acid concentration in solution. In the presence of a complexing anion, the equilibrium will be shifted. Studies have shown that the equilibrium constant for this reaction is 4×10^{-7} in 1M perchloric acid, but is shifted to 2.4×10^{-2} in 1M H_2SO_4[14]. This shift in equilibrium results in a 10^4 fold increase in the concentration of Np^{4+}, which is insoluble and drops out of solution. Unfortunately, sulfate is an essential component of the medium because *D. desulfuricans* is a sulfate reducing bacterium, that derives its energy from the reduction of SO_4^{2-}. Once Np^{4+} is formed, it hydrolyses readily at the near neutral pH of these experiments, forming an insoluble hydroxide according to[1]:

$$Np^{4+} + nH_2O \rightarrow Np(OH)_n^{(4-n)+} + nH^+.$$

Our results, supported by literature precedent, reveal that neptunyl(V) is reduced abiotically under the employed experimental conditions. We have not ruled out some concomitant biotic reduction of neptunyl, however the abiotic pathway is clearly dominant. Attempts to conduct biotic Np redox experiments in a simplified growth medium may provide further insight into the mechanism of Np reduction, but they are unlikely to provide insight into the fate of environmental Np in the presence of *D. desulfuricans* . The experiments discussed herein have demonstrated that the abiotic chemistry is controlling the Np speciation in conditions conducive to *D. desulfuricans* growth.

ACKNOWLEDGMENTS

The authors thank Mark Jensen and Jennifer Linton for technical assistance, and James Sullivan for stimulating conversation. The work at Argonne was supported by the U.S. DOE, BES – Chemical Sciences, under contract W-31-109-ENG-38.

REFERENCES

1   J. A. Fahey, in *The Chemistry of the Actinide Elements*, edited by J. J. Katz, G. T. Seaborg and L. R. Morss (Chapman and Hall, London, 1986), Vol. 1, p. Chapter 6.

2   D. R. Lovley, E. J. P. Phillips, Y. A. Gorby, and E. R. Landa, Nature **350**, 413-416 (1991).

3   D. R. Lovley and E. J. P. Phillips, Environ. Sci. Technol. **26**, 2228-2234 (1992).

4   L. L. Barton, K. Choudhury, B. M. Thomson, K. Steenhoudt, and A. R. Groffman, Radioact. Waste Manag. Enviro. Restor. **20**, 141-151 (1996).

5   D. R. Lovley, Microbio. Rev. **55**, 259-287 (1991).

6   M. R. Antonio, M. Tischler, and D. Witzcak, (submitted, this conference).

7   A. J. Bard and L. R. Faulkner, *Electrochemical Methods: Fundamentals and Applications* (Wiley, New York, 1980).

8   P. G. Hagen and J. M. Cleveland, J. Inorg. Nucl. Chem **28**, 2905 (1966).

9   L. Soderholm, M. R. Antonio, C.Williams, and S. R. Wasserman, Anal. Chem. **71**, 4622-4628 (1999).

10   E. A. Hudson, J. J. Rehr, and J. J. Bucher, Phys.Rev. B **52**, 13815-13826 (1995).

11   A. L. Ankudinov, S. D. Conradson, J. MustredeLeon, and J. J. Rehr, Phys.Rev.B **57**, 7518-7525 (1998).

12   M. Pourbaix, *Atlas of electrochemical equilibria in aqueous solutions* (Pergamon Press, Oxford, 1966).

13   J. C. Sullivan, D. Cohen, and J. C. Hindman, J. Am. Chem. Soc. **1954**, 4275-4279 (1954).

14   B. B. Cunningham and J. C. Hindman, *The Actinide Elements*, London, 1954).

15   D. M. Giaquinta, L. Soderholm, S. E. Yuchs, and S. R. Wasserman, Radiochim. Acta **76**, 113-121 (1997).

16   S. R. Wasserman, L. Soderholm, and D. M. Giaquinta, (submitted this conference).

17   S. Wasserman, L. Soderholm, and U. Staub, Chem. Mater. **10**, 559-566 (1998).

18.   J. C. Hindman, L.B. Magnuson and T. J. LaChapelle, J. Amer. Chem. Soc. **71**, 687-693 (1949).

EXTRACELLULAR IRON-SULFUR PRECIPITATES FROM GROWTH OF
Desulfovibrio desulfuricans

MARK R. ANTONIO*, MONICA LEE TISCHLER**, DANA WITZCAK**
*Argonne National Laboratory, Chemistry Division, Argonne, IL 60439, mantonio@anl.gov
**Benedictine University, Department of Biological Sciences, Lisle, IL 60532

ABSTRACT

We have examined extracellular iron-bearing precipitates resulting from the growth of *Desulfovibrio desulfuricans* in a basal medium with lactate as the carbon source and ferrous sulfate. Black precipitates were obtained when *D. desulfuricans* was grown with an excess of $FeSO_4$. When *D. desulfuricans* was grown under conditions with low amounts of $FeSO_4$, brown precipitates were obtained. The precipitates were characterized by iron K-edge XAFS (X-ray absorption fine structure), ^{57}Fe Mössbauer-effect spectroscopy, and powder X-ray diffraction. Both were noncrystalline and nonmagnetic (at room temperature) solids containing high-spin Fe(III). The spectroscopic data for the black precipitates indicate the formation of an iron-sulfur phase with 6 nearest S neighbors about Fe at an average distance of 2.24(1) Å, whereas the brown precipitates are an iron-oxygen-sulfur phase with 6 nearest O neighbors about Fe at an average distance of 1.95(1) Å.

INTRODUCTION

Sulfate reducing bacteria (SRB) in subsurface oil reservoirs can cause significant production problems, especially in formations that have been subjected to advanced recovery operations such as seawater flooding. In secondary petroleum production activities, waterflooding is the most successful and extensively used technique to recover oil from a field in which the natural pressure of the reservoir is no longer sufficient to force the oil out of the pores of the rock.[1, 2] Although waterflooding operations are of economic benefit—in terms of increased crude oil production—flooding oftentimes indirectly leads to an increase in the hydrogen sulfide, H_2S, content of the reservoir fluids, a condition known as souring.[3] Depending upon the aquifer pH, the dissolved H_2S is in equilibrium with the SH^- and S^{2-} anions.[4] Oil fields with low indigenous levels of H_2S have been shown to produce increasing amounts of H_2S as waterflooding operations expand.[5] This is due to, in part, the presence of SRB in the subsurface reservoir.[3, 6-8]

Regardless of their origin and exact genera, sulfate reducers are potent generators of H_2S. The anaerobic, nutrient rich geothermal environments of waterflooded oil reservoirs are excellent habitats for the growth of SRB.[9, 10] The diversity and distribution of SRB species varies from reservoir-to-reservoir with the subsurface physical conditions, e.g., nutrient concentration, salinity, temperature, pressure, porosity, permeability, pH.[9, 11] The metabolism of SRB involves the eight-electron reduction of sulfate, $[SO_4]^{2-}$. One of the final products of the dissimilatory reduction of sulfate is H_2S. In sulfate-rich environments, the reduction proceeds without accumulation of intermediates, e.g., S(V), S(IV), S(II), S(0), S(I-).

In addition to the undesirable effects of souring and the threat to human health, H_2S poses problems with corrosion. Microbial influenced corrosion (MIC) occurs when SRB/H_2S attack the iron and its alloys found in the subsurface tubing, valves, rods, and other components of the well, to produce iron sulfides.[12-17] Failures of ferrous metal oil-field equipment that occur as a result of MIC are of significant engineering concern and a cause of economic loss in the secondary production of petroleum. Moreover, the subsurface production of H_2S leads to yet another problem—fouling—wherein ultrafine solids clog the pores in the formation and reduce reservoir permeability.[2, 18] Colonies of sulfate reducers give rise to ultrafine extracellular precipitates.[19] The solids precipitation can be severe if the reservoir contains iron, which is ultimately found as ferrous ion, Fe(II), in the anoxic downhole conditions. Regardless of the source of H_2S, reservoir souring in combination with the presence of iron can lead to the formation of a variety of ultrafine iron sulfur, Fe-S, precipitates. This happens when H_2S, SH^-, and S^{2-} react with dissolved Fe(II) in the field aquifer and oil as well as with solid, iron-bearing reservoir minerals.[5] These reactions produce black Fe-S precipitates.[20] The precipitation of extracellular Fe-S fines throughout the reservoir can block the movement of fluids.[2] Because the rate at which oil can be extracted depends upon the porosity and permeability of the rock, any blockage or plugging may lead to the premature decline in production of a field with a large amount of oil in place.

Iron-bearing solids are generally ubiquitous in oil reservoirs as naturally occurring minerals such as sulfides, disulfides, carbonates, clays and clay minerals.[20] Depending upon the subsurface microbiological and geochemical environments, the formation of Fe-S precipitates in souring reservoirs can be understood in terms of either biotic or abiotic processes.[5] The former involves microbiologically-influenced reactions of sulfide with dissolved and solid iron.[21-26] Of interest here is the extracellular MIC and biotically-mediated Fe-S production processes mentioned above, wherein H_2S reacts with iron in the extracellular environment to form fine black precipitates. This is to be contrasted with intracellular biologically-mediated mineralization of Fe-S compounds in magnetotactic bacteria.[27-30] Abiotic processes involve reactions of sulfide produced by any number of inorganic souring mechanisms with iron in the subsurface environment. The suggestion has been made that the ferruginous sulfides produced by biotic microbial mediation and abiotic chemical processes may be distinctly different from one another and from the ferruginous sulfides present in the geochemical environment itself.[16, 30, 31] Verification of this suggestion would have significant technical and economic impact on upstream petroleum production operations.

Although extracellular Fe-S precipitates obtained from the growth of SRB have been studied for some time,[21-24] the exact phase relationships are incompletely known and issues of contemporary interest.[25, 26, 32] A fundamental understanding of the subsurface chemistry of iron and sulfur, in general, and the formation of biotic Fe-S precipitates, in particular, can provide insights about fouling in souring oil reservoirs. In addition to implications in petroleum production operations, biotically-mediated Fe-S precipitates are receiving attention as adsorbants.[33] We have isolated and characterized precipitates that formed in the presence of SRB by reaction of H_2S with ferrous ions in the extracellular environment. The noncrystalline Fe-bearing precipitates obtained from anaerobic growth of *Desulfovibrio desulfuricans* were examined by use of Fe K-edge XAFS, [57]Fe Mössbauer-effect spectroscopy, and powder X-ray diffraction.

EXPERIMENTS

Desulfovibrio desulfuricans (ATCC 29577) was grown anaerobically in unfiltered ATCC culture medium 207 with added resazurin, an E_h indicator. $FeSO_4$ was added in concentrations varying from 0.04 mg/mL to 1.0 mg/mL. Studies were also done with a medium of similar composition in which 2 g/L $MgSO_4$ was used in place of 2 g/L $MgCl_2$. Anaerobiosis was achieved and verified using a modification of benchtop techniques.[34] All manipulations took place under a sterile stream of N_2 gas using needles and syringes to inoculate anaerobic media in serum vials sealed with butyl rubber stoppers. Before incubation, the atmosphere in the vials was replaced by a sterile stream of N_2/CO_2 gas. All cultures were grown at 30°C. The precipitates were worked up under anaerobic conditions under a N_2/CO_2 atmosphere in a Hydrovoid Air Control inert atmosphere box, equipped with gloves. After isolation by filtration, each sample was sealed in place on the filter paper, between layers of Kapton® tape. Zero-field, natural abundance [57]Fe Mössbauer spectra were acquired on the solid samples at 295, 77, and 5 K in a LHe exchange-gas cryostat as described elsewhere.[20] The Mössbauer spectra were recorded in the standard transmission geometry with a [57]Co/Rh source and a Kr/CH_4 detector. Velocity calibration over the range 0 ± 11 mm/s was checked by use of a natural iron foil (25 μm thick). All isomer shifts were quoted with respect to natural iron at room temperature. The instrumental linewidth was found to be 0.28 mm/s for iron foil. The data were folded and fit using the program WMOSS.[35] The isomer shift (δ, mm/s), quadrupole splitting (ΔE_Q, mm/s), linewidths (FWHM, Γ, mm/s) and internal magnetic field (H_{int}, kG) were obtained from the fits. Powder X-ray diffraction data were collected with a Scintag diffractometer using Cu Kα radiation. Iron K-edge transmission XANES (X-ray absorption near edge structure) and EXAFS (extended X-ray absorption fine structure) were collected at the APS using the BESSRC beam line 12-BM-B, equipped with a Si<111> monochromator. The energy was calibrated with the inflection point in the first differential XANES of Fe (4 μm thick), which is at 7112 eV in accordance with previous work.[36] The XAFS was analyzed by conventional methods, as described elsewhere,[37, 38] using EXAFSPAK[39] and FEFF7.02.[40]

RESULTS AND DISCUSSION

Black precipitates associated with the growth of *D. desulfuricans* in media containing both $MgSO_4$ and $FeSO_4$ were isolated. In experiments with $FeSO_4$ as the sole sulfate source, there appeared to be a threshold level of $FeSO_4$ necessary for formation of black precipitates. With $FeSO_4$ concentrations of 0.2 mg/mL and higher, black precipitates were formed with bacterial growth. These black precipitates were extremely air-sensitive, rapidly converting to rust colored solids after short exposure. With $FeSO_4$ concentrations below 0.2 mg/mL, brown precipitates formed with bacterial growth. Powder X-ray diffraction data of both precipitates were essentially featureless, providing no evidence of crystallinity. Neither type of precipitate was attracted to a small permanent magnet.

The Fe K-edge XANES of the black precipitates is shown in Fig. 1. It is consistent with the presence of an iron-sulfur coordination environment, as found in Fe-S minerals such as pyrite, greigite, and pyrrhotite.[37, 38, 41] The Fe XANES of pyrrhotite, containing high-spin Fe(II) in octahedral coordination with S, is shown in Fig. 1 as the dashed line. The absorption edge for the black precipitates is at a higher energy than that for pyrrhotite. This suggests the presence of Fe(III) in the black precipitates. The Fe K-edge XANES of the brown precipitates is also shown in Fig. 1. It is consistent with the presence of an Fe(III)-oxygen coordination environment, as found in Fe-O minerals such as hematite, goethite, ferrihydrite, jarosite, etc.[42-44] Because iron XANES of sulfide minerals can be a misleading diagnostic indicator of Fe valence,[45] we exploited ^{57}Fe Mössbauer spectroscopy to provide insights about the iron valence and magnetism in the black as well as brown precipitates.

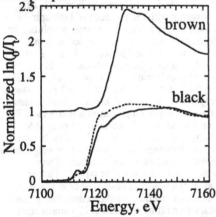

Figure 1. Normalized iron K-edge XANES, $\ln(I_0/I_t)$, recorded at room temperature for the brown (top) and black (bottom) precipitates—solid lines—obtained from growth of *D. desulfuricans*. The Fe XANES for pyrrhotite, $Fe_{1-x}S$, is shown as the dashed line (bottom). The Fe XANES for the brown precipitates is offset for clarity.

The variable-temperature Mössbauer spectra of the black solids are shown in Fig. 2. At room temperature, the spectrum is an asymmetric doublet that is adequately modeled with two overlapping quadrupole split doublets ($\delta_1 = 0.32$, $\Delta E_{Q1} = 0.76$, $\Gamma_1 = 0.39$ mm/s; $\delta_2 = 0.35$, $\Delta E_{Q2} = 1.21$, $\Gamma_2 = 0.35$ mm/s). The isomer shifts, δ, are typical of high-spin Fe(III) with octahedral coordination of S.[46] At 77 K, magnetic splitting is apparent by the decreased intensity of the doublet, which splits into a broad sextet that is typical of a broad distribution of hyperfine fields at iron. At 5 K, the magnetic ordering is complete to reveal another broad spectrum, which is not particularly well modeled with a sum of 2-3 overlapping sextets. Still, the magnetic splitting with H_{int} of approximately 225-265 kG is consistent with the presence of an Fe-S phase. For example, the $Fe_{1-x}S$ (pyrrhotite, mackinawite) and $Fe_{3-x}S_4$ (greigite, smythite) minerals containing Fe(II) and mixed Fe(II)-Fe(III) ions, respectively, as well as amorphous Fe_2S_3 with Fe(III) exhibit similar hyperfine fields of ca. 200-300 kG.[47] At 77 K and above, the Mössbauer spectrum of amorphous

Fe₂S₃ is an asymmetric doublet, and at 4.2 K, the spectrum is a complex combination of broad, overlapping sextets,[48] much like that observed here. Likewise, the 4.2 K Mössbauer spectrum of an amorphous Fe(III) sulfide as a 500-iron atom cluster reveals a mean magnetic hyperfine field of 260 kG[45] that is consistent with the ordering observed for the black precipitates in Fig. 2 at 5 K. However, the Mössbauer spectrum of the black precipitates does not exactly correspond to any one of the known binary iron sulfides. This suggests the possibility that the black precipitates from *D. desulfuricans* contain two or more different Fe-S species, each with two or more different iron sites.

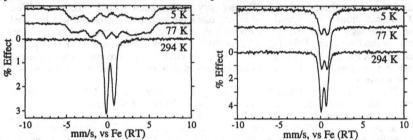

Figures 2 (left) and 3 (right). Variable-temperature, zero-field, natural abundance ^{57}Fe Mössbauer spectra for the black and brown precipitates, respectively.

The brown precipitates isolated from cultures grown with low levels of FeSO₄ were also studied by Mössbauer spectroscopy. The spectra are shown in Fig. 3. At room temperature, the Mössbauer spectrum of these solids is an asymmetric doublet that is adequately modeled with two overlapping quadrupole split doublets ($\delta_1 = 0.39$, $\Delta E_{Q1} = 0.49$, $\Gamma_1 = 0.38$ mm/s; $\delta_2 = 0.41$, $\Delta E_{Q2} = 0.84$, $\Gamma_2 = 0.35$ mm/s). The isomer shift values are consistent with the presence of high-spin Fe(III) with octahedral coordination of O.[49] At 77 and 5 K, there is no evidence for magnetic splitting in the spectra. Rather, at 5 K, the symmetric doublet spectrum is satisfactorily modeled with one quadrupole split doublet ($\delta = 0.49$, $\Delta E_Q = 0.67$, $\Gamma = 0.57$ mm/s). The Mössbauer data of Figs. 2 and 3 indicate that the brown precipitates produced by the bacteria grown with low levels of FeSO₄ are different from the black precipitates produced by the bacteria grown with high levels of FeSO₄. Furthermore, the brown precipitates are different from the intentionally oxidized black precipitates.

Insights about the Fe coordination in the precipitates from *D. desulfuricans* were obtained from the Fe K-edge EXAFS. The primary data and their corresponding Fourier transforms (FTs) are shown as solid lines in Fig. 4. Each FT reveals one intense peak. It is due to the nearest S neighbors about Fe in the black precipitates and the nearest O neighbors about Fe in the brown precipitates. Both FTs reveal a weak, second peak that is attributable to the next nearest neighboring atoms, which were identified as S from best Z fits with O, S, and Fe atoms. The lack of any structurally significant features beyond about 3 Å in the FT data is consistent with the lack of crystallinity of the precipitates as determined by X-ray diffraction. The dashed lines in Fig. 4 illustrate the fits to the primary data. The agreement indicates that the two-shell curve fitting models adequately describe the data. For the black precipitates, curve fitting revealed 6(1) S atoms at an average distance of 2.24(1) Å and 3(1) S atoms at an average distance of 2.84(2) Å. For the brown precipitates, there were 6(1) O atoms at an average distance of 1.95(1) Å and 4(1) S atoms at an average distance of 3.20(3) Å. The average Fe(III)-S₆ distance for the black precipitates is 0.15 Å shorter than that (2.39 Å) for pyrrhotite with Fe(II)-S₆ coordination. This result is consistent with the 0.135 Å decrease in ionic radius from Fe(II) to Fe(III), for high-spin CN=VI.[50] The presence of a distant Fe...S interaction in the brown precipitates suggests the presence of an Fe-O-S phase.

Other researchers have previously described differences in precipitates associated with *D. desulfuricans* based on the amount of available ferrous ion. For example, Booth et al.[12, 13] and King et al.[14, 15] noticed that differences in the corrosive properties of H₂S produced from the metabolism of sulfate reducing bacteria depended on the amount of iron available to precipitate the sulfide. They speculated that when there was insufficient iron to precipitate all of the sulfide produced by the bacteria, the sulfide would form a protective film on metals, preventing corrosion.

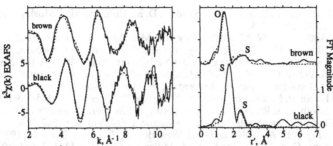

Figure 4. The Fe K-edge $k^3\chi(k)$ transmission EXAFS (left) and the corresponding FT data (right), not corrected for phase shift, for the brown (top curves) and black (bottom curves) precipitates.

CONCLUSIONS

Because of their element-specific, atomic scale perspective of matter, neither Fe XAFS nor Fe Mössbauer can be readily employed to identify bulk phases in the black and brown precipitates obtained from the growth of *D. desulfuricans*. X-ray diffraction is also of little help here because the precipitates showed no evidence of crystallinity. Still, the combined results indicate that the amorphous, nonmagnetic black precipitates produced by *D. desulfuricans* are consistent with the formation of an iron-sulfur phase containing high-spin Fe(III) in octahedral coordination, whereas the brown precipitates are consistent with an iron-oxygen-sulfur phase containing high-spin Fe(III) with octahedral coordination of O.

The presence of ferric ion, Fe(III), in both precipitates is surprising. The oxidation of Fe(II)SO₄ in the culture medium is not due to the presence of O_2 during the growth or work-up procedures. Because motile cells were confirmed in the media, the possibility of O_2 contamination is excluded—*D. desulfuricans* require rigorously anaerobic conditions for growth. Although the source of the oxidant is not known, we suspect that components of the culture medium may be involved. This suggestion is supported by our observation that Np(V) is reduced to Np(IV) in filtered ATCC culture medium 207 without resazurin, both with and without *D. desulfuricans*.[51] In this regard, the Np(V)/Np(IV) and Fe(III)/Fe(II) redox couples have similar formal potentials.[51] The conditions under which our black and brown *D. desulfuricans*-mediated laboratory precipitates were formed are clearly different from those in anoxic subsurface oil reservoirs, where the reducing environment maintains the presence of ferrous ion, Fe(II).

ACKNOWLEDGMENTS

We thank Gail Karet. L. Soderholm, Clayton Williams, and S. Skanthakumar for assistance. The work at ANL was supported by the U.S. D.O.E. Office of Computational and Technology Research-Advanced Energy Projects, under contract W-31-109-ENG-38.

REFERENCES

1   E. C. Donaldson, G. V. Chilingarian, and T. F. Yen, in *Enhanced Oil Recovery, II. Processes and Operations* (Elsevier, Amsterdam, 1989).

2   T. A. Denman and S. Starr, Environ. Sci. Res. **46**, 569-578 (1992).

3   B. Eden, P. J. Laycock, and M. Fielder, (UK Health and Safety Executive - Offshore Technology Report 92 385, 1993).

4   M. A. A. Schoonen and H. L. Barnes, Geochim. Cosmochim. Acta **52**, 649-654 (1988).

5   D. J. Ligthelm, R. B. de Boer, J. F. Brint, and W. M. Schulte, in *Offshore Europe 91 Proc., SPE 23141* (Society of Petroleum Engineers, Richardson, TX, 1991), p. 369-378.

6   T. N. Nazina, A. E. Ivanova, O. V. Golubeva, R. R. Ibatullin, S. S. Belyaev, and M. V. Ivanov, Microbiology, Engl. Transl. **64**, 203-208 (1995).

7   K. O. Stetter, R. Huber, E. Blöchl, M. Kurr, R. D. Eden, *et al.*, Nature **365**, 743-745 (1993).

8   W. J. Cochrane, P. S. Jones, P. F. Sanders, D. M. Holt, and M. J. Mosley, SPE Paper 18368 , 301-316 (1988).

9   K. M. Antloga and W. M. Griffin, Dev. Ind. Microbiol. **26**, 597-610 (1985).

10   J. M. Odom, in *The Sulfate Reducing Bacteria: Contemporary Pespectives*, edited by R. Singleton, Jr. and J. M. Odom (Springer-Verlag, New York, 1993), p. 189-249.

11   D. W. S. Westlake, in *Microbial Enhancement of Oil Recovery - Recent Advances*, edited by E. C. Donaldson (Elsevier, Amsterdam, 1991), p. 257-263.

12   G. H. Booth, P. M. Cooper, and D. S. Wakerley, Br. Corros. J. **1**, 345-349 (1966).

13   G. H. Booth, J. A. Robb, and D. S. Wakerley, in *The Third International Congress on Metallic Corrosion* (Swets-Zeitlinger, Moscow, 1966), Vol. II, p. 542-554.

14   R. A. King, J. D. A. Miller, and D. S. Wakerley, Br. Corros. J. **8**, 89-93 (1973).

15   R. A. King, C. K. Dittmer, and J. D. A. Miller, Br. Corros. J. **11**, 105-107 (1976).

16   M. B. McNeil and B. J. Little, Corrosion **46**, 599-600 (1990).

17   W. Lee, Z. Lewandowski, P. H. Nielsen, and W. A. Hamilton, Biofouling **8**, 165-194 (1995).

18   J. H. Barkman and D. H. Davidson, J. Pet. Technol. , 865-873 (1972).

19   J. M. Galbraith and K. L. Lofgren, Mater. Perform. **26**, 42-49 (1987).

20   M. R. Antonio, G. B. Karet, and J. P. Guzowski, Fuel **79**, 37-45 (2000).

21   L. G. M. Baas Becking and D. Moore, Economic Geology **56**, 259-272 (1961).

22   A. M. Freke and D. Tate, J. Biochem. Microbiol. Technol. Eng. **3**, 29-39 (1961).

23   D. T. Rickard, Stockh. Contr. Geol. **20**, 49-66 (1969).

24   R. O. Hallberg, Neues Jahrb. Mineral., Monatsh., 481-500 (1972).

25   D. A. Bazylinski, Mater. Res. Soc. Symp. Proc. **218**, 81-91 (1991).

26   R. B. Herbert, Jr., S. G. Benner, A. R. Pratt, *et al.*, Chem. Geol. **144**, 87-97 (1998).

27   S. Mann, N. H. C. Sparks, R. B. Frankel, *et al.*, Nature **343**, 258-261 (1990).

28   M. Farina, D. M. S. Esquivel, and H. G. P. Lins de Barros, Nature **343**, 256-258 (1990).

29   D. A. Bazylinski, Chem. Geol. **132**, 191-198 (1996).

30   M. Posfai, P. R. Buseck, D. A. Bazylinski, and R. B. Frankel, Science **280**, 880-883 (1998).

31   M. B. McNeil, J. M. Jones, and B. J. Little, in *Corrosion 91* (NACE, Cincinnati, 1991), Paper No. 580, p. 580/1-580/16.

32   R. Donald and G. Southam, Geochim. Cosmochim. Acta **63**, 2019-2023 (1999).

33   J. H. P. Watson, D. C. Ellwood, Q. Deng, *et al.*, Miner. Eng. **8**, 1097-1108 (1995).

34   W. E. Balch, G. E. Fox, L. J. Magrum, *et al.*, Microbiol. Rev. **43**, 260-296 (1979).

35   T. A. Kent, (WEB Research Company, Eden Prarie, MN, www.webres.com, 1996).

36   I. Song, M. R. Antonio, and J. H. Payer, J. Electrochem. Soc. **142**, 2219-2224 (1995).

37   D. A. Totir, I. T. Bae, Y. Hu, M. R. Antonio, M. A. Stan, and D. A. Scherson, J. Phys. Chem. B **101**, 9751-9756 (1997).

38   D. A. Tryk, S. Kim, Y. Hu, W. Xing, *et al.*, J. Phys. Chem. **99**, 3732-3735 (1995).

39   G. N. George and I. J. Pickering, (http://www-ssrl.slac.stanford.edu/exafspak.html).

40   J. J. Rehr, J. M. de Leon, S. I. Zabinsky, *et al.*, J. Am. Chem. Soc. **113**, 5135-5140 (1991).

41   C. Sugiura, J. Chem. Phys. **80**, 1047-1049 (1984).

42   G. A. Waychunas, M. J. Apted, and G. E. Brown, Phys. Chem. Minerals **10**, 1-9 (1983).

43   H. Yamashita, Y. Ohtsuka, S. Yoshida, *et al.*, Energy Fuels **3**, 686-692 (1989).

44   K. Kaneko, N. Kosugi, and H. Kuroda, J. Chem. Soc., Faraday Trans. 1 **85**, 869-881 (1989).

45   T. Douglas, D. P. E. Dickson, S. Betteridge, J. Charnock, C. D. Garner, and S. Mann, Science **269**, 54-57 (1995).

46   J. B. Goodenough and G. A. Fatseas, J. Solid State Chem. **41**, 1-22 (1982).

47   G. P. Huffman and F. E. Huggins, Fuel **57**, 592-604 (1978).

48   A. H. Stiller, B. J. McCormick, P. Russell, and P. A. Montano, J. Am. Chem. Soc. **100**, 2553-2554 (1978).

49   A. Meagher, V. Nair, and R. Szostak, Zeolites **8**, 3-11 (1988).

50   R. D. Shannon, Acta Cryst. **A32**, 751-767 (1976).

51   L. Soderholm, C. W. Williams, M. R. Antonio, M. L. Tischler, and M. Markos, Mater. Res. Soc. Symp. Proc. **this volume** (2000).

THE STRUCTURE OF ACTINIDE IONS EXCHANGED INTO NATIVE AND MODIFIED ZEOLITES AND CLAYS

STEPHEN R. WASSERMAN*, L. SODERHOLM**, DANIEL M. GIAQUINTA**

*Advanced Photon Source, Argonne National Laboratory, 9700 S. Cass Ave., Argonne IL, 60439
**Chemistry Division, Argonne National Laboratory, 9700 S. Cass Ave., Argonne IL 60439

ABSTRACT

X-ray absorption spectroscopy (XAS) has been used to investigate the structure and valence of thorium (Th^{4+}) and uranyl (UO_2^{2+}) cations exchanged into two classes of microporous aluminosilicate minerals: zeolites and smectite clays. XAS is also employed to examine the fate of the exchanged cations after modification of the mineral surface using self-assembled organic films and/or exposure to hydrothermal conditions. These treatments serve as models for the forces that ultimately determine the chemical fate of the actinide cations in the environment. The speciation of the cations depends on the pore size of the aluminosilicate, which is fixed for the zeolites and variable for the smectites.

INTRODUCTION

Recent studies have examined the effect of organic surface modification and hydrothermal treatments on the structure of various ions, including uranium, iron, and thorium, that had been exchanged into the interlayer of natural clay minerals.[1-4] The surface modification involved the addition of a thin hydrophobic organic film derived from either octadecyltrimethoxy- or octadecyltrichlorosilane, $CH_3(CH_2)_{17}SiX_3$, X = –OCH$_3$, –Cl. These studies demonstrated that reduction and aggregation of the interlayer ions can occur upon hydrothermal treatment of the modified clays. The reactions of the interior metal cations generally follow the reduction potential of the ion and its susceptibility to hydrolysis in aqueous solution. In this paper we extend these experiments to thorium (Th^{4+}) and uranyl (UO_2^{2+}) exchanged into zeolites.

The structures of the zeolites used for these experiments are shown in Figure 1, together with that of the smectite clay used in previous experiments. Both the zeolites and

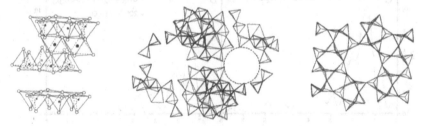

Smectite Clay   **Faujasite**   **ZSM-5**

Figure 1. Structures of a smectite clay and two types of zeolites: faujasite and ZSM-5. The pore sizes of the zeolites are 7.4 and 5.1-5.6 Å, respectively.

clays are aluminosilicates. Substitutions within the framework result in a net negative charge for the aluminosilicate. In both zeolites and smectite clays, aluminum replaces silicon, while in the clays additional substitutions of magnesium for aluminum occur. Electrical neutrality in both classes of materials is maintained through the presence of cations, either within the interlayer of the clay or inside the pores of the zeolites. The cations in the native materials can be exchanged for other ions, such as the Th^{4+} and UO_2^{2+} used in these experiments. For the zeolites, the cation exchange capacity is proportional to the incorporation of aluminum into the framework. In this study two types of zeolites were used. The first, faujasite, has a symmetrical pore structure with a diameter of 7.4 Å. The second, ZSM-5, has a slightly smaller pore, whose diameter varies between 5.1 and 5.6 Å. The fixed sizes of the openings into the interior of the zeolite contrast with the variable interlayer dimensions of the smectite, which depends on the degree of hydration of the clay.

EXPERIMENTAL SECTION

The preparation of ion-exchanged and surface-modified zeolites followed the same procedures previously used for the creation of the corresponding clay minerals.[1-4] Hydrothermal treatments also mirrored the earlier procedures and exposed the zeolites to water at 200 °C within a Teflon bomb. X-ray absorption spectra for uranyl and thorium in faujasite were obtained at the Advanced Photon Source (APS) using BESSRC beamline 12BM. Beamline 4-3 of the Stanford Synchrotron Radiation Laboratory (SSRL) was used

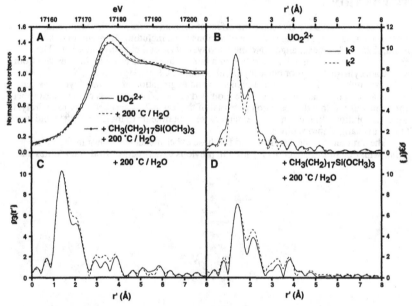

Figure 2. X-ray absorption near edge spectra (XANES) (A) and radial structure functions (B – D, not phase corrected) for uranyl in faujasite. The radial structure functions for each of the three samples shown in A are presented separately in B, C, and D. (1) ion-exchanged zeolite (A ——, B), (2) after hydrothermal processing at 200 °C (A ---·, C) and (3) following formation of a monolayer from octadecyltri-methoxysilane, $C_{18}H_{37}Si(OCH_3)_3$, and hydrothermal treatment (A–●–, D). For each radial structure function, transforms of the $k^3\chi(k)$ (——) and $k^2\chi(k)$ (---·) data are presented. The range for the forward Fourier transforms was $\Delta k = 2.5 - 10.5$ Å^{-1}.

for the samples that incorporated uranyl into ZSM-5. The monochromators at each station were equipped with Si<111> and Si<220> crystals, respectively. Data for the uranyl-exchanged ZSM-5 were collected in fluorescence mode using a Lytle detector whose ion chamber contained krypton gas. A 9-element germanium detector measured the fluorescence from the faujasite samples that contained thorium or uranyl. A 3-absorption-length strontium filter was placed between the sample and the ion chamber for the uranium L_3-edge spectra. No filter was used during the collection of the thorium L_3-edge spectra. At SSRL energy calibration was maintained through acquisition of the near-edge spectrum of UO_2 before or after each spectrum. The experimental data were analyzed with XAMath, a package for XAS analysis based on Mathematica®.[5]

RESULTS

The near-edge x-ray absorption spectra (XANES) and radial structure functions for uranyl exchanged into faujasite are shown in Figure 2. The data from the original exchanged zeolite are similar to those of uranyl in solution, with two axial oxygen ligands at 1.77 Å and 4 to 5 equatorial oxygen atoms at 2.41 Å.[6] Following hydrothermal treatment, the differences in the spectra are relatively minor, suggesting that the uranium remains in the +6

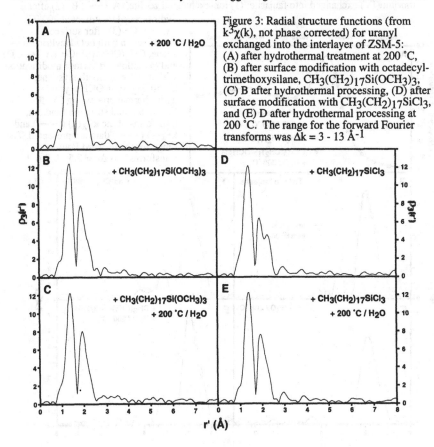

Figure 3: Radial structure functions (from $k^3\chi(k)$, not phase corrected) for uranyl exchanged into the interlayer of ZSM-5: (A) after hydrothermal treatment at 200 °C, (B) after surface modification with octadecyltrimethoxysilane, $CH_3(CH_2)_{17}Si(OCH_3)_3$, (C) B after hydrothermal processing, (D) after surface modification with $CH_3(CH_2)_{17}SiCl_3$, and (E) D after hydrothermal processing at 200 °C. The range for the forward Fourier transforms was $\Delta k = 3 - 13$ Å^{-1}

state, although the equatorial ligands appear somewhat more disordered. After coating the exchanged faujasite with a hydrophobic organic thin film generated using octadecyltrimethoxysilane followed by hydrothermal processing at 200 °C, the absorption edge of uranium shifts to lower energy by 0.7 eV and the structure of the edge changes. In addition, the intensity of the feature in the radial distribution due to the axial ligands is reduced by approximately 25 percent. Since uranium is the only high-Z element in this material, a comparison of the k^2- and k^3-weighted EXAFS should indicate which, if any, of the features in the radial structure function are due to uranium-uranium interactions. No significantly greater intensity is found in the radial structure functions from any of the k^3 EXAFS relative to the k^2 data, thereby demonstrating that the uranium in these samples is isolated from the other uranium ions. These results are consistent with a partial reduction of the uranium in the zeolite to the +4 oxidation state upon addition of the organic coating and aqueous processing at elevated temperature. Similar reactivity has been observed for uranyl in smectite clays.[1,3] In the clay, however, a greater fraction of the uranyl is reduced and aggregation of the uranium into small UO_2 clusters is also observed.

Figure 3 compares the radial structure functions of uranyl exchanged into a ZSM-5

Figure 4: Near-edge spectra and radial structure functions (without phase correction) for thorium(IV) exchanged into faujasite: (1) ion-exchanged zeolite, (A ——— , B) (2) after hydrothermal treatment at 200 °C (A - - -; C), (3) after surface modification with octadecyltrichlorosilane, $CH_3(CH_2)_{17}SiCl_3$ (A –⊖–, D) and (4) following surface modification with octadecyltrimethoxysilane, $CH_3(CH_2)_{17}Si(OCH_3)_3$ and hydrothermal processing (A –■– , E). For each radial structure function, transforms of the $k^3\chi(k)$ (———) and $k^2\chi(k)$ (- - -) data are presented. The range for the forward Fourier transforms was $\Delta k = 2.5 - 9.5$ Å^{-1}

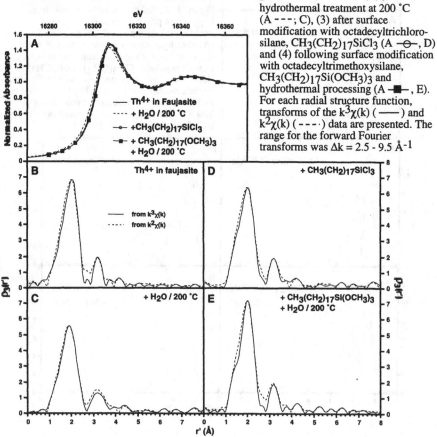

zeolite that contained 2.2 percent aluminum by weight. As found with the uranyl within faujasite, the uranyl structure remains intact after hydrothermal processing of the original exchanged zeolite. This structure is also preserved after addition of octadecyltrimethoxy-silane to the exterior of the zeolite particles. In contrast to faujasite and the earlier observations for uranyl in clay, hydrothermal treatment of the hydrophobic material does not lead to reduction of uranyl to uranium(IV). However, when octadecyltrichlorosilane is used to modify the surface of the zeolite, chloride is found at the equatorial position of the uranyl. This ligand replaces approximately half of the oxygen at the equatorial position. The chloride anion is a by-product of the reaction between the silane and water or hydroxyl groups at the surface of the zeolite. After hydrothermal treatment of this sample, no chloride is found on the uranium within the zeolite. Identical results were obtained from corresponding samples created using a second ZSM-5 zeolite whose aluminum content was 1.7 percent.

The results of similar measurements on thorium(IV) exchanged into faujasite are presented in Figure 4. Both the XANES and EXAFS data indicate that, after addition of a coating using the trimethoxysilane followed by hydrothermal treatment, the structure of the internal thorium is unperturbed. In contrast to the observations with uranyl in the same zeolite, the addition of a film based on the trichlorosilane does not result in a change in the shape or position of the edge, nor is chlorine detected in the coordination sphere of the thorium. However, upon hydrothermal processing of the thorium-exchanged zeolite, a shift in the edge of 0.7 eV to lower energy occurs. A decrease in the observed intensity of the first coordination shell and a broadening of a second peak in the radial distribution function at $r' = 3.2$ Å is also observed. Based on a comparison of the Fourier transforms of $k^2\chi(k)$ and $k^3\chi(k)$, we conclude that the latter feature is probably due to an interaction with the surface of the zeolite and does not reflect aggregation of the thorium. These results may indicate that some hydrolysis of the thorium has occurred.

Attempts to introduce thorium into the ZSM-5 apparently failed. While it is difficult to assess accurately the concentration of an element within a powder sample based on a comparison of edge jumps, we estimate that the thorium content of the exchanged ZSM-5 is less than 5 percent of the corresponding sample created using faujasite. This low concentration precluded the use of XAS to determine the speciation of the thorium that was present. The successful introduction of Th^{4+} into faujasite and our inability to incorporate the ion into ZSM-5 suggests that the cross-sectional size of an aqueous thorium complex is at least that of the pore size of the ZSM-5 but less than that of faujasite.

DISCUSSION

Our results demonstrate that the geometrical structure of zeolites and clays plays a significant role in the reactions of uranyl and thorium(IV) within exchanged and modified aluminosilicates. Within the variable pores of smectite clays, reduction of uranyl occurs readily. The small pores of the zeolites, fixed in size, limit the interactions that lead to changes in oxidation state and prevent subsequent aggregation. For thorium in faujasite, even if hydrolysis of the cation has occurred to some degree, the restricted motion of the cation prevents any coalescence of the actinide species. The differences in reactivity reflect changes in the mobility of the ions, and possibly the reducing agent, within the pores, which results in an entropic difference in the reaction free energy.

The experiments reported here raise two additional issues regarding the properties of uranyl and thorium(IV) within the exchanged and modified zeolites. First, why does the chloride coordinated to the uranyl within ZSM-5 after addition of the trichlorosilane disappear during hydrothermal processing? It is possible that this disappearance reflects loss of uranium from the sample, rather than alteration of the coordination of the actinide ion. In earlier work we demonstrated that, when octadecyltrichlorosilane is used to modify a smectite clay, the uranium can leach from the sample.[1] However, when the coating on the clay is created from the corresponding trimethoxysilane, the uranium remains within the interlayer. For the clays and zeolites, the cations are held within the pores or interlayer by coulombic forces. The replacement of at least two neutral equatorial water molecules by chloride anions results in a neutral or negatively charged complex that can escape, in the presence of bulk water, into solution. Our inability to detect any chloride after hydrothermal treatment can be used to develop hypotheses concerning the presence of mixed aquo-chloro species in these systems.

The second issue in these studies on cation interactions within modified microporous solids is whether the exchanged ions are located in the interior of the aluminosilicate particle or merely adsorbed on the external surface. Our results suggest that the former is the best description for these organic-inorganic composites. The fact that only minor amounts of thorium are found in the ZSM-5 compared to faujasite indicates that pore size affects incorporation, which in turn suggests that the cation passes into the core of the zeolite. The observed differences in the speciation of uranyl within zeolites and clays, after modification of the aluminosilicate surface and hydrothermal treatment, also imply that the cation is in the interior of the aluminosilicate.

CONCLUSION

The results described here demonstrate the variety of reactions that can occur between actinide metals and water and organic materials within microporous aluminosilicates. These extensions of our earlier studies with clays show that the reactivity of actinide cations depends strongly on the structure, and not just the composition, of the mineral interface. The results are suggestive of the complexity of actinide speciation in the environment.

ACKNOWLEDGMENTS

We would like to thank Steven E. Yuchs for aid in the preparation of these samples and S. Skanthakumar for assistance during the acquisition of the XAS data. This work was supported by the U. S. Department of Energy, Office of Basic Energy Sciences-Chemical Sciences (L. S., and D. G.) and Materials Sciences (S. R. W.) under contract W31-109-ENG-38.

REFERENCES

1. Stephen R. Wasserman, Daniel M. Giaquinta, Steven E. Yuchs, and L. Soderholm, in Scientific Basis for Nuclear Waste Management, edited by W. J. Gray and Ines R. Triay (Mater. Res. Soc. Proc. 465, Pittsburgh, PA, 1997) pp. 473-480.

2. Stephen R. Wasserman, L. Soderholm, and Urs Staub, Chem. Mater., 10, 559-566 (1998)

3. D. M. Giaquinta, L. Soderholm, S. E. Yuchs, S. R. Wasserman, Radiochim. Acta, 76, 113-121 (1997).

4. D. M Giaquinta, L. Soderholm, S. E. Yuchs, and S. R. Wasserman, J. Alloys Compounds, 249, 142-145 (1997)

5. S. R. Wasserman. XAMath is available on the World Wide Web at ixs.csrri.iit.edu/database/programs/XAMath.

6. P. G. Allen, J. J. Bucher, D. K. Shuh, N. M. Edelstein, and T. Reich,. Inorg. Chem., 36, 4676-4683 (1997)

EXAFS AND RAMAN STUDIES OF PTMG$_n$:MCl$_2$ AND PTMG/PEG$_n$:MCl$_2$ COMPLEXES (M = Co, Zn)

C.A. Furtado*, A.O. Porto**, G.G. Silva**, R.A. Silva**, M.C. Martins Alves***, P.J. Schilling****, and R. Tittsworth****

*Centro de Desenvolvimento da Tecnologia Nuclear - CDTN/CNEN, C.P. 941, 30123-970, Belo Horizonte, MG, Brazil, clas@urano.cdtn.br
**Instituto de Ciências Exatas, Universidade Federal de Minas Gerais, Belo Horizonte, MG, Brazil
***Laboratório Nacional de Luz Síncrotron – LNLS, Campinas, SP, Brazil
****Center for Advanced Microstructures and Devices – CAMD, Louisiana State University, Baton Rouge, LA, USA

ABSTRACT

Extended x-ray absorption fine structure (EXAFS) and Raman spectroscopy measurements have been performed in a series of liquid polymer electrolytes prepared using poly(tetramethylene glycol) (PTMG), and copolymer poly(tetramethylene glycol/poly(ethylene glycol) (PTMG/PEG), as matrices, and ZnCl$_2$ or CoCl$_2$ as dopants in the concentration range of n = 30 to 90, where n is the molar ratio of Oxigen/Metal cation. EXAFS results have shown the presence of Co-O and Co-Cl coordination shells for PTMG/CoCl$_2$ and PTMG/PEG/CoCl$_2$ systems. Zn-based systems have shown only Zn-Cl bonds in all the concentration range studied. The presence of ZnCl$_2$ and CoCl$_2$ species was confirmed by Raman measurements, by the presence of bands characteristic of Cl-Zn-Cl and Co-Cl stretching modes.

INTRODUCTION

Polymer electrolytes are complex systems obtained from the dissolution of salts into a polymer matrix. They are able to absorb volume variations of electrodes and retain electrochemical and dimensional stability superior to that of liquid electrolytes. These systems are used in electrochemical devices as batteries and capacitors, electrochromic windows and thermoluminescent displays, whose construction depends on the availability of different kinds of specific ions. Considerable progress has been made in the understanding of the conducting process in these systems since the pioneering work of Armand et al.[1]. The interactions between the polymeric chain and the ions are an important parameter related to the behaviour of the ionic conduction mechanism. The polymer has a flexible disordered phase in which ionic species can move through the free volumes existing in the bulk. The ionic transport occurs only in the amorphous region[2]. However, many aspects of the salt solvation are still not very clear[3]. The understanding of the local structure of these systems should allow a better design and development of new materials with higher technological performance.

The local structure around metal cations with atomic numbers $\geq$ 19 can be perfectly investigated with EXAFS, a technique which gives information about the number of neighbours, distances and thermal/static disorder. In the partial-radial distribution function, F(R), the peaks correspond to different atomic shells around the absorber. The Raman technique has been used as

45

a valuable tool to investigate the ion-ion and ion-polymer interaction. Micro-Raman spectroscopy allows the determination of phase separation on a microscale.

Systems based on polyethylene oxide (PEO) and Zn and Co salts have been subject of investigations using these techniques. Linford[4] presented a review of EXAFS studies in PEO based systems. Previous EXAFS results were discussed by Mc Breen et al.[5, 6] and Lathan et al.[7] , and Raman study was obtained by Chintapalli and Frech[8]. We have previously studied liquid[9] and solid[10] electrolytes systems based on PTMG due to the lower glass transition temperature and the higher amorphous character of these matrices in relation to PEO. In this work, EXAFS and Raman spectroscopy measurements have been performed in a serie of liquid polymer electrolytes prepared using poly(tetramethylene glycol) (PTMG) and copolymer poly(tetramethylene glycol - co - ethylene glycol) (PTMG/PEG) as matrices and $ZnCl_2$ or $CoCl_2$ as dopants in the concentration range of n from 30 to 90.

EXPERIMENTAL

The homopolymer PTMG and the copolymer PTMG/PEG are commercial oligomers (Aldrich) with average numeric molecular weight of about 1000 and polydispersivity equal to 2, determined by size exclusion chromatography method. The PTMG/PEG ratio of the copolymer was estimated in 1:0.5 by analysis of ^1H NMR spectra. These polymers are liquid at room temperature. Glass transition temperature, T_g = -90^{0}C, and melting temperature, T_m between 5 and -27^{0}C, for homopolymer and copolymer respectively, were determined by differential scanning calorimetric measurements[9].

Liquid polymer electrolyte samples were prepared by dissolution of $ZnCl_2$ and $CoCl_2.6H_2O$ into the polymeric matrix in the concentration range of n from 30 to 90. In PTMG/$ZnCl_2$ and PTMG/$CoCl_2$ systems, acetone was used as a solvent and PTMG/PEG/salt systems were prepared without solvent. These systems were vacuum-dried at room temperature for 48 hours before use.

EXAFS measurements were performed at Co (7709 eV) and Zn (9659 eV) K edge at CAMD – LSU synchrotron facility running at 1.5 GeV with average current of 150 mA. These data were acquired at room temperature in transmission mode using Nitrogen-filled ion chambers. Monochromatic X-ray beams were obtained using Si (111) and Si (220) double crystal monochromator[11] for cobalt and zinc K edge, respectively. The energy was scanned with a 2 eV and 0.5 eV step in the EXAFS and XANES regions, respectively. Sample mass was chosen to obtain $\Delta\mu t \approx 1$ at the edge jump, where the μ is the absorption coefficient and t is the thickness of the sample. Polymer electrolyte samples were sealed in a steel sample holder with kapton windows.

Raman spectra were recorded in a triple monochromator spectrometer (Dilor XY) equipped with a multi-array CCD detector (Gold). A microscope (Olympus BH-2) was coupled to the spectrometer, allowing a Raman analysis with spatial resolution of about 1 μm (micro-Raman technique). A Krypton laser was used operating in the green line (λ = 514.5 nm) with power of 30 mW.

RESULTS AND DISCUSSION

EXAFS Measurements

EXAFS rough data were treated using Michalowicz chain programs[12]. EXAFS signal was extracted from rough data using standard procedure[13]. Numerical refinement of the EXAFS expression was made using curve wave approximation and theoretical phases (θ_j), and amplitudes (F_j) as calculated by McKale et al.[14].

Fourier-filtered experimental and calculated Fourier transforms ($F(R)$) of k^3-weighted EXAFS signal for some samples are shown in Figures 1 and 2. The structural parameters obtained with fitting procedure are shown in Table I. For each system, in the concentration range mentioned, 2 to 4 compositions were studied. Significant changes were not observed as function of the concentration. Therefore, only the extreme compositions for each system is shown.

For PTMG/PEG/CoCl$_2$ and PTMG/CoCl$_2$, the best fits for theoretical EXAFS functions were obtained when Co-O and Co-Cl coordination shells were considered (Table I). The use of a double shell model was required in the curve fitting procedure for all Co samples.

As seen in Table I, the structural parameters determined for Zn systems are quite different from those obtained for cobalt systems. In the former, only one coordination shell was found related to Zn-Cl interaction. All attempts to fit the data to a double coordination shell consisting of Zn-Cl and Zn-O failed.

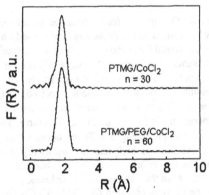

Figure 1 - Experimental (—) and calculated (-----) partial-radial distribution functions ($F(R)$) around Co of PTMG/CoCl$_2$ with n = 30 and PTMG/PEG/CoCl$_2$ with n = 60 obtained at Co K edge.

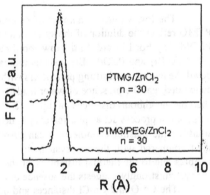

Figure 2 - Experimental (—) and calculated (----) partial-radial distribution functions ($F(R)$) around Zn for PTMG/ZnCl$_2$ with n = 30 and PTMG/PEG/ZnCl$_2$ with n = 30 obtained at Zn K edge.

Table I - Co and Zn coordination shells: number of neighbours, N_j, distances, R_j, Debye-Waller factor, σ_j, and the difference between the experimental and calculated threshold energy, ΔE_0. The R_j and N_j estimated errors are ± 0.02 Å and ± 0.5, respectively.

System		N_j	R_j (Å)	σ_j (Å)	ΔE_0 (eV)
PTMG/CoCl$_2$					
n = 30	Co-O	1.7	2.10	0.110	-13.6
	Co-Cl	1.0	2.23	0.053	-1.7
n = 90	Co-O	0.3	1.76	0.087	-13.6
	Co-Cl	3.1	2.21	0.095	-1.3
PTMG/PEG/CoCl$_2$					
n = 60	Co-O	1.7	2.04	0.009	-14.7
	Co-Cl	0.9	2.29	0.013	-2.2
n = 90	Co-O	1.5	2.06	0.030	-14.3
	Co-Cl	0.9	2.29	0.019	-1.5
PTMG/ZnCl$_2$					
n = 30	Zn-Cl	2.6	2.18	0.076	-1.2
n = 90	Zn-Cl	2.4	2.17	0.068	-0.6
PTMG/PEG/ZnCl$_2$					
n = 30	Zn-Cl	2.4	2.18	0.075	1.02
n = 90	Zn-Cl	2.6	2.18	0.075	-0.5

The low values of number of neighbours and Co-O distances for electrolytes based on PTMG reflect the dilution of the system and that they are similar to the diluted systems based on PEO[4, 15]. For Linford[4], the low coordination number found for metal-O bonds in a series of (PEO)$_8$ZnBr$_2$ and PEO/CoBr$_2$ systems indicates the absence of solvent. This author explored in detail the evolution of fitting parameters with different kinds of drying procedures of samples and concluded that the presence of water leads to a high number of oxygen neighbours (up to 6) and absorber-neighbour distances of up to 2.3 Å. Water and other solvents used in the sample preparation process act as plasticizers promoting an increasing in the polymer chain distances. In addition, the plasticizer molecules can surround the cations producing a different coordination shell compared with the dry system. In our case we can not use high temperature to dry the samples due to the thermal instability of the material. The low coordination number obtained in our system strongly suggests the absence of occluded solvent in the polymer-salt complexes.

The Co-O and Co-Cl distances and number of neighbours obtained for PTMG/CoCl$_2$ are similar to those reported to Linford[4] for the PEO/CoBr$_2$ (n = 30) system, indicating that at low cation concentration PTMG coordinates to metal cations as efficiently as PEO.

The Zn-based systems studied present only Cl as a neigbhour and the Zn-Cl distance in PTMG/ZnCl$_2$ at n = 30 is similar to Zn-Cl distance in PEO/ZnCl$_2$ at n = 8. Bandara et al.[15] observed that for PEO$_n$/MBr$_2$ systems the number of oxygen neighbours decreases as the atomic number increases from cobalt to zinc and reported considerable evidence that the (PEO)$_8$ZnBr$_2$ system contains neutral ZnBr$_2$ species.

Raman spectroscopy

Figs. 3 and 4 show the typical Raman band profile for the polymer/$ZnCl_2$ and polymer/$CoCl_2$ systems at different compositions, in the low frequency region. The spectra were normalized using the intensity of the Raman peak of the pure polymers at $\omega = 840$ cm^{-1} as reference. This peak is associated with the rocking of the C-H bonds and is not affected by the introduction of the salt.

With the introduction of the salt it was observed a band at $\omega = 305$ cm^{-1} with line width $\Delta\omega = 18$ cm^{-1} in the polymer/$ZnCl_2$ systems (Fig. 3), in addition to the polymer bands[9]. This band is assigned to the Cl-Zn-Cl symmetric vibration of ZnA_2 species in solution[16]. Its intensity increases with increasing concentration.

For the polymer/$CoCl_2$ system (Fig. 4), the weak band at approximately 300 cm^{-1} may be associated to Co-Cl stretching in solution[17]. The band localized at $\omega = 515$ cm^{-1} is present at the same frequency in the spectrum of the pure salt $CoCl_2.6H_2O$, indicating the precipitation of salt microcrystals.

It was not possible to identify Raman bands or changes in the spectra between 100 and 1600 cm^{-1} due to cation-polymer chain interaction, probably related to the low concentration range studied. These results characterize the weak interaction between Zn^{2+} and Co^{2+}-polymer chain in systems based on PTMG. Raman results confirm the EXAFS results previously obtained, clearly showing that for the Zn^{2+}-polymer system there is no interaction between the metallic ion and the polymeric chain. Our results are in accord with those of other authors. Raman studies of Chintapalli and Frech[8] shown that in liquid $(PPO)_n ZnBr_2$ (n = 20, 40 ,60) system, the formation of $ZnBr_2$ clusters and ionic pairs is very frequent due to the weak interaction between the cation and the polymer chain. Mendolia et al.[18] measured conductivities of PTMG/$CoBr_2$ and PEG/$CoBr_2$ systems and observed low conductivity values and the predominance of tetrahedral neutral species $CoBr_2L_2$ (L = ligand) in PTMG samples in contrast with PEG systems. They attributed this fact to the stability of the five-membered chelate ring (-Co-O-CH_2CH_2-O-) formed in PEG/cobalt systems. On the other hand, in PTMG/cobalt systems, a weaker seven-membered chelate ring (-Co-O-CH_2-CH_2-CH_2-CH_2-O-) is formed.

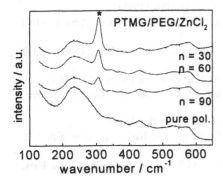

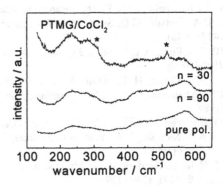

Figure 3 - Normalized Raman spectra for the PTMG/PEG/$ZnCl_2$ system at different compositions

Figure 4 - Normalized Raman spectra for the PTMG/$CoCl_2$ system at different compositions

CONCLUSION

EXAFS and Raman spectroscopic studies for the PTMG/CoCl$_2$, PTMG/PEG/CoCl$_2$, PTMG/ZnCl$_2$ and PTMG/PEG/ZnCl$_2$ systems show that, even in diluted solution, these complexes exist mostly as undissociated ZnCl$_2$ and CoCl$_2$ species due to the formation of the less effective seven-membered chelate ring for PTMG systems. This weaker cation-polymer interaction prevents the displacement of the anions from the first coordination sphere of the metal ion and as a result neutral species predominate. CoCl$_2$.6H$_2$O shows a slight higher tendency to associate with ether oxygen and EXAFS measurements were able to detect the Co- polymer chain coordination.

ACKNOWLEDGEMENTS

This work has been supported by the Brazilian Agencies FAPEMIG, CNPq and PADCT. Part of this work were developed at DCM – beamline at CAMD – LSU - USA. Special thanks are given to Prof. M. A. Pimenta for the use of the Raman Laboratory at UFMG – MG - Brazil.

REFERENCES

1. M.B. Armand, J.M. Chabagno and M.J. Duclot, *Fast Ion Transport in Solids*, edited by. P. Vashista, J.N. Mundy and G.K.Shenoy (Elsevier North Holland, New York, 1979), p. 131.
2. C. Berthier, W. Gorecki, M. Minier, M. Armand, J.M. Chabagno and P. Rigaud, Solid State Ionics **11**, 91 (1983).
3. W.H. Meyer, Adv. Mater. **10** (6), 439 (1998).
4. R.G. Linford, Chemical Society Reviews, *1995*, 267.
5. J. McBreen, X.Q. Yang, H.S. Lee, Y. Okamoto, J. Electrochem. Soc. **143** (10), 3198 (1996).
6. J. McBreen, X.Q. Yang, H.S. Lee, Y. Okamoto, Electrochim. Acta **40** (13-14), 2115 (1995).
7. R.J. Lathan, R.G. Linford, R. Pynenburg, W.S. Schilindwein, Electrochim. Acta **37** (9), 1529 (1992); R.J. Lathan, R.G. Linford, W.S. Schilindwein, Faraday Discuss. Chem. Soc. **88**, 103 (1989).
8. S. Chintapalli and R. Frech, Electrochim. Acta **40** (13-14), 2093 (1995).
9. C.A. Furtado, G.G. Silva, M.A. Pimenta and J.C. Machado, Electrochim. Acta **43** (10-11), 1477 (1998).
10. C.A. Furtado, G.G. Silva, J.C. Machado, M.A. Pimenta, R.A. Silva, J. Phys. Chem. **103** (34), 7102 (1999).
11. M.C. Correa, H. Tolentino, A. Craievich and C. Cusatis, Rev. Sci. Inst. **63**, 896 (1992).
12. A.Michalowicz, Notice d'Utilisation des Programmes EXAFS pour le Macintosh, Fortran, 77 (1989).
13. B.K. Teo, *EXAFS: Basic Principles and Data Analysis* (Springler, Berlin, 1986).
14. A.G. McKale, B.W. Veal, A.P. Paulikas, S.K. Chen and G.S. Knapp, J. Amer. Chem. Soc. **110**, 3763 (1988).
15. H. M. N. Bandara, W. S. Schindwein, R. J. Latham and R. G. Linford, J. Chem. Soc. Faraday Trans. **90** (23), 3549 (1994).
16. D.F.C. Morris, E.L. Short, D.N. Waters, J. Inorg. Nucl. Chem. **25**, 975 (1963).
17. G.W. Watt, D.S. Klett, Inorg. Chem. **3** (5), 782 (1964).
18. M.S. Mendolia and G.C. Farrington, Electrochim. Acta **37** (9), 1695 (1995).

CHARGE TRANSFER AND LOCAL STRUCTURE IN THERMOELECTRIC GERMANIUM CLATHRATES

A.E.C. PALMQVIST*, B.B. IVERSEN**, L.R. FURENLID***, G.S. NOLAS****,
D. BRYAN*****, S. LATTURNER*****, G.D. STUCKY*****

*Department of Applied Surface Chemistry, Chalmers University of Technology, SE-412 96 Göteborg, Sweden, anders.palmqvist@surfchem.chalmers.se
**Department of Chemistry, University of Aarhus, DK-8000 Aarhus C, Denmark
***Department of Radiology, University of Arizona, Tucson, AZ 85721
****Marlow Industries Inc., Dallas, TX 75238-1645
*****Department of Chemistry, University of California, Santa Barbara, CA 93106

ABSTRACT

Germanium clathrates have recently received attention as potential highly efficient thermoelectric materials based on the phonon glass electron crystal (PGEC) concept [1-4]. A combined EXAFS and XANES study has been performed in order to investigate the local structure and the degree of charge transfer between the guest atoms and the framework atoms. Analysis of the Sr, Ga, and Ge K-edge XANES spectra of a $Sr_8Ga_{16}Ge_{30}$ clathrate reveals that the atoms are close to neutrally charged and that the degree of charge transfer is low in agreement with recent theoretical predictions and charge density distribution measurements [3,5].

INTRODUCTION

Thermoelectricity is a unique phenomenon of energy conversion suitable for power generation and cooling applications. In general thermoelectric (TE) materials are used in applications where advantages such as small size, simplicity, no moving parts, low pollution risk and high reliability outweigh their current higher cost and lower efficiency compared to the conventional power generators and compressor-based coolers [6].

The parameter that determines the dependence of device efficiency upon material properties is the dimensionless thermoelectric figure of merit $ZT = TS^2\sigma/\kappa$. In this equation S is the Seebeck coefficient, σ and κ are electrical and thermal conductivity, respectively, where the latter consists of electronic κ_e and lattice κ_l components. To develop a TE device of high efficiency, the material should have large values of both S and σ and a small value of κ. The state of the art materials have ZT's of around one, and some well known examples are the TAGS alloys $((AgSbTe_2)_{1-x}(GeTe)_x)$, Bi_2Te_3 and the Bi-Sb alloys. Recently TE materials have received a lot of attention spurred by theoretical and experimental developments as well as by growth in commercial markets for the TE technology. New materials concepts based on structures that have phonon glass electron crystal (PGEC) properties, e.g. skutterudites and clathrate structures have been reported and show great promise for high thermoelectric performance [1,2,7]. The PGEC concept can be realized in materials that have crystal structures with low values of κ_l, achieved by the efficient scattering of phonons by anharmonically vibrating atoms in the structure, and an electronic conductivity that is not affected by this atomic "rattling".

Mat. Res. Soc. Symp. Proc. Vol. 590 © 2000 Materials Research Society

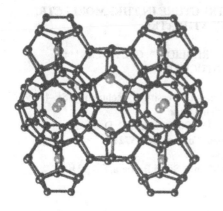

Figure 1.
The structure of clathrate $Sr_8Ga_{16}Ge_{30}$, where dark atoms are Ga or Ge, and light atoms are Sr. The two cage types (dodecahedron and tetrakaidecahedron) are present in a 2:6 ratio in the unit cell.

The germanium based clathrate Type I with a composition of $Sr_8Ga_{16}Ge_{30}$ was first reported in 1986 [8]. The structure consists of two types of cages, the 20-atom dodecahedron and the 24-atom tetrakaidecahedron, which are present in a 2:6 ratio in the unit cell as shown in Figure 1. The cage frameworks are built up from the germanium and gallium atoms, and the alkaline earth atoms reside inside the cages as weakly bound guests. There is a large number of elemental compositions that can form this structure since frameworks of all but the lightest and the heaviest of the elements in groups IIIB, IVB, and VB have been made with guests from all but the lightest and heaviest elements of groups IA, IIA, and VIIB [8-13]. Clathrates containing transition metal elements from groups IB and IIB have also been prepared [14]. There are for example clathrates with the compositions $A_8B_{44}\square_2$ (A=IA, B=IVB, $\square$=vacancy) [9,10], $A_8B_{16}C_{30}$ (A=IIA, B=IIIB, C=IVB) [8], $A_8B_8C_{38}$ (A=VIIB, B=VB, C=IVB) [11], $A_8B_3C_{32}D_{11}$, $A_8B_4C_{30}D_{12}$, and $A_8B_{12}C_{14}D_{20}$ (A=VIIB, B=IIIB, C=IVB, D=VB) [11], $A_8B_8C_{38}$ (A=IIA, B=IIB, C=IVB) [12], and even compositions like $A_8B_6C_{40}$ (A=IIA, B=IB, C=IVB) [13], and $A_8B_{15.5}C_{30.5}$ (A=IIA, B=IB, C=VB) [14]. Common to most reported compositions of clathrates is the sum of the number of valence electrons of the elements present. Based on this empirical fact, it is tempting to assume that, in order for the structures to form, it is necessary that the guest atoms in the cages donate or accept valence electrons to or from the framework atoms so that all the framework atoms have four valence electrons.

A number of studies have shown that the weakly bound guest atoms exhibit extreme motion inside the oversized cavities. This motion is contributing to the effective reduction of the thermal conductivity of the clathrate materials, which is important for obtaining a high ZT. However, ZT will not improve if the "rattling" behavior of the guest atoms also lowers the electrical conductivity. If there is a large degree of charge transfer between the guest and framework atoms, it is likely that the rattling affects the electrical conductivity so that the clathrates are less interesting candidates as good TE materials [3]. Nevertheless several clathrate compositions have been reported with ZT around 0.2-0.3, which constitutes some of the most promising non-optimized structures discovered so far. Recent theoretical calculations surprisingly show, that contrary to common belief, the Sr guest atoms in $Sr_8Ga_{16}Ge_{30}$ seem to be close to neutral, and that the electrical conduction takes place through the framework atoms and not through the Sr "wires" [3]. In this paper we examine experimentally the valence states of the guest and host atoms, and the local structure of some promising germanium clathrates based on X-ray absorption spectroscopy (XANES and EXAFS) studies.

EXPERIMENTAL

Structure and Composition

The solid state synthesis of the compounds studied in this work has been previously described elsewhere [15]. The structure and composition of the crystals have been determined by both single crystal and powder X-ray diffraction at room temperature on a Bruker SMART CCD single crystal diffractometer and a Scintag powder diffractometer. The composition has also been determined by inductively coupled plasma atomic emission spectroscopy ICP-AES.

X-ray Absorption

X-ray Absorption Near Edge Structure (XANES) and Extended X-ray Absorption Fine Structure (EXAFS) measurements were carried out at beamline X10C at the National Synchrotron Light Source (NSLS), Brookhaven National Laboratory. The K-edges of Ga, Ge and Sr were studied in transmission mode on powdered samples evenly smeared onto a scotch tape which was folded into 8 layers of sample to ensure an even sample distribution without pinholes. Elemental as well as oxidized standards (sealed under argon in aluminized Mylar film if air sensitive) were used as references for the photon energy scale at the absorption edge and as reference compounds for the local structure analysis of the different atoms. The data were analyzed using the MacXAFS3.6 software package [16], which is based on the UWXAFS codes and includes the FEFF 3.1 program for calculation of EXAFS spectra of reference compounds. The energy of the absorption edges was determined at the maximum value of the derivative of the absorption curve.

RESULTS & DISCUSSION

XANES Analysis

Sr K-edge

Figure 2 shows the Sr K-edge X-ray absorption edge of the clathrate $Sr_8Ga_{16}Ge_{30}$ compared to those of elemental Sr and $Sr(OH)_2 \cdot 8H_2O$. The Sr in the clathrate has an absorption edge energy which is ca. 1eV higher than that of elemental Sr, and ca. 7eV lower than that of Sr(II) in the hydroxide. The values of the absorption edges are given in Table I.

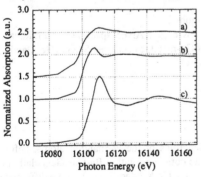

Figure 2.
Sr K-edge XANES spectra of a) Sr-metal and b) clathrate $Sr_8Ga_{16}Ge_{30}$ compared to a reference of c) $Sr(OH)_2 \cdot 8H_2O$. The intensity of the absorption spectra has been normalized.

Table I. Absorption edge energies at derivative maximum of clathrate $Sr_8Ga_{16}Ge_{30}$ compared to elemental and oxidized standards of Sr, Ga and Ge.

Compound	Sr K-edge (eV)	Ge K-edge (eV)	Ga K-edge (eV)
$Sr_8Ga_{16}Ge_{30}$	16100	11102	10366
Metallic standard	16099 (Sr metal)	11102 (Ge metal)	10365 (Ga metal)
Oxidized standard	16107 ($Sr(OH)_2 \cdot 8H_2O$)	11107 (GeO_2)	10370 (Ga_2O_3)

Ge K-edge

The X-ray absorption edge of the clathrate $Sr_8Ga_{16}Ge_{30}$ at the Ge K-edge is given in Figure 3 and compared with those of elemental Ge and GeO_2. It is clear that the oxidation state of the Ge in the clathrate resembles more that of the elemental Ge than that of GeO_2. The value of the absorption edge energy is the same as that of elemental Ge, and 5eV lower than that of GeO_2 as given in Table I indicating a neutral oxidation state of the Ge atoms.

Ga K-edge

Figure 4 shows the X-ray absorption at the Ga K-edge of the clathrate $Sr_8Ga_{16}Ge_{30}$ compared to elemental Ga and Ga_2O_3. Also in this case, the oxidation state of the atoms in the clathrate shows to be closer to that of the elemental form when comparing absorption edge energies. The energy of the clathrate´s edge is 1eV higher than that of the Ga metal and 4eV lower than for Ga_2O_3 as shown in Table I indicating a slightly positive Ga oxidation state.

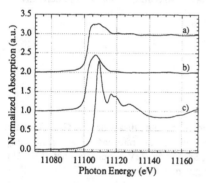

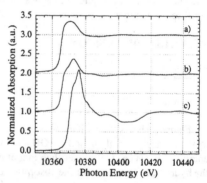

Figure 3
Ge K-edge XANES spectra of a) a reference of Ge-metal, b) clathrate $Sr_8Ga_{16}Ge_{30}$, and c) GeO_2.

Figure 4
Ga K-edge XANES spectra of a) a reference of Ga-metal, b) clathrate $Sr_8Ga_{16}Ge_{30}$, and c) Ga_2O_3.

EXAFS Analysis

Ge K-edge

Figure 5 shows the radial distance distribution functions of the germanium atoms, before correction for phase shifts, in the $Sr_8Ga_{16}Ge_{30}$ clathrate compared with a $Ba_8Ga_{16}Ge_{30}$ clathrate and Ge-metal as calculated from the EXAFS analysis of the Ge K-edge. There is some difference

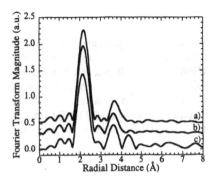

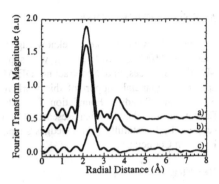

Figure 5
k^2 weighted Fourier transforms without phase shift corrections giving the radial distance distribution functions centered around germanium in the clathrates a) $Ba_8Ga_{16}Ge_{30}$, and b) $Sr_8Ga_{16}Ge_{30}$, compared with c) Ge-metal as determined by EXAFS analysis of the Ge K-edge.

Figure 6
k^2 weighted Fourier transforms without phase shift corrections giving the radial distance distribution functions centered around gallium in the clathrates a) $Ba_8Ga_{16}Ge_{30}$, and b) $Sr_8Ga_{16}Ge_{30}$, compared with c) Ga-metal as determined by EXAFS analysis of the Ga K-edge.

between the Sr clathrate and the Ba clathrate, especially on the second coordination sphere, which is at a 0.15 Å shorter distance in the Sr sample than in the Ba sample. The first coordination sphere of the Sr sample is at a slightly shorter distance than that of the Ba sample.

Ga K-edge

The radial distance distribution functions centered around the Ga atoms, before correction for phase shifts, resulting from the EXAFS analysis of the Ga-K edge of the two clathrate samples and the Ga-metal are shown in Figure 6. The Sr containing sample has a 0.15 Å shorter distance in this case, too, to the second coordination sphere compared to the Ba clathrate, whereas the distance to the first coordination sphere is virtually the same in the two samples.

CONCLUSIONS

The degree of charge transfer between host and guest atoms in the thermoelectric $Sr_8Ga_{16}Ge_{30}$ clathrate has been studied using XANES analysis. The analysis of the Sr K-edge suggests that the Sr atoms keep their valence electrons to a large extent and have an oxidation state that is essentially zero since the absorption edge energy is closer to that of elemental strontium than to that of $Sr(OH)_2 \cdot 8H_2O$. This low degree of charge transfer is confirmed by the XANES analysis of the Ga and Ge K-edges of the same compound. The Ge atoms in the clathrate have the same absorption edge energy as Ge-metal, whereas the Ga atoms seem very slightly positively charged. Radial distance distribution functions centered on the Ge and Ga atoms were obtained by EXAFS analysis of the $Sr_8Ga_{16}Ge_{30}$ clathrate and compared with those of a $Ba_8Ga_{16}Ge_{30}$ clathrate. It was found that the first coordination spheres of both Ga and Ge atoms are very similar in the two compounds, but that the second coordination sphere is at a 0.15 Å longer distance for both the Ga and Ge atoms in $Ba_8Ga_{16}Ge_{30}$ compared to $Sr_8Ga_{16}Ge_{30}$.

ACKNOWLEDGEMENTS

The authors would like to acknowledge the assistance of the beam personnel and the use of beamline X10C at NSLS, which was supported by the U.S. Dept. of Energy, Division of Materials Sciences, under contract no. DE-AC02-98CH10886. GDS, DB, SL, and AECP greatly appreciate financial support of this work from the Office of Naval Research. AECP thanks STINT, the Swedish Foundation for International Cooperation in Research and Higher Education, and the Swedish Research Council for Engineering Sciences (TFR) for financial support. BBI gratefully acknowledges support from the Carlsberg Foundation and the Danish Research Councils.

REFERENCES

1. G.A. Slack in *CRC Handbook of Thermoelectrics*, edited by D.M. Rowe, CRC Press, Boca Raton, 1995, pp. 407-440.
2. G.S. Nolas, J.L. Cohn, G.A. Slack, S.B. Schujman, Appl. Phys. Lett., **73**, 178-180 (1998).
3. N.P. Blake, L. Møllnitz, G. Kresse, H. Metiu, J. Chem. Phys., **111**, 3133-3144 (1999).
4. B.B. Iversen, A.E.C. Palmqvist, D.E. Cox, G.S. Nolas, G.D. Stucky, N.P. Blake, H. Metiu, J. Solid State Chem., *in press*.
5. B.B. Iversen, A. Bentien, A.E.C. Palmqvist, D. Bryan, S. Latturner, G.D. Stucky, N. Blake, H. Metiu, G.S. Nolas, D. Cox, in *Applications of Synchrotron Radiation Techniques to Materials Science V*, edited by S.R. Stock, D.L. Perry, S.M. Mini, MRS 1999 Fall Meeting Symposium Proceedings.
6. D.M. Rowe in *CRC Handbook of Thermoelectrics*, edited by D.M. Rowe, CRC Press, Boca Raton, 1995, pp. 1-5.
7. D. Mandrus, B.C. Sales, V. Keppens, B.C. Chakoumakos, P. Dai, L.A. Boatner, R.K. Williams, J.P. Thompson, T.W. Darling, A. Migliori, M.B. Maple, D.A. Gajewski, E.J. Freeman, Mat. Res. Soc. Symp. Proc. Vol. **478**, p. 199 (1997).
8. B. Eisenmann, H. Schäfer, R. Zagler, J. Less-Comm. Met., **118**, 43-55 (1986).
9. J.S. Kasper, P. Hagenmuller, M. Pouchard, C. Cros, Science, **150**, 1713-1714 (1965).
10. J.-T. Zhao, J.D. Corbett, Inorg. Chem., **33**, 5721-5726 (1994).
11. v. H. Menke, H.G. von Schnering, Z. Anorg. Allg. Chem. **395**, 223-238 (1973); H.G. von Schnering, H. Menke, Z. Anorg, Allg. Chem. **424**, 108-114 (1976).
12. A. Czybulka, B. Kuhl, H.-U. Schuster, Z. Anorg. Allg, Chem. **594**, 23-28 (1991).
13. G. Cordier, P. Woll, J. Less-Comm. Met., **169**, 291-302 (1991).
14. J. Dünner, A. Mewis, Z. Anorg. Allg. Chem., **621**, 191-196 (1995).
15. J.D. Bryan, V.I. Srdanov, G.D. Stucky, D. Schmidt, Physical Review B, **60**, 3064-3067 (1999).
16. MacXAFS3.6 Software Package, C.E. Bouldin, W.T. Elam, L. Furenlid, Physica B, **208&209**, 190-192 (1995).

A HIGH-RESOLUTION ANGLE-RESOLVED PHOTOEMISSION STUDY OF THE TEMPERATURE DEPENDENT ELECTRONIC STRUCTURE OF THE PENTATELLURIDE ZrTe5

D. N. McIlroy*, Daqing Zhang*, Bradley Kempton*, J. Wharton*, R. T. Littleton**,
T. M. Tritt**, and C. G. Olson†
*Department of Physics, University of Idaho, Moscow, ID 83844-0903
**Department of Physics, Clemson University, Clemson, SC 29634-1911
†Ames Laboratory, Iowa State University, Ames, IA 50011

ABSTRACT

The temperature dependent band structure of the pentatelluride ZrTe5 has been examined using the technique of high-resolution angle-resolved photoemission in conjunction with synchrotron radiation. Specifically, the band dispersion along the X-Γ-X high symmetry axis was mapped at 20K (T<Tc) and 170K (T>Tc), where Tc≃ 160K. One electron band and two hole bands within 0.5eV of the Fermi level have been identified. The dispersion of the bands indicates that they have extremely small electron and hole effective masses. The hole bands were observed to distort and shift in energy upon raising the sample temperature above Tc, indicative of a surface distortion.

I. INTRODUCTION

Beginning in the mid-1980's with the discovery of high temperature superconductivity in the cuparates, low dimensional (2-D) and quasi-low dimension materials have received considerable attention. The majority of these systems are oxides that exhibit novel behavior such as superconductivity [1-7], colossal magnetoresistance (CMR) [8-12], and charge density wave [13-16] formation [13-16]. The aforementioned behavior is typically accompanied by a metal-nonmetal or semimetal-nonmetal transition associated with either the formation of gaps [1-3,5,6,7,12] or changes in the widths of gaps [8,9]. Many low dimensional systems are inherently unstable, which is related to correlation and exchange effects, and consequently experience structural instabilities such as Peierls distortions [13,14], Jahn-Teller distortions [16], or CDW distortions [15]. The technique of photoemission has been instrumental in furthering our understanding of nonmetal to metal transitions of low dimensional and quasi-low dimensional systems [1-24], as well as the fundamental physics driving the transitions.

The pentatellurides (ATe5, A = Zr, Hf, etc.) are a family of quasi two-dimensional materials. The crystal lattice of the pentatellurides is orthorhombic and belongs to the space group Cmcm. The unit cell contains 24 atoms and the general structure of the pentatellurides can be viewed as planes comprised of linear chains, where the planes are weakly held together by Van der Waals forces. In a similar vein as the low dimensional systems previously discussed, the pentatellurides exhibit novel transport behavior expected of low dimensional materials. In particular, the resistivity of these materials is highly anisotropic depending on the direction of conduction, i.e. along the chains, perpendicular to the chains, or perpendicular to the planes [25,26]. In addition, an anomalous peak has been observed in the resistivity as a function of temperature, reminiscent of the perovskites [9], and suggests that the pentatellurides may exhibit a nonmetal to metal transition. While this in itself is intriguing, these materials also exhibit promising thermoelectric properties [25,26]. A temperature dependent transition in their thermopower from electrons as the primary charge carrier to holes as the primary carrier corresponds to the anomalous behavior in the resistivity. The resistivity and thermopower of ZrTe5 as a function of temperature is presented in Fig. 1. It is clear from Fig.1 that the two effects are coupled. It has been proposed that the anomalies in the transport data shown in Fig.1 are a consequence of charge density wave formation [25,26]. However, to date all efforts to observe charge density wave formations have been unsuccessful.

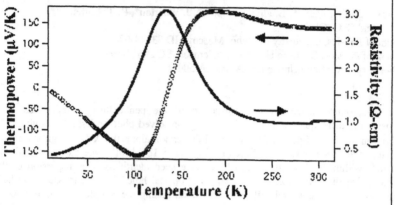

Figure 1. The absolute thermopower (α) and relative resistivity, R(T)/R(300K) of ZrTe$_5$ as a function of temperature.

In order to understand the transport properties of pentatellurides we have performed temperature dependent high-resolution angle-resolved photoemission spectroscopy (ARPES) on ZrTe$_5$. The goal of this work is to probe the microscopic electronic structure, i.e. the density of states, band structure, etc., and either identify the presence of CDW formation or identify another mechanism responsible for the transport properties of ZrTe$_5$. This work has focused primarily on the density of states in the vicinity of the Fermi level since it is the near Fermi region of the band structure that dictates the bulk transport properties. In this paper we present the results of these ARPES experiments, as well as conclusions related to the transport properties of pentatellurides based on the ARPES data.

II. EXPERIMENTAL DETAILS AND RESULTS

The ARPES measurements were conducted on single crystal ZrTe$_5$ whiskers, where the details of crystal growth are described elsewhere [27]. The measurements were performed at the Ames/Montana beamline at the Synchrotron Radiation Center at the University of Wisconsin-Madison. The samples were attached to a coldfinger cooled by a closed-cycle helium refrigerator with an absolute temperature of 20K, which in turn was inserted into a ultra-high vacuum chamber with a base pressure of 3×10^{-11} Torr. The samples were cleaved *in situ* exposing the a-b plane (defined as the a-c plane in refs. 25 and 26). The a-b planes are the cleavage planes, which bond via Van der Waals forces. In addition, the chain structures discussed earlier lie in the a-b planes. The energy distribution curves (EDC) of ZrTe$_5$ in the vicinity of the Fermi level was acquired with monochromatic photons of an energy hv = 18 eV. All photoemission spectra were normalized to the photon flux and referenced to the Fermi level of a sputtered Pt foil in electrical contact with the sample. The instrumental resolution (ΔE) was 50 meV and determined from the width of the Fermi level of the Pt foil acquired at 20K. The angular resolution of the spectrometer was $\pm1°$. Photoemission spectra were acquired along the Γ-X (a-dir.) and the Γ-X' (b-dir.) high symmetry axis by rotating the analyzer at 20K and

170K, although, only the data acquired along the Γ-X direction will be discussed in this paper for the sake of brevity.

In Fig.2 we present the EDC's of ZrTe₅ acquired about the zone center along the X-Γ-X high symmetry axis in 1° steps at 20K (T<T$_c$; T$_c$ ≃ 160K) and 170K (T>T$_c$), respectively.

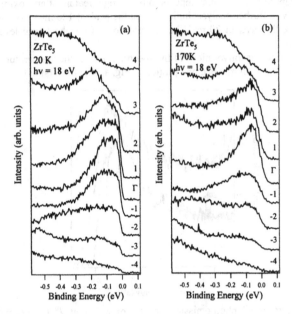

Figure 2. Angle–resolved photoemission spectra of ZrTe₅ about Γ along the Γ-X high symmetry direction at (a) 20 K and (b) 170 K, where the numbering on each spectrum represents the position of the analyzer relative to the surface normal.

Examination of Fig. 2 reveals an asymmetry in the intensity of the spectral features about the zone center (Γ). This has been attributed to cross section effects, i.e. selection rules, associated with the polarization of the synchrotron radiation relative to the sample orientation [28]. This effect has also been observed in the X'-Γ-X' direction, as well as for other pentatelluride samples examined to date by the authors. The EDC acquired at 20K at the Γ point in Fig. 2(a) consists of a single 250 meV wide peak, yet is quickly seen to broaden into multiple peaks within only a few degrees off normal. Deconvolution of the data was performed using Gaussian peaks rather than Voigt functions since the authors were only concerned with peak positions rather than lifetime broadening. Attempts to deconvolute the spectra in Fig. 2 with two Gaussian peaks on top of an integrated background yielded poor fits. However, better success was obtained by using three Gaussians. This deconvolution of the band structure in the vicinity of the Fermi level into three states is consistent with Fermi surface measurements by Kamm et al. [25], where three distinct states were identified utilizing the Shubnikov-de Haas effect. Based on the valence of Zr ([Kr]4d²5s²) and Te (([Kr]4d¹⁰5s²5p⁴) we can conclude that the states at the Fermi level consist of hybridized 4d bands of Zr and 5p bands of Te.

A comparison of the two EDC's at Γ in Fig. 2 reveal a narrowing of the convoluted peak width from approximately 250 meV at 20K to 150 meV at 170K. To emphasize the effects of temperature on the band structure, spectra acquired at Γ+2° at 20K and 170K have been plotted together in Fig. 3. What was once a shoulder at 70 meV and a peak at 120 meV at T= 20K has collapsed to form what appears to be a single feature at approximately 100 meV at T = 170K. However, further analysis indicates that the peak at 120 meV has shifted towards the Fermi level while the peak at 70 meV has shifted down from the Fermi level, indicative of a surface distortion.

Based upon the deconvolutions of the spectra in Fig. 2, the temperature dependent band structure of $ZrTe_5$ has been mapped and plotted in Fig. 4. The dispersion of the band closest to

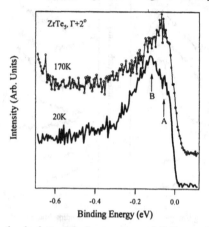

Figure 3. Angle-resolved photoemission spectra of $ZrTe_5$ at 2° off normal along the Γ-X direction acquired at 20K and 170K, respectively.

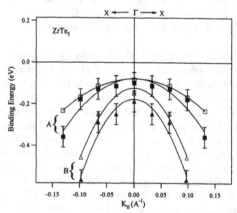

Figure 4. The dispersion of the two hole bands of $ZrTe_5$ based on the deconvolution of the spectra in Fig.2. The closed symbols correspond to data acquired at T=20K and the open symbols for data acquired at 170K. The solid lines represent the parabolic fits of the bands.

the Fermi level has been omitted from Fig. 4 due to difficulties in quantifying the dispersion. The results of the fits of the spectra in Fig. 4 for this band produce a dispersionless peak that quickly decreases in intensity away from the zone center. The authors are of the opinion that the decreasing intensity of the upper band indicates that it is an electron band that quickly disperses across the Fermi level. The other two bands are clearly hole bands based on their downward dispersion away from the zone center. We have concluded that the band closest to the Fermi level is an electron band and the two higher binding energy bands (labeled A and B in Fig. 4) are hole bands. These assignments contradict the previous assignments of these states of $ZrTe_5$ by Kamm et al. [25] as two electron bands and one hole band, based on the Shubnikov-de Hass effect.

The effects of temperature on the band structure in Fig. 4 are quite pronounced and can be summarized as follows: Upon raising the temperature from 20K to 170K the band labeled A shifts to higher binding energy while the band labeled B shifts to lower binding energy. In addition, the dispersion of both bands decreases above the transition temperature, i.e. larger hole effective masses. The observed distortion of the band structure in Fig. 4 suggests the presence of a structural distortion. A CDW would result in a doubling of the unit cell, which would correspond to a reduction of the surface Brillouin zone (SBZ) by half. If this were the case significantly larger dispersions for the two hole bands due to the reduction of the SBZ would be observed. Since the distortion of the bands across the temperature dependent transition does not correspond to a doubling of the surface unit cell, a CDW model must be ruled out in favor of a Jahn-Teller [16] or Peierls-like distortion [13,14]. Further studies of the surface structure of $ZrTe_5$ across the temperature dependent transition are in order, as well as detailed calculations of the band structure. While the authors have suggested that a structural distortion may occur, care must be taken since the observed effects could also be attributed to changes in electron correlation across the transition.

III. CONCLUSIONS

The temperature dependent band structure of the quasi two-dimensional pentatelluride $ZrTe_5$ has been determined by high-resolution angle-resolved photoemission spectroscopy. The density of states in the vicinity of the Fermi level consists of two hole bands and one electron band, contrary to earlier assignments. Distortions of the hole bands, as well as band shifts, were observed upon raising the sample temperature above T_c ($\simeq$ 160K). The flattening of the bands indicate that the effective masses of the holes increase with elevated temperatures. Taken in conjunction with the band shifts, it has been concluded that a surface distortion occurs. The experimental results preclude CDW formation. Consequently, it has been suggested that the indirectly observed distortions may be better described in terms of Jahn-Teller or Peierls-like distortions.

ACKNOWLEDGEMENTS
The work conducted by D.N.M was partially supported by the NASA Idaho Space Grant Consortium and a grant from the Research Council of the University of Idaho. This work was based on research conducted at the Synchrotron Radiation Center, University of Wisconsin-Madison, which is supported by the NSF under Award No. DMR-95310009.

REFERENCES
[1]. J. E. Demuth, B. N. J. Persson, F. Holtzberg, and C. V. Chandrasekhar, Phys. Rev. Lett. 64, 603 (1990).
[2]. M. R. Norman, H. Ding, M. Randeria, J. C. Campuzano, T. Yokoya, T. Takeuchi, T. Mochiku, K. Kadowaki, P. Guptasarma, and D. G. Hinks, Nature 392, 157 (1998).

[3].   Yonghong Li, Jin Lin Huang, and Charles M. Lieber, Phys. Rev. Lett. 68, 3240 (1992).

[4].   D. L. Mills, R. B. Phelps, and L. L. Kesmodel, Phys. Rev. B 50, 6394 (1994).

[5].   H. Romberg, M. Alexander, N. Nücker, P. Adelmann, and J. Fink, Phys. Rev. B 42, 8768 (1990).

[6].   J. W. Allen, C. G. Olson, M. B. Maple, J.-S. Kang, L. Z. Liu, J.-H. Park, R. O. Anderson, W. P. Ellis, J. T. Markert, Y. Dalichaouch, R. Liu, Phys. Rev. Lett. 64, 595 (1990).

[7].   Z. X. Shen, D. S. Dessau, B. O. Wells, D. M. King, W. E. Spicer, A. J. Arko, D.Marshall, L. W. Lombardo, A. Kapitulnik, P. Dickinson, S. Doniach, J. DiCarlo, A. G. Loeser, C. H. Park, Phys. Rev. Lett. 70, 1553 (1993).

[8].   D. N. McIlroy, J. Zhang, S.-H. Liou, and P. A. Dowben, Phys. Lett. A 207, 367 (1995).

[9].   D. N. McIlroy, C. Waldfried, Jiandi Zhang, J.-W. Choi, F. Foong, S.-H. Liou, and P. A. Dowben, Phys. Rev. B 54, 17438 (1996).

[10].   J.-H. Park, C. T. Chen, S.-W. Cheong, W. Bao, G. Meigs, V. Chakarian, and Y. U. Idzerda, J. Appl. Phys. 79, 4558 (1996).

[11].   J.-H. Park, C. T. Chen, S.-W. Cheong, W. Bao, G. Meigs, V. Chakarian, and Y. U. Idzerda, Phys. Rev. Lett. 76, 4215 (1996).

[12].   D. S. Dessau, T. Saitoh, C.-H. Park, Z.-X. Shen, P. Villella, N. Hamada, Y. Moritomo, and Y. Tokura, Phys. Rev. Lett. 81, 192 (1998).

[13].   K. E. Smith, Annual Reports C, 115 (1993).

[14].   Klaus Breuer, Cristian Stagerescu, Kevin E. Smith, Martha Greenblatt, and Kandalam Ramanujachary, Phys. Rev. Lett. 76, 3172 (1996).

[15].   G.-H. Gweon, J. D. Denlinger, J. A. Clack, J. W. Allen, C. G. Olson, E. DiMasi, M. C. Aronson, B. Foran, and S. Lee, Phys. Rev. Lett. 81, 886 (1998).

[16].   M. Nakamuru, A. Sekiyama, H. Namatame, A. Fujimori, H. Yoshihara, T. Ohtani, A. Misu, and M. Takano, Phys. Rev. B 49, 16191 (1994).

[17].   Y. Haruyama, S. Kodaira, Y. Aiura, H. Bando, Y. Nishihara, T. Maruyama, Y. Sakisaka, and H. Kato, Phys. Rev. B 53, 8032 (1996).

[18].   Zhaoming Zhang, Shin-Puu Jeng, and Victor E. Henrich, Phys. Rev. B 43, 12004 (1991).

[19].   Jesper Nerlov, Qingfeng Ge, and Preben J. Moller, Surf. Sci. 348, 28 (1996).

[20].   Robert. J. Lad and Victor E. Henrich, Phys. Rev. B 38, 10860 (1988).

[21].   Kevin E. Smith and Victor E. Henrich, Phys. Rev. B 38, 5965 (1988).

[22].   W. P. ellis, A. M. Boring, J. W. Allen, L. E. Cox, R. D. Cowan, B. B. Pate, A. J. Arko, and I. Lindau, Sol. Stat. Commun. 72, 725 (1989).

[23].   Klaus Breuer, David M. Goldberg, Kevin E. Smith, Martha Greenblatt, and William McCarroll, Sol. Stat. Commun. 94, 601 (1995).

[24].   I. H. Inoue, I. Hase, Y. Aiura, A. Fujimori, K. Morikawa, T. Mizokawa, Y. Haruyama, T. Maruyama, and Y. Nishihara, Physica C 235-240,1007 (1994).

[25].   G. N. Kamm, D. J. Gillespie, A.C. Ehrlich, T. J. Wieting, and F. Levy, Phys. Rev. B 31, 7617 (1985).

[26].   G. N. Kamm, D. J. Gillespie, A.C. Ehrlich, D. L. Peebles, and F. Levy, Phys. Rev. B 35, 1223 (1987).

[27].   R. T. Littleton IV, T. M. Tritt, C. R. Feger, J. Kolis, M. L. Wilson, M. Marone, J. Payne, D. Verebeli, and F. Levy, Appl. Phys. Lett. 72, 2056 (1998).

[28].   For a review see for example, D. E. Eastman and F. J. Himpsel. "Photoelectron Spectroscopy," *Encyclopedia of Physics*, Second Ed. , 908-912, New York, VCH Publishers, 1991.

X-RAY ABSORPTION CHARACTERIZATION OF DIESEL EXHAUST PARTICULATES

A. J. NELSON*, J. L. FERREIRA*, J. G. REYNOLDS* S.D. SCHWAB** and J. W. ROOS**
*Lawrence Livermore National Laboratory, Livermore, CA 94550
**Ethyl Corporation, Richmond, VA 23217

ABSTRACT

We have characterized particulates from a 1993 11.1 Detroit Diesel Series 60 engine with electronic unit injectors operated using fuels with and without methylcyclopentadienyl manganese tricarbonyl (MMT) and overbased calcium sulfonate added. X-ray photoabsorption (XAS) spectroscopy was used to characterize the diesel particulates. Results reveal a mixture of primarily Mn-phosphate with some Mn-oxide, and Ca-sulfate on the surface of the filtered particulates from the diesel engine.

INTRODUCTION

Methylcyclopentadienyl manganese tricarbonyl (MMT) is an effective means of increasing octane quality of gasoline. [1] Comprehensive automotive testing also shows that vehicles using MMT (0.03125 gram manganese per gallon) in gasoline had 6% lower CO emissions and 20% lower NO_x emissions than automobiles operating on base gasoline. [2] In another application, it is proposed that MMT with overbased calcium alkylbenzene sulfonate (OCABS) be employed as a fuel additive package to reduce diesel smoke and particulate emissions. [3] OCABS are also used to impart detergency in engine oils and fuel combustion catalysts. [4,5]

Identification of the manganese and calcium species formed by the decomposition of MMT and OCABS in diesel engines is of practical importance and of environmental interest. These species end up either on surfaces in the exhaust system of an automobile or are emitted into the atmosphere.

This paper reports on an investigation to determine the surface composition of particulates emitted from vehicles operating on fuel containing various concentrations of MMT and calcium sulfonate using X-ray absorption spectroscopy (XAS).

EXPERIMENTAL

Particulate samples from diesel engine test runs using fuels with and without methylcyclopentadienyl manganese tricarbonyl (MMT) and overbased calcium alkylbenzene sulfonate added were characterized. The fuels used were Colonial diesel (389 ppm S) and Howell diesel (460 ppm S). The particulate samples were collect on filter paper. The test runs are summarized in Table I. It should be noted that engine operation time for Run 1716 had a duration five times that of the other test runs resulting in 5X exposure of the filter.

63

Model compounds were analyzed by XAS and XPS to provide a basis for quantitative analysis of the unknown particulates. The model compounds CaO, $CaCO_3$, $CaSO_4$, $Ca_2P_2O_7$, $Ca(H_2PO_4)_2$, $Ca(NO_3)_2$, $CaCl_2$, MnO, Mn_2O_3, Mn_3O_4, MnO_2, $Mn(OCH_3)_2$, $MnPO_4$, $Mn_5(PO_4)_2[PO_3(OH)]_2 \cdot 4H_2O$ (Mn-Phos) (Alfa Aesar), $Mn(CH_3 COCH=COCH_3)_3$ ($Mn(acac)_3$), MnS (Aldrich) and $MnSO_4 \cdot H_2O$ (Mallinckrodt) were used as received. The powder samples were pressed into indium foil and attached to the sample holder, which was then introduced into the ultra-high vacuum (UHV) chamber for analysis. Similarly, small pieces of the particulate on filter samples were cut and mounted with indium foil on the sample holder prior to their introduction into the analysis chamber.

Table I. Summary of Parameters for Diesel Particulate Test Run Samples

Run Number	Mn (mg Mn/liter)	Ca (mg Ca/liter)	Fuel	Engine Operation Time (min.)
1716	3	18	Colonial	100
1599	50	0	Howell	20
1632	0	50	Howell	20
1605	8	0	Howell	20
1616	0	8	Howell	20
1658	6	36	Howell	20
1591	18	0	Howell	20
1626	0	18	Howell	20

Manganese 2p and calcium 2p core-level XAS were performed on these particulates and the series of Mn and Ca compound standards. This technique probes empty, or unfilled 3d electronic states of manganese and provides information on the local chemical environment. For 2p X-ray absorption, the dipole allowed transitions are 2p → 3d and 2p → 4s, with transitions to the 3d states dominanting. The observed behavior has been interpreted as the combination of crystal field effects with splittings due to multipole 2p–3d and 3d–3d interactions. Crystal field effects, measured by the cubic crystal field parameter, are equal for both the initial and final states. Also, the 2p core hole spin-orbit coupling splits the absorption spectrum into a $2p_{3/2}$ (L_3) and a $2p_{1/2}$ (L_2) part. In addition, it has been observed that the branching ratio of the L_3 edge intensity to the total line strength depends on the oxidation state of the Mn and Ca. Specifically for Mn, this branching ratio decreases in the order Mn(II) → Mn(III) → Mn(IV). [6,7] This analysis was performed at the 8-2 beam line at the Stanford Synchrotron Radiation Laboratory (SSRL) by scanning the photon energy of the incoming monochromatic synchrotron radiation through the Mn 2p and Ca 2p core-level edge (L-edge) while monitoring the total electron yield. The spectra were calibrated arbitrarily with 640 eV for the absorption maximum in the Mn model compounds and ~347 eV for the Ca model compounds. [8]

RESULTS

XAS – Model Compounds

Figure 1 presents the Mn 2p core level XAS of the Mn model compounds. Analysis of these series of model compounds provides standard Mn L-edge spectra of several formal valences, specifically, Mn(II) in $MnSO_4$, MnO and Mn-Phos, Mn(III) in $MnPO_4$, and a mixture of Mn(II) and Mn(III) in Mn_3O_4. All spectra show two absorption regions, the Mn $2p_{3/2}$ (L_3) at ~640 eV and Mn $2p_{1/2}$ (L_2) at ~650 eV. Characteristic features of the absorption regions depends upon the ligand environments and the oxidation states. For example, Mn_3O_4 shows a sharp and broad L_3 region and a very broad L_2 region reflecting the oxide environment and the mixture of Mn(II) and Mn(III) oxidation states, while Mn-Phos and $MnSO_4$ show a relatively sharp L_3 region reflecting the single oxidation state Mn(II). Also note the differences in the energy

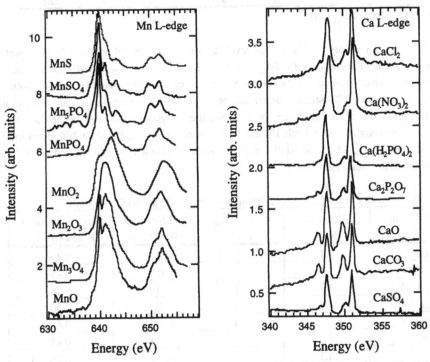

Figure 1. Mn L-edge photoabsorption spectra of model compounds.

Figure 2. Ca L-edge photoabsorption spectra of model compounds.

separation between the L_3 and L_2 peaks and their relative intensities as well as the shoulder on the low energy side of the L_3 peak for the phosphate and sulfate. Table II summarizes the pertinent Mn L-edge spectral characteristics.

Table II. Summary of the Mn L-edge Spectral Features for Mn Model Compounds.

Compound	Formal Mn Valency	$\Delta E(L_3 - L_2)$ (eV)	L_3 Linewidth FWHM (eV)	Branching Ratio $I(L_3)/I(L_3 + L_2)$
MnO	II	12.0	4.2	0.60
Mn_3O_4	II, III	12.0	4.2	0.62
Mn_2O_3	III	11.0	3.8	0.62
MnO_2	IV	10.2	5.3	0.61
$MnPO_4$	III	11.8	1.0	0.78
Hureaulite[a]	II	11.8	1.3	0.78
$MnSO_4$	II	11.8	1.1	0.82
MnS	II	11.8	1.6	0.82

a) $Mn_5(PO_4)_2[PO_3(OH)]_2 \cdot 4H_2O$.

Figure 2 presents the Ca 2p core level XAS of the Ca model compounds. All spectra show two areas of absorption, Ca $2p_{3/2}$ (L_3) at ~348 eV and Ca $2p_{1/2}$ (L_2) at ~351 eV. Also, smaller associated absorptions are seen as either shoulders or resolved satellite peaks for both L_3 and L_2 in all spectra. Quantitative determination of the model compound Ca species relies on energy separations of the small satellite peak and the associated main peak of the Ca L-edge spectra as are summarized in Table III. Since Ca only occurs as Ca(II), subtle differences in the peak shapes, $\Delta E(L_{3a} - L_{3b})$, linewidths and the branching ratio will be utilized to determine Ca speciation in the diesel particulates.

Table III. Summary of the Ca L-edge Spectral Features for Ca Model Compounds.

Compound	$\Delta E(L_3 - L_2)$ (eV)	$\Delta E(L_{3a} - L_{3b})$ (eV)	L_3 Linewidth FWHM (eV)	Branching Ratio $I(L_3)/I(L_3 + L_2)$
$CaSO_4$	3.4	0.8	0.6	0.46
CaO	3.4	1.2	0.7	0.45
$CaCO_3$	3.4	1.1	0.7	0.46
$Ca_2P_2O_7$	3.2	1.1	0.7	0.50
$Ca(H_2PO_4)_2$	3.2	0.8	0.7	0.48
$Ca(NO_3)_2$	3.2	0.8	0.8	0.43
$CaCl_2$	3.3	1.0	0.8	0.46

XAS Analysis – diesel particulates

The Mn XAS spectra for the diesel particulate samples are presented in Figure 3. Table IV summarizes the spectral characteristics. From visual comparison, these spectra best match the Mn L-edge spectrum for Mn_2O_3. The L_3 linewidth suggests a mixture of species, most likely containing oxide. The L_3–L_2 energy separation suggests pure oxides or a mixture of oxides and other species such as phosphate and sulfate. The branching ratios suggest a mixture of phosphate and/or sulfate with oxides.

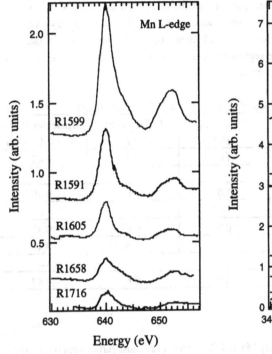

Figure 3. Mn L-edge photoabsorption spectra of diesel particulates.

Figure 4. Ca L-edge photoabsorption spectra of diesel particulates.

Table IV. Summary of the Mn L-edge Spectral Features for Diesel Particulates

Sample	$\Delta E(L_3 - L_2)$ (eV)	L_3 Linewidth FWHM (eV)	Branching Ratio $I(L_3)/I(L_3 + L_2)$
Run 1716	12.0	3.2	0.73
Run 1599[†]	12.0	3.2	0.77
Run 1632[‡]	–	–	–
Run 1605[†]	12.0	2.7	0.79
Run 1616[‡]	–	–	–
Run 1658	12.0	3.2	0.67
Run 1591[†]	12.0	2.8	0.79
Run 1626[‡]	–	–	–

[†]no Ca added to fuel, [‡]no Mn added to fuel

67

The Ca XAS spectra for the diesel particulate samples are presented in Figure 4. Table V summarizes the spectral characteristics. Visual comparison of the Ca L_3 spectral features for the diesel particulates with the Ca L_3 spectral features for the Ca model compounds would seem to indicate that the Ca speciation has several possibilities and that the species may be different for some of the samples. The L_3 linewidth suggests in all cases $CaSO_4$. The L_3–L_2 energy separation suggests phosphate and nitrate (chloride is most likely not occurring because of the lack of a chloride source). The L_{3a}–L_{3b} satellite separation suggests sulfate, phosphate, and nitrate. The branching ratio suggests phosphate and sulfate.

Table V. Summary of the Ca L-edge Spectral Features for the Diesel Particulates

Sample	$\Delta E(L_3 - L_2)$ (eV)	$\Delta E(L_{3a} - L_{3b})$ (eV)	L_3 Linewidth FWHM (eV)	Branching Ratio $I(L_3)/I(L_3 + L_2)$
Run 1716	3.5	1.0	0.5	0.47
Run 1599[†]	–	–	–	–
Run 1632[‡]	3.3	1.0	0.6	0.49
Run 1605[†]	–	–	–	–
Run 1616[‡]	3.2	0.9	0.6	0.47
Run 1658	3.2	1.0	0.6	0.48
Run 1591[†]	–	–	–	–
Run 1626[‡]	3.3	0.9	0.6	0.47

[†]no Ca added to fuel, [‡]no Mn added to fuel

CONCLUSIONS

We have applied XAS to identify Mn and Ca species in exhaust particulates from a diesel engine operated using fuel with MMT and overbased calcium alkylbenzene sulfonate. These compounds are reported to improve combustion characteristics leading to reduced diesel smoke and particulate emissions.

We have compared the energy separation of the XAS L_3 and L_2 edges, the L_3 linewidth, and the branching ratio of these diesel samples with the spectral characteristics of the Mn and Ca model compounds to determine the surface composition of the samples. Results for the diesel particulate samples indicate that the Mn is present as a mixture of phosphate and/or sulfate with oxides and that the Ca is primarily present as $CaSO_4$, consistent with the higher operating temperatures. A high sulfur content fuel and diesel operating conditions would probably yield primarily sulfates. However, Ethyl simulations indicate that $MnSO_4$ is not stable at higher temperatures.

Acknowledgments

This work was performed under the auspices of the U.S. Department of Energy under Contract W-7405-ENG-48. The XAS work was performed at the Stanford Synchrotron Radiation

Laboratory, which is supported by the U.S. Department of Energy, Office of Basic Energy Sciences, under Contract No. DE-AC03-76SF00515.

REFERENCES

1. J. E. Faggan, J. D. Bailie, E. A. Desmond and D. L. Lenane, SAE Paper 750925 (1975), Warrenville, PA: Society for Automobile Engineering.

2. D. P. Hollrah and A. M. Burns, Oil & Gas Journal, March 11, 1991, pp. 86–90.

3. G.R. Wallace, Hydrocarbonaceous Fuel Composition and Additives Thereof, US Pat 5944858.

4. G.G. Pritzker, Natl. Petroleum News **37**, 793 (1945).

5. J.F. Marsh, Colloidal lubricant additives, Chem. Ind. (London, 1987) p. 470.

6. S.P. Cramer, F.M.F. deGroot, Y. Ma, C.T. Chen, F. Sette, C.A. Kipke, D.M. Eichhorn, M.K. Chan, W.H. Armstrong, E. Libby, G. Christou, S. Brooker, V. McKee, O.C. Mullins and J.C. Fuggle, J. Am. Chem. Soc. **113**, 7937 (1991).

7. M.M. Grush, J. Chen, T.L. Stremmler, S.J. George, C.Y. Ralston, R.T. Stibrany, A. Gelasco, G. Christou, S.M. Gorun, J.E. Penner-Hahn and S.P. Cramer, J. Am. Chem. Soc. **118**, 65 (1996).

8. Zhanfeng Yin, M. Kasrai, G.M. Bancroft, K. Fyfe, M.L. Colaianni and K.H. Tan, Wear **202**, 192 (1997).

XANES ANALYSIS OF BCC/FCC TWO-PHASE BINARY ALLOYS

P.J. SCHILLING[a] , R.C. TITTSWORTH[b], E. MA[c], and J.-H. HE[c]
[a] Mechanical Engineering Department, University of New Orleans (UNO), New Orleans, LA 70148, USA, pschilli@uno.edu
[b] Center for Advanced Microstructures and Devices, Louisiana State University, Baton Rouge, LA 70806, USA
[c] Department of Materials Science and Engineering, Johns Hopkins University, Baltimore, MD 21218, USA

ABSTRACT

A critical factor in the characterization of two-phase binary alloy systems is the determination of the phase fractions and compositions of the two coexisting solid solutions for any given overall composition of the two-phase mixture. In some systems, for example nanocrystalline alloys formed by high-energy ball-milling, these parameters are difficult to attain by traditional techniques like X-ray diffraction. A new technique has been developed to obtain these quantities indirectly from X-ray absorption near edge structure (XANES) data collected at the two relevant absorption edges, with the formulation of this technique presented here in detail. The technique has been tested using Fe-Ni fcc and bcc standards and the results indicate that the method is accurate to within 5%. This method has been applied to two-phase (f.c.c. and b.c.c.) binary alloys formed by ball-milling of $Cu_{100-x}Fe_x$ (x = 50-80).

INTRODUCTION

In order to adequately characterize two-phase binary alloy systems, it is necessary to determine the phase fractions and compositions of each phase. For systems in equilibrium, these parameters can be obtained from the appropriate binary phase diagram. In non-equilibrium systems, these parameters are sometimes difficult to obtain. This is the case for nanocrystalline Fe-Cu alloys formed by high energy ball milling [1]. One method for such phase analysis is through the quantitative analysis of X-ray diffraction (XRD) data. However, in the Fe-Cu system, while peak areas in XRD patterns can be used to estimate phase fractions, compositions of the phases cannot be obtained [1,2]. A new approach has been developed to address this problem based on X-ray absorption near edge structure (XANES) spectroscopy. Linear combination fitting of a XANES spectrum collected at the absorption edge of a particular element can give the fractional distribution of that element among the constituent phases. From the results of such fitting at both edges (corresponding to the two elements in the binary system), along with the overall compostion, the phase fractions and compositions can be extracted. Experimental results for the fcc/bcc Fe-Cu system [3] demonstrate the potential for the technique, but the conceptual foundation and formulation of the relevant relations have not been presented previously. A full explanation of this procedure is presented here, including a detailed derivation of the relations used to extract the phase fractions and compositions from the XANES fitting results. Results of a study on standard samples in the Fe-Ni system to establish the accuracy of the technique are summarized, followed by a brief discussion of the experimental results obtained from Fe-Cu alloys prepared by high-energy ball-milling [3].

Mat. Res. Soc. Symp. Proc. Vol. 590 © 2000 Materials Research Society

XANES ANALYSIS

XANES spectroscopy is a powerful element-specific probe of local electronic and crystallographic environment, and is particularly sensitive to the geometrical arrangement of nearest and even more distant neighbors to the absorbing atom [4-6]. A common analysis technique for multi-component systems involves the simulation of an experimental XANES spectrum as a linear combination of the spectra from the separate constituents using a non-linear least-squares procedure[7,8]. The technique described here is based on linear combination fitting of XANES spectra at the absorption edges of the two constituent elements in a binary system. From these fitting results, phase fractions and compositions can be extracted for a two-phase alloy. A few simple relations can be used to calculate the phase fractions and compositions from the fitting results, and their derivations will be presented. While these relations hold in theory for all systems, the application of the technique hinges on the ability to perform linear combination fitting of the XANES spectra as a sum of the spectra of the two phases, and therefore will only work if the features in the XANES spectra of the two phases differ significantly. This point will be addressed further with respect to particular examples (fcc/bcc alloys in the Fe-Ni and Fe-Cu systems), after presentation of the conceptual basis for the technique. The technique is based on the assumption of a two-phase system. If there is a third phase (crystalline or amorphous) present, this formulation obviously does not apply.

The explanation of the technique depends strongly on clarity in the nomenclature and symbolism used, as well as a clear understanding of exactly what information is directly obtained from the XANES fitting. An important fact is that the XANES measurement is element specific – a XANES at the K edge of element A probes these absorbers specifically, and represents a sum of contributions from all absorbing A atoms. If we assume a two-phase ($\alpha + \beta$) system, the XANES spectrum collected at the K edge of element A may be simulated as a linear combination of the element-A XANES signatures for these two phases. Again, in order to apply the technique, the XANES spectra of the α and β phases must be very different. The fitting thus gives the fraction of element-A atoms in each phase (e.g. number of element-A atoms in α)/(total number of element-A atoms). Similarly, fitting XANES data from the K edge of the other element in the binary alloy (element B) will give the distribution of the element-B atoms between the α and β phases. It is important to distinguish this element specific information from, for example, what is obtained from quantitative XRD, where the XRD peaks represent the overall fractions of α and β phases.

Thus, we will define a two-phase binary system consisting elements A and B, forming phases, α and β. We assume the overall composition is known:

X_A the mole fraction of A atoms
      (the total number of A atoms / the total number of atoms)

X_B the mole fraction of B atoms
      (the total number of A atoms / the total number of atoms)

From least squares fitting of XANES spectra as linear combinations of standard spectra for phases α and β at the absorption edge of element A, we obtain:

f_a^A the fraction of A atoms which are in the α phase

(the number of A atoms in the α phase / the total number of A atoms)

f_β^A the fraction of A atoms which are in the β phase

(the number of A atoms in the β phase / the total number of A atoms)

From similar fitting at the absorption edge of element B, we obtain values for:

f_α^B the fraction of B atoms which are in the α phase

f_β^B the fraction of B atoms which are in the β phase

Thus, for each sample we know the overall composition and (from XANES fitting) the distribution of each element between the two phases. This represents a complete characterization of the two-phase system, but not in the terms normally used – phase fractions and compositions of each phase. These terms will be represented as:

f_α , f_β the phase fractions for the system

X_A^α , X_B^α the composition of the α phase

X_A^β , X_B^β the composition of the β phase

The relations to obtain these desired parameters from the overall composition and XANES results can be derived based on conservation of matter. The following parameters will be used:

N the total number of atoms (A and B)

N_A the total number of A atoms

N^α the number of atoms (A and B) in the α phase

N_A^α the number of A atoms in the α phase

The total number of A atoms can be written in terms of the total number of atoms and the overall composition as:

$$N_A = NX_A \tag{1}$$

Next, we will write an expression for f_α^A in terms of the parameters just defined. Since f_α^A represents the number of A atoms in the α phase divided by the total number of A atoms, we can write:

$$f_\alpha^A = \frac{N_A^\alpha}{N_A} = \frac{N_A^\alpha}{NX_A} \tag{2}$$

This can be rewritten as:

$$N_A^\alpha = f_\alpha^A X_A N \tag{3}$$

Similarly:

$$N_A^\beta = f_\beta^A X_A N \; ; \quad N_B^\alpha = f_\alpha^B X_B N \; ; \quad N_B^\beta = f_\beta^B X_B N \tag{4}$$

73

The phase fraction for the α phase (the number of atoms in the α phase divided by the total number of atoms) can now be written as:

$$f_\alpha = \frac{N^\alpha}{N} = \frac{N_A^\alpha + N_B^\alpha}{N} = \frac{f_\alpha^A X_A N + f_\alpha^B X_B N}{N} \qquad (5)$$

or:

$$f_\alpha = f_\alpha^A X_A + f_\alpha^B X_B \qquad (6)$$

The composition of the α phase, written as the mole fraction of A atoms, is the number of A atoms in the α phase divided by the total number of atoms in the α phase:

$$X_A^\alpha = \frac{N_A^\alpha}{N^\alpha} = \frac{f_\alpha^A X_A N}{f_\alpha N} = \frac{f_\alpha^A X_A N}{\left(f_\alpha^A X_A + f_\alpha^B X_B\right)N} \qquad (7)$$

or:

$$X_A^\alpha = \frac{f_\alpha^A X_A}{f_\alpha^A X_A + f_\alpha^B X_B} \qquad (8)$$

Equations (6) and (8) are the relations we desire. These give values for the phase fraction and phase composition in terms of the parameters obtained by linear combination fitting of the XANES spectra and the overall elemental composition. The set of desired parameters can be completed using the following similar relations:

$$f_\beta = f_\beta^A X_A + f_\beta^B X_B \qquad \text{(or } f_\beta = 1 - f_\alpha \text{)} \qquad (9)$$

$$X_B^\alpha = \frac{f_\alpha^B X_B}{f_\alpha^A X_A + f_\alpha^B X_B} \qquad \text{(or } X_B^\alpha = 1 - X_A^\alpha \text{)} \qquad (10)$$

$$X_A^\beta = \frac{f_\beta^A X_A}{f_\beta^A X_A + f_\beta^B X_B} \qquad (11)$$

$$X_B^\beta = \frac{f_\beta^B X_B}{f_\beta^A X_A + f_\beta^B X_B} \qquad \text{(or } X_B^\beta = 1 - X_A^\beta \text{)} \qquad (12)$$

APPLICATION TO BCC/FCC SYSTEMS

As stated earlier, the relations above are theoretically valid for any system, the limiting factor is the ability to successfully perform the linear combination fitting of the XANES spectra. For a particular two-phase binary system, this requires (1) that the XANES spectra for the two phases differ significantly enough to accommodate fitting, and (2) that the XANES spectrum for a particular phase does not change appreciably as a function of composition over the compositional range of this phase. This second requirement means that it is possible to use a single standard XANES spectrum for each phase in the linear combination fitting.

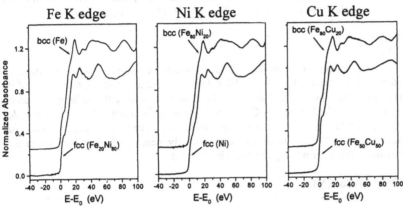

Figure 1. Comparison of XANES spectra for bcc and fcc phases including Fe, Ni, and Cu, collected at the corresponding K absorption edge.

Our results indicate that these two criteria are met in binary alloys consisting of bcc/fcc mixtures in the Fe-Ni and Fe-Cu systems. This is based on the significant difference between the characteristic XANES spectra for these two phases at the Fe, Ni, and Cu K edges. Representative spectra are presented in Figure 1 to demonstrate this. Furthermore, comparisons of spectra for each phase (bcc and fcc) indicate that the changes in the spectra as a function of phase composition are minor in comparison the these large differences among phases [3,9].

As a test of the applicability of this method to such systems, we have measured combinations of bcc and fcc Fe-Ni alloys. This system was chosen as a test because suitable standards (with known phase fractions and compositions) could be obtained [9]. XANES spectra were obtained at the CAMD synchrotron facility, and the linear combination fitting was performed using the WinXAS software package [10]. The fitting results were inserted into equations (6), (8), and (11) to obtain the phase fraction for the bcc phase, the mole fraction of Fe in the bcc phase, and the mole fraction of Fe in the fcc phase, respectively. The results are summarized in Table I, which compares the known phase fraction and compositions to those obtained from the data analysis. The results demonstrate that the phase fractions and phase compositions can be determined to within 5 %.

Based on this positive result, the technique has been further applied to XANES data collected in the ball-milled Fe-Cu system, which forms a two-phase

Table I.
Pseudo-two-phase Fe-Ni alloy standards.

Sample	True Values			Values Derived from XANES Fitting		
	f_{bcc}	X_{Fe}^{bcc}	X_{Fe}^{fcc}	f_{bcc}	X_{Fe}^{bcc}	X_{Fe}^{fcc}
1	0.48	0.8	.20	0.49	0.79	0.21
2	0.55	0.8	.64	0.60	0.78	0.65
3	0.76	0.8	0	0.75	0.81	0.00
4	0.22	1	.20	0.24	0.95	0.19
5	0.26	1	.64	0.30	0.97	0.63
6	0.08	1	.20	0.10	1.0	0.18
7	0.11	1	.64	0.16	0.96	0.62

bcc/fcc mixture in the compositional range of ~ 50% to 80% Fe [3]. The XANES analysis indicates that throughout this region the two coexistent phases have almost identical composition, consistent with that of the overall mixture. This is in contrast to the normal breakdown of a two-phase system at equilibrium in which the phase compositions of each phase are constant throughout the region and the phase fractions follow the lever rule.

SUMMARY

A new technique has been developed to obtain phase fractions and compositions in two-phase binary alloy systems, base on XANES data collected at the absorption edges of the two constituent elements. For bcc/fcc alloys, Fe-Ni standards indicate the technique is accurate to within ~ 5%. In the ball-milled Fe-Cu system, this analysis indicates that the system does not follow the lever rule. This is not necessarily unexpected, since this is a positive heat-of-mixing (non-equilibrium) system. The technique is currently being applied to ball-milled Fe-Ni to determine whether a lever rule type scenario is observed.

ACKNOWLEDGMENT

The authors wish to thank Joe Wong for useful discussions and for providing the Fe-Ni alloy standards. This work was supported by the State of Louisiana through the Center for Advanced Microstructures and Devices (CAMD), and the National Science Foundation, Grant No. DMR-9896379.

REFERENCES

1. L.B. Hong and B. Fultz, *Acta Mater.* **46**, 2937 (1998).
2. L.B. Hong and B. Fultz, *J. Appl. Phys.* **79**, 3946 (1996).
3. P.J. Schilling, J.H. He, R.C. Tittsworth, and E. Ma, *Acta Mater.*, **47**, 2525 (1999).
4. A. Bianconi, in X-Ray Absorption: Principles, Applications, Techniques of EXAFS, SEXAFS, and XANES, edited by D. C. K. a. R. Prins (John Wiley & Sons, New York, 1988), Vol. 92, p. 573.
5. G.E. Brown, Jr., G. Calas, G. A. Waychunas, and J. Petiau, in Spectroscopic Methods in Mineralogy and Geology, edited by F. C. Hawthorne (Mineralogical Society of America, Chelsea, Michigan, 1988), Vol. 18, p. 431.
6. J. Stöhr, NEXAFS Spectroscopy (Springer-Verlag, New York, 1992).
7. G. S. Waldo, R. M. K. Carlson, J. M. Moldowan, K.E. Peters, and J.E. Penner-Hahn, *Geochimica et Cosmochimica Acta* **55**, 801 (1991).
8. G. N. George, M. L. Gorbaty, S. R. Kelemen, M. Sansone, *Energy & Fuels* **5**, 93 (1991).
9. P.J. Schilling and R.C. Tittsworth, *J. Synchrotron Rad.*, **6**, 497-499 (1999).
10. T. Ressler, *J. Physique IV France*, **7**, C2-269 (1997).

SPECTROSCOPIC CHARACTERISATION OF
MIXED TITANIA-SILICA XEROGEL CATALYSTS

MA Holland[+], DM Pickup, G Mountjoy, SC Tsang[*], GW Wallidge[#], RJ Newport, ME Smith[#].
School of Physical Sciences, University of Kent at Canterbury, CT2 7NR, UK.
+ contact author: mah3@ukc.ac.uk
* Centre for Catalysis, Department of Chemistry, University of Reading, RG6 6AD, UK.
# Department of Physics, University of Warwick, Coventry, CV4 7AL, UK.

ABSTRACT

The synthesis of high surface area $(TiO_2)_{0.18}(SiO_2)_{0.82}$ xerogels has been achieved using the sol-gel route. Heptane washing was used before the drying stage to minimise capillary pressure and hence preserve pore structure and maximise the surface area. The as-prepared xerogels were tested for their catalytic activity using the epoxidation of cyclohexene with *tert*-butyl hydrogen peroxide (TBHP) as a test reaction. Surface areas up to 450 m^2g^{-1} were achieved with excellent selectivities and reasonable percent conversions. SAXS data has identified that heptane washing during drying, in general, results in a preservation of the pore structure, and produces more effective catalysts with higher surface areas and larger pore diameters. FT-IR spectroscopy has revealed that the catalytic activity is dependant upon the number of Si-O-Ti linkages, inferring intimate mixing of the precursors at the atomic level. XANES data reveals the presence of reversible 4/6-fold Ti sites that are thought to be 'active' catalytic sites. The most effective catalyst was produced with a calcination temperature of 500 °C, and a heating rate of 5 °Cmin^{-1}.

INTRODUCTION

The sol-gel synthesis of materials offers a number of inherent advantages over traditional preparation techniques. A suitable choice of the precursors allows the production of high-purity, atomically mixed, homogeneous materials. Recently, sol-gel derived titania-silica oxides have been exploited for their potential use as catalysts and catalytic supports; they are particularly suitable as epoxidation catalysts [1,2].

The textural characteristics of sol-gel derived titania-silica oxides are strongly dependant on the processing of the gel. The sol-gel method offers the possibility of preparing solid materials with high surface areas and controlled pore volumes and pore size distributions. Drying a newly prepared gel under ambient conditions often induces high capillary pressures, due to differential strain between the pore liquid and the gel network, which leads to a collapse of the porosity. Gels produced in this way are termed *xerogels*. Previously, our group has extensively studied the structure of xerogels produced using routine drying [3,4,5]. In this work we use an alternative method, by initially ageing our xerogel samples for 10 days, after which the samples were then washed in n-heptane for a further 10 days before drying was started. This solvent was chosen as it has a low surface tension, γ, which reduces the capillary pressure, P_c, during drying, according to Equation 1, and hence helps preserve pore structure.

$$P_c = \frac{-2\gamma\cos\theta}{r} \qquad (1)$$

r is the pore radius and θ the contact angle. The resulting xerogel samples were heat treated and studied using several structural techniques. The catalytic activity of the samples was assessed using the epoxidation reaction of cyclohexene with *tert*-butyl hydroperoxide (TBHP). This enables a comprehensive investigation of the correlation between structural characteristics as a function of varying synthesis conditions, and the resultant catalytic performance.

METHOD

Sample preparation

Samples were prepared with the following precursors: tetraethoxyorthosilicate (TEOS, Aldrich, 98%), titanium (IV) isopropoxide (Ti(OPr^i)$_4$, Aldrich, 97%), and isopropyl alcohol (IPA, Aldrich, 98%). The IPA acts as a mutual solvent, and HCl was used as a catalyst to promote the hydrolysis and condensation reactions. During mixing of the precursors, the different rates of hydrolysis of the silicon and titanium alkoxides can cause phase separation, producing Ti-rich and Si-rich regions, which reduces the catalytic usefulness of the material. We circumvent this problem using the method of Yoldas [6] which involves partially hydrolysing the Si alkoxide prior to mixing it with the Ti alkoxide. Before mixing with the Ti alkoxide, the TEOS was pre-hydrolysed for 2 hours using TEOS:IPA:H$_2$O in a 1:1:1 ratio, in the presence of an acid catalyst (pH of 1). The Ti(OPr^i)$_4$ and water were then added, such that the ratio, R, of water to alkoxide equals 2. With this ratio, complete hydrolysis can only occur by utilising water released from initial condensation.

All gels were prepared with the nominal composition (TiO$_2$)$_{0.18}$(SiO$_2$)$_{0.82}$. The resulting wet gels were separated into two equal quantities, and were initially left to age for 10 days. After this period, half of the yield was washed for two 5 day periods using n-heptane. Following this, the residual n-heptane was decanted off and both samples were left to dry slowly (using a pinhole in the top of the otherwise sealed container to reduce the drying rate). The resultant xerogels were then finely ground, and heat treatments were performed at temperatures ranging up to a maximum of 750 °C. Heating rates of 2 and 5 °C min^{-1} were used, with each set-point temperature maintained for 2 h.

Experiment and data analysis - FTIR, UV-VIS, NMR

Infrared spectra were recorded in diffuse reflectance mode using a Biorad FTS175C spectrometer controlled by Win-IR software. Samples were diluted (1:10 by weight) in dry KBr and scanned in the range 4000-400 cm^{-1} with a resolution of 4 cm^{-1}. The region between 1450-650 cm^{-1} is characteristic of a silica network [5,7] and is of particular interest to us, as evaluation of this region provides a semi-quantitative estimate of the Si-O-Ti connectivity [8], or so-called Ti dispersion. The normalised and corrected intensity of the vibration due to Si-O-Ti bonding (Ti dispersion), $I_{Si-O-Ti}$, can be obtained from

$$I_{(Si-O-Ti)} \approx (I_{(Si-O-Ti,Si-OH)_{\sim 960 cm^{-1}}} - kI_{(-OH)_{\sim 3590 cm^{-1}}})/I_{(Si-O-Si)_{\sim 1090 cm^{-1}}} \qquad (2)$$

The ^{17}O MAS NMR spectra were acquired on a Chemagnetics CMX300 Infinity spectrometer. The spectra were collected at 40.7 MHz under MAS at typically 15.1 kHz, with a recycle delay of 1-2.5 s, using a Chemagnetics 4 mm double-bearing probe. A rotor-synchronised 90°-τ-180° echo sequence was applied with a short τ delay of 1 rotor period (66 μs) to overcome problems of probe ringing, thereby allowing the spectra to be phased correctly. The spectra are dominated by a peak, with an isotropic chemical shift, δ, at ~0 ppm and shows structure at negative shift due to a 2nd order quadrupole interaction. This resonance has spinning sidebands from the MAS process at -325 and 350 ppm. The signal at ~220 ppm is attributed to

Si-O-Ti linkages [9]. The resonance near 0 ppm is assigned to Si-O-Si linkages with a contribution from Si-OH groups, hence it is not a 2nd order quadrupolar lineshape.

Experiment and data analysis - EXAFS, XANES, SAXS

Titanium K-edge X-ray absorption spectra were collected at Station 7.1 at the Daresbury Laboratory SRS using a Si (111) crystal monochromator and 50% harmonic rejection. EXAFS and XANES spectra were collected at the Ti K-edge (4966 eV) in transmission mode at room temperature. The XANES spectra were processed according to our earlier work [3] to obtain the absorbance, $\mu t = \ln(I_t/I_0)$, and then pre- and post-edge backgrounds were fitted to obtain normalised absorbance, $1+\chi(E) = (\mu t(E) - \mu t_{pre})/(\mu t_{post} - \mu t_{pre})$. A single Lorentzian was used to model the dominant pre-edge peak, i.e. the XANES peak of interest, and additional Lorentzians were used to model the remainder of the absorption edge. The EXAFS data may be introduced using the following heuristic formula

$$\chi(k) = AFAC\sum_j \frac{N_j}{kR_j^2} |f(\pi,k,R_j)| e^{-2R_j/\lambda(k)} e^{-2\sigma_j^2 k^2} \sin(2kR_j + 2\delta(k) + \psi(k,R_j)) \qquad (3)$$

where $\chi(k)$ is the magnitude of the X-ray absorption fine structure as a function of the photoelectron wave vector k. $AFAC$ is the proportion of electrons that undergo an EXAFS-type scatter and has been fixed at 0.9 for this analysis. N_j is the co-ordination number and R_j is the interatomic distance for the jth shell. The Debye-Waller factor is $A = 2\sigma^2$. $\delta(k)$ and $\psi(k, R_j)$ are the phase shifts experienced by the photoelectron, $f(\pi, k, R_j)$ is the amplitude of the photoelectron backscattering and $\lambda(k)$ is the electron mean free path [4]. The programs EXCALIB, EXBACK and EXCURV98 were used to extract the EXAFS signal and analyse the data to obtain the structural parameters, N, R and A.

The SAXS experiments were carried out on Station 8.2 of the SRS, Daresbury Laboratory, U.K. X-rays of wavelength $\lambda=1.54$ Å were used, with four sets of collimating slits to reduce slit scattering to a negligible level. Samples were contained between two thin sheets of kapton. The data analysis was used primarily to extract information about pore structures using the Guinier Equation

$$I(Q) = I_0 \exp(-\frac{R_g^2 Q^2}{3}) \qquad (4)$$

A power law background was used to take account of the behaviour at low Q due to scattering from the powder/void interface [10].

Experiment and data analysis - catalytic test reaction

Measured amounts, 0.2 g of the xerogel catalysts, prepared as above, and a magnetic stirrer were placed inside a 100 cm³ three neck round bottom flask fitted with a thermometer, reflux condenser and a tapped dropping funnel. The apparatus was flushed with argon for 30 minutes before the addition of 10.0 cm³ dichloromethane, 1.26 cm³ pure cyclohexene (>99 %, Sigma), and 1.0 cm³ TBHP, (5.5 M in decane) via the dropping funnel where 2:1 molar ratio of cyclohexene to TBHP is maintained. The mixture was then refluxed with continual stirring at about 45 °C for 24 h. The mixture was then allowed to cool; the supernatant was collected, centrifuged at 4000 rpm for 4 minutes, and analysed.

RESULTS AND DISCUSSION

The as-prepared xerogels, show good selectivities and reasonable percent conversions overall, as shown in Table I. The surface areas achieved for the xerogels are also quite high, up to ~450 m²g⁻¹. The most effective catalyst was that heat treated to 500 °C at a heating rate of 5 °Cmin⁻¹. Heptane washing, in general, preserves the pore structure of the xerogel and results in increased catalytic performance.

The SAXS data reveal a difference in pore structure between heptane washed and non-washed samples, as shown in Figure 1. The shape of the SAXS in the region between ~0.03 Å⁻¹-0.14 Å⁻¹ can be regarded as a representation of the pore distribution for a given sample. For both

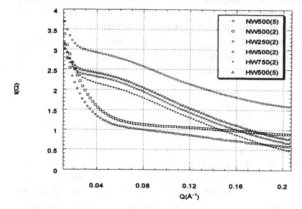

FIGURE 1. SAXS plots for six xerogel samples. Pore distribution immediately observed for heptane washed xerogel samples

Sample	Washing	R_{HT}[a]	HeatT[b]	SArea[c]	Conv%[d]	$S_{epox\%}$[e]	$S_{alc\%}$[f]	$S_{ket\%}$[g]
HW250(2)	Heptane	2	250	430	16.8	96.1	2.5	1.4
HW500(2)	Heptane	2	500	365	15.9	76.1	8.1	15.9
HW750(2)	Heptane	2	750	120	7.8	91.7	4.3	4.0
HW500(5)	Heptane	5	500	320	26.2	97.8	1.7	0.4
NW500(2)	None	2	500	182	8.7	>99	<0.5	<0.5
NW500(5)	None	5	500	25	19.5	91.6	3.9	4.5

TABLE I. Preparation conditions and catalytic performance for the six xerogel samples.[a] Rate of heat treatment (°Cmin⁻¹). [b]Final temperature(°C). [c]Surface area (m²g⁻¹). [d]Catalytic conversion%. [e]Epoxidation%. [f]Alcohol%. [g]Ketone%. Note that the errors are ±10 m²g⁻¹ in SA and ±0.1 % in catalytic activities.

Sample	R_{Ti-O}[a]	N_{Ti-O}[b]	Ti[4]irrev[c]	I(Si-O-Ti)[d]	d_p[e]	I_p[f]
HW250(2)	1.845	3.4	2	68	50	3.80
HW500(2)	1.850	2.7	14	77	54	1.40
HW750(2)	1.830	3.5	24	35	51	1.75
HW500(5)	1.845	2.5	15	97	54	1.05
NW500(2)	1.830	2.9	14	44	14	0.75
NW500(5)	1.845	2.9	21	91	21	0.30

TABLE II. Structural characterisation for the six xerogel samples. [a]R_{Ti-O} represents Ti-O bond distance (Å). [b]N_{Ti-O} represents Ti co-ordination. no. [c]Ti[4]irrev. represents relative amount of fixed 4-fold Ti sites. [d]Intensity of Si-O-Ti linkages. [e]d_p represents mean pore diameter(Å). [f]I_p represents intensity of pore scattering relative to background. Note that the errors are ±0.005 in R_{Ti-O}, ±0.5 in N_{Ti-O}, ±2 in Ti[4]irrev., ±2 in I(Si-O-Ti), ±2 Å in d_p and ±0.05 in I_p.

of the non-washed samples, the SAXS plots in this region are concave, indicative of little or no distinctive pore distribution. Fitting of the Guinier equation reveals small pore sizes, with an average diameter of only ~20 Å. In the case of the heptane washed samples, the capillary pressure is reduced during drying, resulting in preservation of the pore structure. The resultant SAXS plots support this observation, showing a smoothly humped curve in the region ~0.03 Å⁻¹ - 0.14 Å⁻¹, indicative of a significant pore distribution. Fitting of the Guinier equation for the heptane washed samples yields much larger characteristic pore sizes, with an average diameter of ~50 Å, as shown in Table II. Finally, from the calculations of the intensity of pore scattering relative to the background, which is representative of the relative abundance of pores, we conclude that there are more pores (by an order of magnitude) for the heptane washed samples compared to the non-washed samples.

The EXAFS results yield average Ti-O distances in the range 1.83-1.85 Å for all samples. These distances are all close to 1.82 Å (the distance associated with 4-fold Ti, compared with that of 1.95 Å for 6 fold Ti sites) and hence, we can conclude that the majority of the Ti is 4-fold in all of the samples.

This conclusion that the titania and silica are atomically in all the samples is supported by the ^{17}O MAS NMR spectra (see Figure 2) which show the presence of resonances due to Ti-O-Si bonding and an absence of those due to Ti-O-Ti bonding.

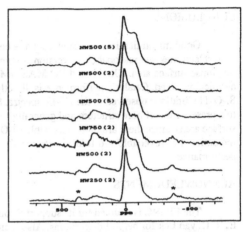

FIGURE 2. ^{17}O MAS NMR spectra of six xerogel samples. (*- Spinning sidebands)

The FT-IR spectra (e.g. see Figure 3) were used to estimate the relative intensity of the Si-O-Ti linkages, from the area of the band that occurs at *ca.* 960cm⁻¹. Table II displays the intensity of Si-O-Ti linkages. A direct correlation is observed between the relative intensity of Si-O-Ti linkages and the catalytic activity; a similar conclusion has been arrived at by other authors [8] suggesting that Si-O-Ti linkages are associated with the 'active sites'. It should be noted that the quantity $I_{(Si-O-Ti)}$ derived from the FT-IR is not a simple measure of Ti mixing. It also includes the effects of surface area because it is normalised by dividing by $I_{(Si-O-Si)}$. The latter quantity will be reduced for samples with high surface area because they have less Si(OSi)₄ units and the silica network is less rigid.

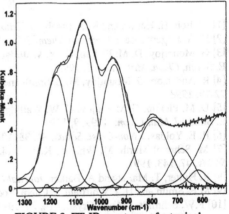

FIGURE 3. FT-IR spectrum of a typical xerogel sample, including deconvolution of bands using Gaussian curve fitting.

The structural parameters derived from the XANES data, are shown in Table II. The XANES data allow us to infer the relative abundance of the following types Ti sites: (i) irreversible 4-fold Ti, which are not readily altered by interaction with H_2O ligands, and are considered 'catalytically inactive', (ii) reversible 4/6 fold Ti sites [3] which readily interact with H_2O ligands, and are considered to be responsible for the catalytic activity. The presence of the latter is implied by the sometimes low values of $Ti^{[4]}irrev$ in Table II. Since we know from the EXAFS and the NMR results that these samples are atomically mixed, the variation in irreversible 4-fold Ti sites must be due to the presence of reversible 4/6 fold Ti sites. (Note, however, that the present XANES data is meaningful only in a relative sense, and absolute values for the proportions of irreversible and reversible 4-fold Ti sites cannot reliably be generated.)

CONCLUSIONS

Good mixed titania-silica catalysts have been synthesised using the sol-gel route, with high selectivites and reasonable conversion percentages, using heptane washing techniques to maximise surface area. EXAFS and ^{17}O MAS NMR studies show all the samples to be atomically mixed. Catalytic performance is directly proportional to the normalised intensity of Si-O-Ti vibrations observed in the FT-IR spectra. SAXS data confirm that heptane washing leads to a preservation of pore structure and generally results in more effective catalysts with higher surface areas, since there are more accessible Si-O-Ti surface species. XANES data gives an indication that 'reversible' 4/6-fold Ti sites are also important in determining catalytic performance.

ACKNOWLEDGMENTS

The EPSRC is thanked for its support through grant GR/L28647. We also, wish to thank E. R. H. van Eck for helpful discussions. Also thanked, are Dr N. J. Terrill and Dr K. C. Cheung for help with X-ray experiments.

REFERENCES

[1] M. Itoh, H. Hattori and K. J. Tanabe, *J. Catalysis*, 225, **35**, 1974.

[2] K. A. Jorgensensen, *American Chem. Soc.*, 89, **3**, 1989.

[3] G. Mountjoy, D. M. Pickup, G. W. Wallidge, R. Anderson, J. M. Cole, R. J. Newport and M. E. Smith, *Chem. Mater.*, 11, **5**, 1999.

[4] R. Anderson, G. Mountjoy, M. E. Smith and R. J. Newport, *J. Non-Cryst. Solids*, 232-234, **72-79**, 1998.

[5] D. M. Pickup, G. Mountjoy, G. W. Wallidge, R. Anderson, J. M. Cole, R. J. Newport and M. E. Smith, *J. Mat. Chem.*, 1299, **9**, 1999.

[6] B. E. Yoldas, *J.Non-Cryst. Solids*, 81, **38**, 1980.

[7] M. Schraml-Marth, M. Walther, K. L. Wokaun, B. E. Handy and A. Baiker, *J. Non-Cryst. Solids*, 47, **143**, 1991..

[8] R. Hutter, T. Mallat and A. Baiker, *Journal Catal.*, 177-189, **153**, 1995.

[9] M.E. Smith and H. J. Whitfield, J. Chem. Soc., Chem. Commun., 723, 1994.

[10] P.W. Schmidt, J. Appl. Cryst., 414, **24**, 1991.

SYNCHROTRON-BASED STUDIES OF TRANSITION METAL INCORPORATION INTO SILICA-BASED SOL-GEL MATERIALS

G. Mountjoy*, D.M. Pickup*, M.A. Holland*, G.W. Wallidge**, R.J. Newport*, M.E. Smith**
*School of Physical Sciences, University of Kent at Canterbury, Canterbury CT2 7NR, U.K.
**Department of Physics, University of Warwick, Coventry CV4 7AL, U.K.

ABSTRACT

Previous structural studies on titania- and zirconia-silica xerogels have shown the occurrence of homogeneous mixing at low metal content, and phase separation at high metal content. The use of additional, complementary, synchrotron-based methods can contribute to a fuller structural description of these materials. We present new X-ray absorption near edge structure (XANES) and SAXS results for $(ZrO_2)_x(SiO_2)_{1-x}$ xerogels and compare them with previous results for $(TiO_2)_x(SiO_2)_{1-x}$ xerogels. Significant differences between $(TiO_2)_x(SiO_2)_{1-x}$ and $(ZrO_2)_x(SiO_2)_{1-x}$ xerogels are observed in the affects of heat treatment on the coordination of homogeneously mixed metal atoms, and in the development of phase separated metal oxide regions.

INTRODUCTION

Mixed titania-silica, $(TiO_2)_x(SiO_2)_{1-x}$, and zirconia-silica, $(ZrO_2)_x(SiO_2)_{1-x}$, materials with low metal content, i.e. x<0.5, are potentially useful in a number of technological applications [1,2], including catalysis [3,4]. These materials are also interesting from a structural point of view because the silica network is tetrahedral, but Ti and Zr prefer coordinations greater than 4 [5]. In comparison to high temperature routes, preparation by the sol-gel process [6] has the advantages of using liquid precursors and occurring at low temperatures. However, the structures of xerogels are strongly dependent on the details of preparation and heat treatment.

This paper concerns acid-catalysed $(TiO_2)_x(SiO_2)_{1-x}$ and $(ZrO_2)_x(SiO_2)_{1-x}$ xerogels. Our group has studied the same samples using SAXS [7], diffraction, IR, NMR and X-ray absorption spectroscopies [8,9,10,11 and references therein]. In general, metal atoms are homogeneously incorporated into the silica network at low concentrations, i.e. x<0.2, and substantial phase separation of metal oxide occurs at high concentrations, i.e. x~0.4. This paper focuses on two complementary methods which augment the structural description of these materials. X-ray absorption fine structure (EXAFS) and near edge structure (XANES) probe the local atomic environment of metal atoms, and conversely, small angle X-ray scattering (SAXS) probes mesoscopic-scale inhomogeneities due to phase separation. Here we present new SAXS results for $(ZrO_2)_x(SiO_2)_{1-x}$ xerogels, and compare them with $(TiO_2)_x(SiO_2)_{1-x}$ xerogels [7].

We have previously reported Ti and Zr K-edge EXAFS results for titania- [8,9] and zirconia-silica [10] xerogels (respectively). Table 1 shows selected results for metal-oxygen coordination number N, distance R, and Debye-Waller term $A=2\sigma^2$. In reference compounds, Ti with 4-, 5- and 6-fold coordination has Ti-O distances of 1.81, 1.70/1.99 and 1.96Å respectively [12]. The EXAFS results indicate that for x<0.2, Ti has coordination of 4 in heat treated samples, but >4 in unheated samples. In samples with x~0.4, Ti has a coordination closer to 6. Zr-O coordinations in reference compounds are shown in Table 2. The EXAFS results in Table 1 indicate that for x~0.4, Zr has coordination of 7. In samples with x~0.1, the coordination is characterised by a split Zr-O shell, somewhat similar to cubic ZrO_2. However, there remains some ambiguity in the interpretation of the EXAFS results. Here we present new XANES results for $(ZrO_2)_x(SiO_2)_{1-x}$ xerogels, and compare them with $(TiO_2)_x(SiO_2)_{1-x}$ xerogels [7].

Mat. Res. Soc. Symp. Proc. Vol. 590 © 2000 Materials Research Society

Table 1: Metal-oxygen coordination from EXAFS [8,10] in xerogels.

x	heat treat. (°C)	metal-oxygen coordination		
		N	R (Å)	A (Å^2)
(TiO$_2$)$_x$(SiO$_2$)$_{1-x}$				
0.41	none	6.2(10)	1.87(1)	0.05(1)
	750	5.0(6)	1.90(1)	0.022(5)
0.18	none	5.5(10)	1.84(1)	0.03(1)
	750	4.4(4)	1.81(1)	0.016(3)
(ZrO$_2$)$_x$(SiO$_2$)$_{1-x}$				
0.4	none	7.9(5)	2.14(1)	0.023(2)
	750	7.4(5)	2.12(1)	0.028(2)
0.1	none	3.1(4)	2.00(1)	0.006(1)
		6.4(13)	2.25(1)	0.03(1)
	750	2.4(8)	1.98(1)	0.007(2)
		4.4(17)	2.15(4)	0.04(2)

Table 2: Zr-O coordination in reference compounds [13].

compound	Zr-O coord.		symmetry
	N	R (Å)	
ZrSiO$_4$	4	2.13	tetrahedral
	4	2.27	
tetrag. ZrO$_2$	4	2.08	tetrahedral
	4	2.38	
cubic ZrO$_2$	3	2.04	distort. tetrah.
	4	2.28	
mono. ZrO$_2$	7	2.16	disordered
Zr(OH)$_4$	7	2.14	disordered
zirconolite	7	2.16	disordered
BaZrO$_3$	6	2.09	octahedral
Zr propoxide	2	1.96	distort. octah.
	4	2.18	

EXPERIMENT

Sample Preparation

The (TiO$_2$)$_x$(SiO$_2$)$_{1-x}$ samples were prepared using a two-step hydrolysis procedure with acid catalysis (HCl, pH=1). Firstly, tetraethoxyorthosilicate (TEOS) was prehydrolysed for 2hrs by mixing with water and isopropanol in approximately equimolar ratios. Secondly, Ti isopropoxide, and then water, were added dropwise while stirring, such that the final ratio of alkoxide to water equals 2. The resulting gels were dried in ambient conditions and then under vacuum to produce xerogels. Heat treatment consisted of heating at 5°C/min followed by 2hrs at constant temperature. The (ZrO$_2$)$_x$(SiO$_2$)$_{1-x}$ samples were prepared in the same way, except that the solvent was propanol, and the metal alkoxide was Zr propoxide diluted 1:5 in propanol.

X-ray absorption spectroscopy

X-ray absorption spectroscopy experiments were carried out at the Ti and Zr K-edges on stations 8.1 and 9.2 (respectively) of the SRS, Daresbury Laboratory, U.K. Here we describe the Zr K-edge XANES experiments. Samples of suitable and uniform thickness were prepared by grinding and pressing into pellets. Spectra were collected at the Zr K-edge (17998eV) in transmission mode. At the Zr K-edge, the monochromator resolution is ~5eV, and the limited lifetime of the core-hole causes a broadening of ~2eV [14]. Absorption was measured using standard ion chambers. The XANES spectra were processed in the usual way to obtain normalised absorbance as a function of x-ray energy E [14]. Energies are reported relative to the main inflection point. Experiments at the Ti K-edge (4966eV) were similar [11], except that the energy resolution was ~1eV, the monochromator energy was calibrated during each experiment, and energies are reported relative to the first inflection point of Ti metal.

Small angle x-ray scattering

Here we describe the SAXS experiments for (ZrO$_2$)$_x$(SiO$_2$)$_{1-x}$ xerogels (the SAXS experiments for (TiO$_2$)$_x$(SiO$_2$)$_{1-x}$ xerogels were similar [7]). The SAXS experiments were carried out on station 8.2 of the SRS, Daresbury Laboratory, U.K. X-rays of wavelength λ=1.54Å were used with 4 sets of collimating slits and a camera length of ~4.0m. The angular scale of scattering is given by $Q=4\pi\sin(\theta/2)/\lambda$, where θ is the scattering angle. Absorption was measured using standard ion chambers. The SAXS was recorded using a quadrant detector. The SAXS detector response was calibrated using a radioactive ^{55}Fe source. The SAXS intensity $I(Q)$ was obtained using standard data reduction techniques.

RESULTS

<u>X-ray absorption spectroscopy</u>

The K-edge XANES spectra for transition metal oxides show characteristic features [15]. The shape of the main absorption peak represents transitions to p-type continuum states and "shape resonances" of the metal atom environment. Modelling these using multiple scattering calculations is a very complex endeavour. In many cases, useful qualitative information can be obtained by comparison with reference compounds representing standard coordinations.

The XANES spectra may contain pre-edge peak(s) representing transitions to low-lying states with pd mixing. Such transitions are disallowed for completely centrosymmetric metal atom sites, but increase in intensity as the degree of centrosymmetry decreases. The prominence of pre-edge peaks varies for different elements. Ti is an element that exhibits very strong pre-edge peaks, as shown in Figure 1. Furthermore, the height of the pre-edge peak increases from 6- to 5- to 4-fold coordination (denoted $Ti^{[6]}$, $Ti^{[5]}$, $Ti^{[4]}$) due to decreasing centrosymmetry as the coordination changes from octahedral to square pyramidal to tetrahedral, respectively [12]. This approach is put on a quantitative basis by fitting Lorentzians to the pre-edge peak to obtain its relative position and height. Studies of reference compounds have shown that reliable information about Ti coordination can be obtained, as shown in Figure 2 [12].

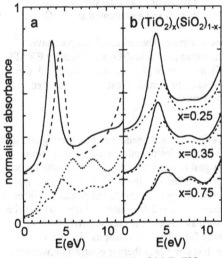 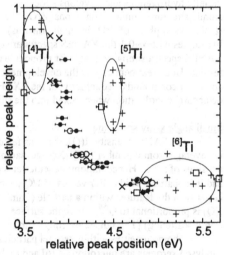

Figure 1: Ti K-edge XANES of (a) $BaTiO_4$, Na_2SiTiO_5, anatase, $ZrTiO_4$ (top to bottom), and (b) $(TiO_2)_x(SiO_2)_{1-x}$ xerogels with x=0.25, 0.35 and 0.75 unheated (dashed) and heat treated at 750°C (solid) [11].

Figure 2: Height and position of pre-edge peak in reference compounds (+ and □), $(TiO_2)_x(SiO_2)_{1-x}$ xerogels prepared with isopropoxide (o) and acetyl-acetone (•) [11], and titania-silica glasses (×) [12,16].

85

We have used quantitative characterisation of the pre-edge peak to investigate Ti coordination in $(TiO_2)_x(SiO_2)_{1-x}$ xerogels [11]. Typical XANES spectra are shown in Figure 1, and the relative peak positions and heights are shown in figure 2. The results show that xerogels contain a mixture of $Ti^{[6]}$ and $Ti^{[4]}$. Unheated samples contain $Ti^{[6]}$ which is isolated in samples with low x, and in phase separated TiO_2 regions in samples with high x. In samples with low x, heat treatment causes the conversion of isolated $Ti^{[6]}$ to $Ti^{[4]}$, with the loss of hydroxyl and water groups and the substitution of Ti for Si in the silica network. This effect is reduced as x increases because more of the Ti is in phase separated regions, remaining in the form of $Ti^{[6]}$.

To obtain information from the XANES spectra of $(ZrO_2)_x(SiO_2)_{1-x}$ xerogels, we compare them with reference compounds, as shown in Figure 3. In samples with x=0.4, the XANES spectra is initially similar to that in $Zr(OH)_4$ and with heat treatment it develops some similarity to that in tetragonal ZrO_2. This implies an amorphous ZrO_2 phase which is initially hydrogenous and develops a local atomic structure similar to tetragonal ZrO_2 (the stable crystal phase of ZrO_2 in small domains).

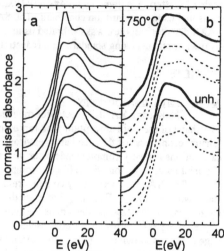

Figure 3: Zr K-edge XANES of (a) $ZrSiO_4$, tetragonal, cubic and monoclinic ZrO_2, $Zr(OH)_4$, zirconolite [13], $BaZrO_3$, Zr propoxide (top to bottom), and (b) $(ZrO_2)_x(SiO_2)_{1-x}$ xerogels with x=0.1(short dash), 0.2 (long dash), 0.3 (thin solid) and 0.4 (thick solid), unheated and heat treated at 750°C.

In samples with x=0.1 the XANES spectra are quite different. The unheated samples have XANES spectra similar to that in Zr propoxide rather than the cubic ZrO_2 suggested by EXAFS results. Heat treatment causes the development of a much more prominent pre-edge peak than in any other compounds. Together with the EXAFS results, this implies that Zr has distorted octahedral coordination when mixed into the silica network.

Small angle x-ray scattering

The SAXS intensity, I(Q), contains features due to inhomogeneities in scattering strength, i.e. average atomic number [17]. Xerogel samples are powdered, and typically have packing densities of 50%. Hence, inhomogeneities at the largest length scales are due to powder particles and inter-particle voids. For values of 1/Q which are smaller than the particle size (~μm), but larger than the features within a particle (~nm), the scattering intensity due to particle surfaces is I(Q) is proportional to Q^{d-6}, where the surface "dimension" d is 2 for smooth sharp, surfaces (i.e. Porod scattering) [17], but >2 for rough surfaces, such as observed in clay minerals [18].

Inhomogeneities within xerogel particles arise for at least two reasons. Firstly, acid-catalysed xerogels are microporous [6] and on length scales ≤2nm there is contrast between the silica-based network and pores. Secondly, mixed titania- and zirconia-silica xerogels may contain regions of phase separated metal oxide [e.g. 9,10]. Such phase separated regions may be approximated as independent regions of finite spatial extent, for which I(Q) is proportional to $\exp(-R_g^2Q^2/3)$, where R_g is the radius of gyration [17]. Such scattering causes a shoulder in I(Q) at $1/Q \sim R_g$.

Figure 4 shows the SAXS intensity for $(TiO_2)_x(SiO_2)_{1-x}$ xerogels, with x=0, 0.08, 0.18 and 0.41 [7]. In the samples with x=0 and 0.08 there is no phase separation of TiO_2 (^{17}O NMR spectra show no OTi_n configurations [9], and X-ray absorption spectroscopy shows $Ti^{[4]}$ [11]). The I(Q) show strong Q^{-n} scattering at low Q which is expected due to the surfaces of xerogel powder particles. Interestingly, the values of n are <4, corresponding to rough surfaces, with estimated dimension d=2.35±0.1. The particle surfaces are rough because they are perforated by micropores which permeate the silica-based network. At high Q the I(Q) show a plateau which changes slightly with heat treatment [7]. This feature is due to inhomogeneities within the bulk of the xerogel particles due to micropores.

Having established the SAXS features for homogeneous xerogel samples, we now consider those for phase separation of TiO_2. This occurs in samples with x=0.41 (^{17}O NMR spectra show OTi_n configurations [9], and X-ray absorption spectroscopy shows $Ti^{[6]}$ [11]). The I(Q) clearly have additional intensity in the region of $0.04<Q<0.1\text{Å}^{-1}$. With heat treatment the additional intensity becomes more pronounced and moves to lower Q values, corresponding with the formation and growth of anatase crystals. Note that the x=0.18 sample initially has Ti homogeneously mixed in the silica network, but phase separation begins to occur at 750°C due to the reduced solubility of Ti (note that there is no crystallisation at this stage) [7].

Figure 5 shows the SAXS intensity for $(ZrO_2)_x(SiO_2)_{1-x}$ xerogels. The sample with x=0.1 has Zr homogeneously mixed in the silica network (^{17}O NMR spectra show no OZr_n configurations [10]), and the I(Q) is the same as for SiO_2 and $(TiO_2)_{0.08}(SiO_2)_{0.92}$ xerogels. In contrast, the sample with x=0.4 contains substantial phase separated ZrO_2 (^{17}O NMR spectra shows OZr_n configurations [10]). This is apparent in the I(Q) for the unheated sample with x=0.4, and to a lesser extent x=0.3, which have additional intensity in the region $0.06<Q<0.1\text{Å}^{-1}$. The I(Q) for the unheated $(ZrO_2)_{0.4}(SiO_2)_{0.6}$ and $(TiO_2)_{0.41}(SiO_2)_{0.59}$ xerogels are very similar. However, after heat treatment the I(Q) for the $(ZrO_2)_{0.4}(SiO_2)_{0.6}$ xerogel shows no strong feature representing phase separation. This indicates that the separate ZrO_2 phase is much less

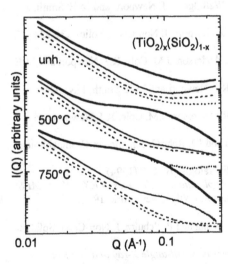
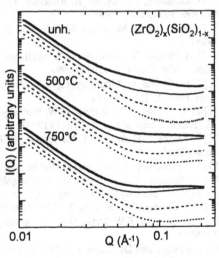

Fig. 4: SAXS I(Q) for $(TiO_2)_x(SiO_2)_{1-x}$ xerogels with x=0 (short dash), 0.08 (long dash), 0.18 (thin solid) and 0.41 (thick solid)

Fig. 5: SAXS I(Q) for $(ZrO_2)_x(SiO_2)_{1-x}$ xerogels with x=0.1 (short dash), 0.2 (long dash), 0.3 (thin solid) and 0.4 (thick solid).

susceptible to consolidation and eventual crystallisation than the separate TiO_2 phase (the former does not crystallise at 750°C, but the latter begins to crystallise at 500°C). The separate amorphous ZrO_2 is hence stabilised by the surrounding silica network.

CONCLUSIONS

The XANES spectra for $(ZrO_2)_x(SiO_2)_{1-x}$ xerogels with x=0.4, in which there is phase separated ZrO_2, indicates the local atomic structure develops with heat treatment from being like that in $Zr(OH)_4$ to like that in tetragonal ZrO_2. The XANES spectra for xerogels with x=0.1, in which Zr is homogeneously mixed into the silica network, indicates that the Zr coordination is initially similar to that in Zr propoxide and remains octahedral after heat treatment. This contrasts with $(TiO_2)_x(SiO_2)_{1-x}$ xerogels with x<0.2, in which Ti coordination is initially octahedral but becomes tetrahedral after heat treatment. The SAXS intensities show that inhomogeneities due to phase separation of metal oxide in both $(TiO_2)_{0.4}(SiO_2)_{0.6}$ and $(ZrO_2)_{0.4}(SiO_2)_{0.6}$ are initially very similar. However, whereas TiO_2 regions undergo consolidation and subsequent crystallisation, this does not happen to ZrO_2 regions which are stabilised by the surrounding silica network.

REFERENCES
[1] P.C. Shultz and H.T. Smyth, in *Amorphous Materials*, edited by Douglas EW and Ellis B (Wiley, London, 1972).
[2] M. Nogami, J. Non-Cryst. Solids **69**, 415 (1985).
[3] M. Itoh, H. Hattori, and K.J. Tanabe, J. Catalysis **35**, 225 (1974).
[4] J.B. Miller and E.I. Ko, J. Catalysis **159**, 58 (1996).
[5] R. Gill, *Chemical Fundamentals of Geology* (Unwin Hyman, London, 1989).
[6] C.J. Brinker and G.W. Scherer, *Sol-gel science: the Physics and Chemistry of Sol-gel Processing* (Academic Press, San Diego, 1990)
[7] G. Mountjoy, J.S. Rigden, R. Anderson, G.W. Wallidge, R.J. Newport, and M.E. Smith, J. Mat. Res. submitted (1999).
[8] R. Anderson, G. Mountjoy, M.E. Smith and R.J. Newport, J. Non-Cryst. Solids **232-234**, 72 (1998).
[9] D.M. Pickup, G. Mountjoy, G.W. Wallidge, R. Anderson, J.M. Cole, R.J. Newport and M.E. Smith, J. Mat. Chem. **9**, 1299 (1999).
[10] D.M. Pickup, G. Mountjoy, G.W. Wallidge, R.J. Newport and M.E. Smith, Phys. Chem. Chem. Phys. **1**, 2527 (1999).
[11] G. Mountjoy, D.M. Pickup, G.W. Wallidge, R. Anderson, J.M. Cole, R.J. Newport and M.E. Smith, Chem. Mater. **11**, 1253 (1999).
[12] F. Farges, G.E. Brown, A. Navrotsky, H. Gan and J.J. Rehr Geochimica et Cosmochimica acta **60**, 3023 (1996).
[13] F. Farges, G.E. Brown, and D. Velde, Am. Mineralogist **70**, 838 (1994).
[14] A. Bianconi, in *X-ray absorption: principles, applications, techniques of EXAFS, SEXAFS and XANES*, edited by Koninbsberger D.C. and Prins R. (Wiley, New York, 1987).
[15] L.A. Grunes Phys. Rev. B. **27**, 2111 (1983).
[16] R.B. Greegor, F.W. Lytle, D.R. Sandstrom, J. Wong and P. Schultz, J. Non-Cryst. Solids **55**, 27 (1983).
[17] LA Feigin and DI Svergun, in *Structure analysis by small-angle X-ray and neutron scattering* (Plenum Press, New York, 1987)
[18] P.W. Schmidt, J. Appl. Cryst. **24**, 414 (1991)

Scattering and Diffraction

An Inelastic Nuclear Resonant Scattering Study of Partial Entropies of Ordered and Disordered Fe₃Al

B. FULTZ*, W. STURHAHN**, T. S. TOELLNER** and E. E. ALP**
* Div. Engineering and Applied Science, 138–78, California Institute of Technology, Pasadena, California 91125, btf@hyperfine.caltech.edu
** Advanced Photon Source, Argonne National Laboratory, Argonne, Illinois 60439

ABSTRACT

Inelastic nuclear resonant scattering spectra were measured on alloys of Fe_3Al that were chemically disordered, partially-ordered, and $D0_3$-ordered. The phonon partial DOS for ^{57}Fe atoms were extracted from these data, and the change upon disordering in the partial vibrational entropy of Fe atoms was obtained. By comparison to previous calorimetry measurements, it is shown that the contribution of the Fe atoms to the vibrational entropy is a factor of 10 smaller than that of the Al atoms. With the assistance of Born - von Kármán model calculations on the ordered alloy, it is shown that differences in the vibrational entropy originate primarily with changes in the optical modes upon disordering. The phonon DOS of ^{57}Fe was found to change systematically with chemical short range order in the alloy. It is argued that changes in the vibrational entropy originate primarily with changes in the chemical short-range order in the alloy, as opposed to long-range order.

INTRODUCTION

Vibrational Entropy

Phase diagrams of materials have been the subject of extensive experimental and theoretical research. In recent years, free energies of solid phases have been calculated by elegant combinations of methods for calculating the electronic energy with the local density approximation and methods using cluster approximations for calculating the *configurational* entropy [1,2]. The change in *vibrational* entropy during a solid state phase transition is often thermodynamically important, however [3-8]. It is now accepted that vibrational entropy must be considered in alloy thermodynamics. Recent work has focused on the reasons underlying the differences in vibrational entropy of different states of materials [6-10].

Knowledge of the phonon density of states (DOS) is an important step towards the understanding of vibrational entropy. For alloys, however, the total phonon DOS masks important details. The different atomic species in an alloy are not expected to have identical motions. The partial phonon DOS, defined below, describes the spectrum of motions of the individual atoms in the alloy. Knowledge of the motions of individual atoms, rather than the overall vibrational spectrum, provides more of the information needed to identify reasons for the differences in vibrational entropy of alloy phases. In this paper we use knowledge of the ^{57}Fe phonon partial DOS in Fe_3Al to show that the changes in vibrational entropy upon ordering are caused primarily by changes in the motions of Al atoms. Furthermore, we show that much of this change in the motion of Al atoms occurs in those phonons described as optical modes of the alloy.

Mat. Res. Soc. Symp. Proc. Vol. 590 © 2000 Materials Research Society

Inelastic Nuclear Resonant Scattering

To date the most detailed information on phonons in solids has been obtained from studies on coherent inelastic neutron scattering by single crystals of pure elements or ordered compounds. The momentum and energy transfers between the neutron and the crystal are used to identify both the frequencies and wavevectors of phonon excitations in the solid. Unfortunately, for many interesting materials it is impossible to obtain single crystals for coherent inelastic neutron scattering experiments. This is especially true for materials far from thermodynamic equilibrium, highly disordered alloys for example, where the methods for preparing single crystals will destroy the interatomic disorder. Furthermore, interpreting the phonon frequencies from neutron groups measured on single crystals of disordered alloys is typically performed with a virtual crystal approximation, which is known to be unreliable [9]. For such materials it is appropriate to measure the spectrum of vibrational excitations, without information on the phonon wavevectors. An experimental advance towards more detailed data is to measure the phonon partial DOS for the individual atoms in an alloy.

A method using synchrotron radiation for measuring the phonon partial DOS in solids has been developed recently [11,12]. The method evolved from studies on nuclear forward scattering of synchrotron radiation, which has been increasingly practical since 1985 [13]. (Reference [14] is a recent review of nuclear forward scattering.) Undulator-based third generation synchrotron sources have invigorated these applications of the Mössbauer effect. The Mössbauer effect is, by definition however, purely elastic scattering. Nevertheless, outside the narrow energy window for the Mössbauer effect itself, it is now possible to observe nuclear transitions that are accompanied by phonon creation or annihilation. This method of inelastic nuclear resonant x-ray scattering is similar to incoherent inelastic neutron scattering, where an elastic line is surrounded by inelastic intensity that can be used to obtain the phonon DOS of a material.

PRINCIPLES

Partial Vibrational Entropy

In the classical limit, vibrational entropy is related to the phonon DOS, $g(\varepsilon)$, as

$$S_{vib} = -3k_B \int_0^\infty g(\varepsilon) \ln\left(\frac{\varepsilon}{\varepsilon_0}\right) d\varepsilon \qquad (1)$$

where ε is the phonon energy, and ε_0 is a constant that will vanish when differences in entropy are calculated. The phonon DOS, $g(\varepsilon)$, is a spectrum of phonon frequencies, each of energy ε_j. In discrete form it is

$$g(\varepsilon) = \frac{1}{3N} \sum_j^{3N} \delta(\varepsilon - \varepsilon_j) \qquad (2)$$

Substituting Eq. 2 into 1 gives

$$S_{vib} = -\frac{k_B}{N} \int_0^\infty \sum_j^{3N} \delta(\varepsilon - \varepsilon_j) \ln\left(\frac{\varepsilon}{\varepsilon_0}\right) d\varepsilon \qquad (3)$$

For each phonon mode in the alloy, denoted, "j", we have a polarization vector, $\mathbf{e}_{\xi,j}$, that describes the displacement and phase of the atom, ξ, in the unit cell [15]. To ensure consistent amplitudes for all phonons, there must be a normalization of the polarization vectors for all atoms in the unit cell

$$\sum_{\xi} \left| \mathbf{e}_{\xi,j} \right|^2 = 1 \tag{4}$$

We therefore write:

$$S_{vib} = -\frac{k_B}{N} \sum_{\xi} \int_0^{\infty} \sum_j^{3N} \left| \mathbf{e}_{\xi,j} \right|^2 \delta(\varepsilon - \varepsilon_j) \ln\left(\frac{\varepsilon}{\varepsilon_0}\right) d\varepsilon \tag{5}$$

To separate the contributions of individual atoms to the vibrational entropy, it is convenient to define a "phonon partial DOS" as

$$g_{\xi}(\varepsilon) = \frac{1}{3N} \sum_j^{3N} \left| \mathbf{e}_{\xi,j} \right|^2 \delta(\varepsilon - \varepsilon_j) \tag{6}$$

Recognizing the expression of Eq. 6 in Eq. 5, we see that Eq. 5 is a sum of partial entropy contributions, S_{ξ}, from each atom, ξ, in the unit cell

$$S_{\xi} = -3k_B \int_0^{\infty} g_{\xi}(\varepsilon) \ln\left(\frac{\varepsilon}{\varepsilon_0}\right) d\varepsilon \tag{7}$$

The total vibrational entropy is the sum of partial entropies obtained from the phonon partial DOS's

$$S_{vib} = \sum_{\xi} S_{\xi} \tag{8}$$

Data Analysis

The phonon partial DOS of ^{57}Fe, $g_{Fe}(\varepsilon)$, is obtained from the inelastic nuclear resonant scattering spectra by correcting for the thermal occupancies of phonon modes and multiphonon scattering. The measured scattering intensity, $I(Q,\varepsilon)$, is a sum of resonant absorption processes, each accompanied by a different number of phonon creations or annhilations, $I_n(Q,\varepsilon)$. (Here ε is the energy loss, and Q is the momentum transfer, which is 7.3 Å^{-1} for the 14.41 keV transition in ^{57}Fe.) We use the definitions [16]

$$\gamma_0 \equiv \int_0^{\infty} \coth\left(\frac{\varepsilon}{2k_B T}\right) \frac{g_{Fe}(\varepsilon)}{\varepsilon} d\varepsilon \tag{10}$$

$$2W = \frac{(\hbar Q)^2}{2M} \gamma_0 \tag{11}$$

$$A_1(\varepsilon) \equiv \frac{g_{Fe}(\varepsilon)}{\gamma_0 \varepsilon \left(1 - \exp\left(\frac{\varepsilon}{k_B T}\right)\right)} \tag{12}$$

$$A_n(\varepsilon) \equiv \int_{\infty}^{\infty} A_1(\varepsilon - \varepsilon') A_{n-1}(\varepsilon') d\varepsilon \qquad (13)$$

The $2W$ is exponentiated to form the Debye-Waller factor, $A_1(\varepsilon)$ is the single scattering profile, and $A_n(\varepsilon)$ is the scattering profile for the creation of n phonons.

Extracting the phonon partial DOS, $g_{Fe}(\varepsilon)$, is the goal of our data analysis. We can do this easily with Eq. 12 once we know the single scattering profile, $A_1(\varepsilon)$. This requires that the experimental spectra are corrected for multiphonon scattering. For fixed values of Q and temperature, the energy dependence of the various phonon contributions, $I_n(Q,\varepsilon)$, is

$$I_n(Q,\varepsilon) = \frac{(2W)^n}{n!} e^{-2W} A_n(\varepsilon) \qquad (14)$$

The measured scattering intensity, $I(Q,\varepsilon)$, is the sum of intensities from all phonon processes

$$I(Q,\varepsilon) = \sum_n I_n(Q,\varepsilon) \qquad (15)$$

Our problem is to use the measured $I(Q,\varepsilon)$ to isolate the single scattering profile, $A_1(\varepsilon)$. A convenient approach was mentioned by M. Y. Hu, et al. [17]. This method automatically corrects for the instrument resolution, $Z(\varepsilon)$, as obtained from the zero-loss peak. This resolution function, $Z(\varepsilon)$, is convoluted with the scattering from the specimen

$$I(Q,\varepsilon) = Z(\varepsilon) * \left(e^{-2W} + \sum_n I_n(Q,\varepsilon) \right) \qquad (16)$$

The first term in the braces is the zero loss peak, reduced by the Debye-Waller factor. Taking the Fourier transform, $\mathcal{F}[]$, of Eq. 16 simplifies the convolutions of Eq. 13, which become multiplications in Fourier space.

$$\mathcal{F}[I(Q,\varepsilon)] = \mathcal{F}[Z(\varepsilon)] e^{-2W} \left(1 + \sum_n \frac{(2W)^n}{n!} (\mathcal{F}[A_1(\varepsilon)])^n \right) \qquad (17)$$

The term in braces on the right side of Eq. 17 is recognized as the expansion of an exponential function

$$\mathcal{F}[I(Q,\varepsilon)] = \mathcal{F}[Z(\varepsilon)] e^{-2W} \left(e^{2W \mathcal{F}[A_1(\varepsilon)]} \right) \qquad (18)$$

To isolate the single scattering profile, $A_1(\varepsilon)$, we take the logarithm of Eq. 18, rearrange, and perform the inverse transformation, $\mathcal{F}^{-1}[]$. If we neglect the normalization of the single scattering profile, and if we delete the spike at zero energy (which is suppressed by the thermal correction of Eq. 12 anyway), we obtain

$$A_1(\varepsilon) = \mathcal{F}^{-1}\left[\ln\left(\frac{\mathcal{F}[I(\varepsilon)]}{\mathcal{F}[Z(\varepsilon)]} \right) \right] \qquad (19)$$

Data analysis therefore proceeds as follows. After subtraction of a constant background of order 0.01 Hz, the zero loss profile, $Z(\varepsilon)$, is identified in the measured spectrum. This elastic peak is subtracted from the data, and Eq. 19 is used to obtain the single scattering profile, $A_1(\varepsilon)$. The single scattering profile is easily converted into the phonon partial DOS, $g_{Fe}(\varepsilon)$, with Eq. 12. This data analysis procedure is perfomed with the software package PHOENIX, written by W. Sturhahn, which provides other analyses of the inelastic scattering and tests the reliability of the experimental spectra.

EXPERIMENTAL

The measurements were performed at the 3-ID undulator beamline of the Advanced Photon Source. A premonochromater consisting of water-cooled diamond (111) crystals in a nondispersive setting produced a beam of 14.413 keV synchrotron radiation with an energy bandwidth of 1.2 eV. Further monochromatization to an 875 μeV bandwidth was provided by a high-resolution monochromater [18], comprising two asymmetically-cut silicon (975) crystals in a dispersive geometry. The photon flux incident on the foil sample within this bandwidth was 6×10^8 Hz. The energy alignment of the high-resolution monochromater was tuned around the nuclear resonance in steps of 300 μeV. An avalanche photodiode with an active area of 2 cm^2 was mounted 4 mm above the specimen and used to detect K-shell internal conversion x-rays of 6.4 keV. To suppress the x-ray fluorescence processes excited by the synchrotron flash, data were acquired approximately 30 ns after the flash. All measurements were performed at room temperature.

Measurements were performed on three foils of ^{57}Fe$_3$Al of 3 μm thickness. They were prepared by arc-melting a 50 mg ingot under an argon atmosphere, followed by piston-anvil quenching into a thin foil and cold rolling. A JEOL Superprobe 733 electron microprobe was used to check for chemical heterogeneities, and to determine that the chemical composition of the foil was Fe-26.0 at.% Al. One specimen was used directly in this state, and is denoted the "disordered" sample. A second foil of ^{57}Fe$_3$Al was annealed at 473 K for 1 h to induce partial chemical order, and is denoted the "partially-ordered" sample. The third foil of ^{57}Fe$_3$Al was annealed at 773 K for 6 days, followed by 723 K for 40 days, and is denoted the "ordered" sample. The states of chemical long-range order (LRO) of the three samples of ^{57}Fe$_3$Al were determined by x-ray diffractometry and by conversion electron Mössbauer spectrometry. Some x-ray diffraction patterns are presented in Fig. 1. Only the "ordered" sample shows any D0$_3$ LRO, as evidenced by intense (1/2 1/2 1/2) and (100) x-ray superlattice diffractions. The diffraction pattern from the partially-ordered sample may show some broad intensity from about 28 - 37 degrees that could be diffuse scattering originating with chemical short-range order (SRO).

Fig. 1 X-ray diffraction patterns (Co Kα radiation, Debye-Scherrer optics) from the three specimens of ^{57}Fe$_3$Al, scaled for equal intensities of the (110) fundamental diffraction at 52°.

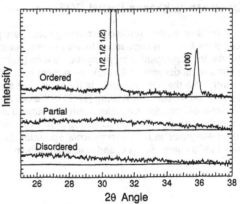

Conversion electron Mössbauer spectrometry was used to measure the chemical short range order in the samples (Fig. 2). A perfect state of DO_3 chemical order would have Fe atoms with only 0 or 4 Al neighbors (the 4[b] and 8[c] sites in Wyckoff notation) in a 1:2 ratio. The excess of Fe atoms with 3 Al neighbors in our ordered sample indicates a LRO parameter between 0.7 and 0.8 (where perfect DO_3 order would be 1.0). The disordered sample shows a much broader distribution of local chemical environments, peaking at 2 1nn Al atoms as expected for a random solid solution. For the disordered alloy, the probability of a finding an Fe atom with the number, n, of 1nn Al atoms follows approximately the binomial distribution [19], although there was a small excess of Fe atoms with 4 Al neighbors. The partially-ordered sample shows an HMF distribution that is intermediate between the ordered and disordered alloy.

Fig. 2 a. Conversion electron Mössbauer spectra from the specimen of bcc ^{57}Fe and the three specimens of DO_3-ordered ^{57}Fe$_3$Al.
b. HMF distributions of the ^{57}Fe$_3$Al specimens, extracted by the method of [20]. Peaks in the HMF distribution are labeled with numbers indicating the different numbers of 1nn Al neighbors about the ^{57}Fe atom.

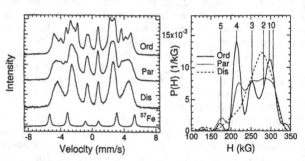

Local vibrational amplitudes of the Fe atoms and Al atoms with respect to their 1nn atoms were measured with extended electron energy loss fine structure spectrometry (EXELFS) [9,21,22], performed at temperatures ranging from 100 K to 420 K. The Al K-edge and Fe $L_{2,3}$-edge electron energy-loss spectra were acquired with a Gatan 666 parallel-detection magnetic prism spectrometer attached to a Philips EM 430 transmission electron microscope. Processing of the spectra to obtain EXELFS oscillations and radial distribution functions (RDF's) followed procedures described elsewhere [5,9,21,22].

RESULTS

Experimental Phonon Partial DOS

Inelastic nuclear resonant scattering spectra are presented in Fig. 3. The Mössbauer elastic peak at $\varepsilon = 0$ meV is surrounded by the inelastic contribution from nuclear resonant photon absorptions accompanied by phonon excitation (right), and phonon absorption (left). The vibrational modes around 43 meV are present in the disordered alloy, but show a distinct growth in intensity with the chemical order in the alloy. The non-zero intensity at energies above 50 meV is primarily multiphonon scattering [16,23]. We corrected for the multiphonon scattering by the Fourier-logarithm deconvolution method described above, and by an iterative procedure [24]. For comparison, the two phonon scattering profile calculated with the iterative procedure is shown in Figure 4 together with the experimental spectrum. These spectra were converted into phonon partial DOS curves for ^{57}Fe, and results are shown at the bottom of Fig. 5.

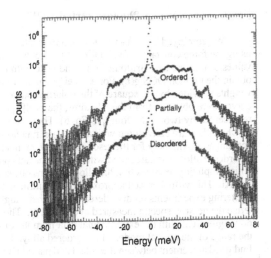

Fig. 3. Conversion x-ray intensity versus photon energy from the disordered, partially-ordered, and ordered alloys near the elastic (Mössbauer) resonance set at 0 meV. Curves for the partially-ordered and ordered samples are shifted vertically by factors of 10 and 100, respectively.

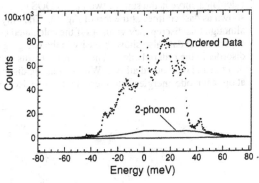

Fig. 4. Experimental spectrum of 3, together with the 2-phonon contribution to the scattering.

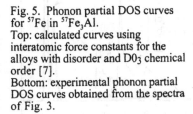

Fig. 5. Phonon partial DOS curves for ^{57}Fe in ^{57}Fe$_3$Al.
Top: calculated curves using interatomic force constants for the alloys with disorder and D0$_3$ chemical order [7].
Bottom: experimental phonon partial DOS curves obtained from the spectra of Fig. 3.

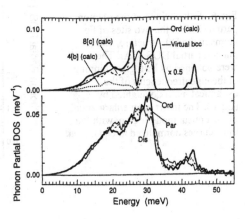

Born - von Kármán Phonon Partial DOS

We employed a Born - von Kármán model to calculate the phonon densities of states (DOS) using the force constants of van Dijk [25]. The dynamical matrix was diagonalized for about 10^7 values of **k** in the first Brillouin zone, and histogram binning of the eigenfrequencies was used to obtain the phonon DOS $g(\varepsilon)$. For each eigenmode, four other histograms were prepared with weights provided by the square of the polarization vectors, $|e_{\xi,j}|^2$, so the phonon partial DOS curves were obtained from the Al atoms, the Fe 4[b] atoms, and the Fe 8[c] atoms (the latter were represented by two of the four ξ in Eq. 6). These results are shown in Fig. 6, and at the top of Fig. 5 (where they are convoluted with a Gaussian function of 850 μeV full-width-at-half-maximum to account approximately for the resolution of the monochromator). The partial DOS curve for the disordered alloy was calculated with force constants obtained from bcc dispersion curves [26], which implicitly assumes that the alloy contains one species of atom with the average mass of Fe and Al. This virtual crystal approximation is intrinsic to the analysis of coherent inelastic neutron scattering experiments on disordered alloys when single phonon frequencies are used to parameterize the neutron spectra measured at constant Q. The virtual crystal calculation has particularly poor agreement with the experimental data from the disordered alloy for energies above 34 meV in the range of the optical modes of the ordered alloy. On the other hand, for DO_3-ordered Fe_3Al we find good agreement between the calculated partial DOS and our experimental data (where the calculated curve is the sum of two partial DOS curves for the 4[b] and 8[c] crystallographic sites, shown as dashed lines at the top of Fig. 5). The only distinct discrepancy occurs around 26 meV, although this discrepancy vanishes if the calculated curves are broadened by about 2 meV. The experimental data also show intensity in the energy gap around 34 - 38 meV owing to some disorder in the ordered alloy. The intensity in this energy gap becomes larger for the partially-ordered and disordered alloys. With increasing disorder there is a decrease in intensity of the DOS at optical mode energies, and these modes also decrease in energy.

Fig. 6 Calculations of the phonon partial DOS of the two sites of Fe atoms in the DO_3 structure, and the phonon partial DOS of Al. The results were obtained by histogram binning of the eigenfrequencies of the dynamical matrix, weighted by $|e_{\xi,j}|^2$ (Eq. 6). The insert is an enlargement of the optical mode region, with the three curves normalized to equal area.

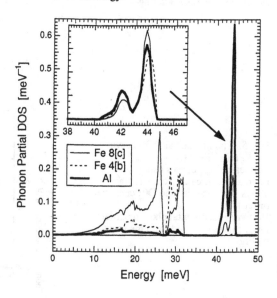

VIBRATIONAL ENTROPY

Comparison with Calorimetry

With Eq. 7 we use the phonon partial DOS curves, $g_{Fe}^D(\varepsilon)$ and $g_{Fe}^O(\varepsilon)$, to obtain $\Delta S_{Fe}^{vibr} \equiv S_{Fe}^D - S_{Fe}^O$. The data of Fig. 5 provide $\Delta S_{Fe}^{vibr} = (+0.01 \pm 0.005)$ k_B (Fe atom)$^{-1}$. Data from two other sets of runs with a lower resolution 5 meV monochromator gave $\Delta S_{Fe}^{vibr} = -0.02$ k_B (Fe atom)$^{-1}$. These data show that ΔS_{Fe}^{vibr} is small.

The difference in total vibrational entropy of disordered and ordered Fe$_3$Al, $\Delta S^{vibr} \equiv S^D - S^O$, is much larger than this small contribution from Fe atoms. This total entropy difference, $\Delta S^{vibr} \equiv S^D - S^O$, is the sum of contributions from the Fe atoms and the Al atoms (c.f. Eq. 8). The present results on S_{Fe}^D and S_{Fe}^O provide direct information on the roles of Fe atoms, but the role of Al atoms can be deduced because the total vibrational entropy difference, ΔS^{vibr}, has been measured previously by calorimetry, and is $\Delta S^{vibr} = (+0.1 \pm 0.03)$ k_B atom^{-1} [6]. Since the partial entropies of Fe and Al are added to obtain ΔS^{vibr}, we use Eq. 8 in the form: $\Delta S^{vibr} = 0.75 \Delta S_{Fe}^{vibr} + 0.25 \Delta S_{Al}^{vibr}$, where ΔS_{Fe}^{vibr} and ΔS_{Al}^{vibr} are the change in entropy upon disordering per Fe and per Al atom. We find $\Delta S_{Al}^{vibr} = 0.37$ k_B atom^{-1}. Evidently the difference in vibrational entropy of disordered and ordered Fe$_3$Al is due primarily to changes in the vibrations of Al atoms.

Comparison with EXELFS

Previously reported EXELFS data [2,9] also show that the Al atoms are primarily responsible for the difference in vibrational entropy of ordered and disordered Fe$_3$Al. In this study the temperature dependence of the EXELFS oscillations was measured for both the Fe and Al K-edges at temperatures from 97 to 400 K. By comparing the 1nn oscillations in the Fourier transform of the EXELFS data, changes in the mean-squared relative displacements (MSRD) measured from the Al K-edge (Al MSRD) and the Fe L$_{2,3}$-edge (Fe MSRD) were determined relative to the lowest-temperature datum. The MSRD data were fit to predictions of the Einstein model [27], with the lowest-temperature MSRD being free to vary. Einstein temperatures so obtained were 377 K (+28 K, −26 K) for Al atoms in disordered Fe$_3$Al, 490 K (+74 K, −54 K) for Al atoms in ordered Fe$_3$Al, 391 K (+18 K, −15 K) for Fe atoms in disordered Fe$_3$Al, and 431 K (+40 K, −31 K) for Fe atoms in ordered Fe3Al.

There was a strong reduction in the Al MSRD upon ordering, but a markedly smaller reduction in the Fe MSRD with ordering. The EXELFS measurements provide a relative displacement between Fe and Al atoms and their entire 1nn shell, which is a mixture of Fe and Al neighbors. Nevertheless, the backscattering cross section of Al is much smaller than that of Fe, so the EXELFS data are sensitive primarily to the relative displacements between the central atom and the Fe atoms in the 1nn shell. Using the expressions

$$\Delta S_{Fe}^{D-O} = 3k_B \ln\left(\frac{\theta_{Fe}^O}{\theta_{Fe}^D}\right) \qquad \Delta S_{Al}^{D-O} = 3k_B \ln\left(\frac{\theta_{Al}^O}{\theta_{Al}^D}\right) \qquad (20a,b)$$

where, for example, θ_{Fe}^O is the is the EXELFS-determined Einstein temperature for Fe in the ordered alloy, we find that and $\Delta S_{Al}^{D-O} = (0.81 \pm 0.41)$ k_B atom^{-1} and $\Delta S_{Fe}^{D-O} = 0.20 \pm 0.17$ k_B atom^{-1}. The EXELFS results indicate a big Al contribution to the entropy, even though the error is large. The $\Delta S_{Al} = 0.37$ k_B atom^{-1} deduced from the inelastic nuclear resonant x-ray scattering plus calorimetry is consistent with the $\Delta S_{Al}^{D-O} = (0.81 \pm 0.41)$ k_B atom^{-1} from the EXELFS. The error bars on the Fe EXELFS results, ΔS_{Fe}^{D-O}, are so large that we cannot reliably extract any numerical value. These EXELFS data from Fe are clearly inadequate for determining the very small $\Delta S_{Fe}^{D-O} =$

0.01 k_B atom^{-1} measured by the inelastic nuclear resonant x-ray scattering, but are not inconsistent with it.

Optical Modes

We can make another independent estimate of ΔS_{Al}^{D-O}. We deduce the optical modes of the Al phonon partial DOS from the measured Fe phonon partial DOS, with guidance from a Born − von Kármán model of the lattice dynamics of ordered Fe$_3$Al. From the lattice dynamics of D0$_3$-ordered alloys we know that a total of 1/4 of the modes must be associated with the optical modes around 43 meV. This results from the fact that 3 out of 12 dispersion curves for the D0$_3$ structure are associated with the optical modes [26]. We also know that 1/4 of the modes are associated with Al atoms, since 1/4 of the atoms in the alloy are Al. By integration of the experimental $g_{Fe}(\varepsilon)$ of the ordered alloy, we find that a fraction of 0.08 of the intensity of the phonon partial DOS of ^{57}Fe lies in the region of the optical modes around 43 meV. For the optical modes to account for 25% of the total DOS, 76% of the phonon partial DOS of Al atoms must therefore be associated with the optical modes. From the Born − von Kármán model of the ordered alloy (Fig. 6) we find a similar number of 80% with the force constants of Robertson [26], and 70% with the force constants of Kentzinger, et al. [28]. Most of the Al phonon partial DOS is associated with the optical modes.

We also know that for an ordered alloy, the energy ranges of the optical modes for the Al and Fe vibrations are identical. This is shown for the ordered alloy in the insert in Fig. 6, which also shows that the shapes of the phonon partial DOS curves for Fe and Al atoms are similar. Our experimental data show that with disorder, the optical modes shift downwards in energy, and some states appear in the gap. We make the assumption that the shapes of the phonon partial DOS curves of Fe and Al are the same, at least in the region of the optical modes. We tested if the change upon disordering of the optical modes, measured for Fe in Fig. 5, but scaled by the factor of 0.76/0.08 for Al, can account for the difference in vibrational entropy measured by calorimetry. The result of applying Eq. 7 to this synthesized Al phonon partial DOS is $\Delta S_{Al}^{u-v} = 0.20$ k_B/Al atom^{-1}, or 0.05 k_B/atom in the alloy. This is of the correct order to account for the change in vibrational entropy of the alloy upon disordering.

Again, it seems reasonable that the contribution of Fe atoms to the difference in vibrational entropy of order and disordered Fe$_3$Al is small, and the main contribution is from Al atoms. Furthermore, it appears that much or most of this change in the Al phonon partial DOS originates with changes in the optical modes. It is interesting that theoretical calculations have not yet detected significant softening of the optical modes of disordered Ni$_3$Al or disordered Fe$_3$Al, but find instead a simple broadening of the modes [29-32], or even a stiffening of them [33].

Atom Arrangements that Control the Phonon DOS

The results of Fig. 5 are useful for determining the spatial range of chemical order that affects the phonon DOS. Two of the samples have chemical SRO, but only one has LRO. Figures 3 and 5 show that a substantial change in phonon partial DOS occurs when only a small amount of SRO is present in the alloy. For the three samples, their fractional differences in chemical SRO seem comparable to their fractional changes in $g_{Fe}(\varepsilon)$. It is clear that the ^{57}Fe phonon partial DOS in Fe$_3$Al depends much more strongly on SRO than on LRO.

An analysis of why the chemical SRO should have an important effect on the vibrational entropy of the order-disorder transition was presented previously [10]. A general relationship between the phonon DOS and chemical short-range order may be expected from the slopes of phonon dispersion curves. Flat dispersion curves provide a high density of phonon states, especially when they include Brillouin zone boundaries. They also provide a slow group velocity of sound. Slow propagation of a wavepacket of energetic phonons requires that much of the energy of lattice vibrations is associated with localized atom movements. We expect localized atom movements to be affected strongly by SRO. We should expect SRO to have a major effect on the phonon DOS because of the high density of phonon states associated with these localized atom movements. We note that this interpretation is at odds with interpretations of vibrational entropies of order-disorder transformations based on a Debye model with a constant velocity of sound.

SUMMARY

The phonon partial densities of states and the partial vibrational entropies of the different atoms in the unit cell were defined, and shown useful for elucidating reasons for differences in vibrational entropies of alloy phases. Inelastic nuclear resonant scattering spectra were measured on alloys of Fe_3Al that were chemically disordered, partially-ordered, and $D0_3$-ordered. The relationship between the phonon partial DOS for ^{57}Fe atoms and inelastic nuclear resonant x-ray spectra was explained, and used to extract phonon partial DOS spectra from the three alloys of Fe_3Al. By comparison to previous calorimetry measurements, it is shown that the contribution of the Fe atoms to the vibrational entropy is an order of magnitude smaller than that of the Al atoms. This seems consistent with previous lower-quality EXELFS measurements on ordered and disordered Fe_3Al. With the assistance of Born - von Kármán model calculations on the ordered alloy, it is argued that change in the vibrational entropy upon disordering originates primarily with changes in the dynamics of Al atom motions in the optical modes. The phonon DOS of ^{57}Fe was found to change systematically with chemical short range order in the alloy. It was argued that changes in the vibrational entropy should originate primarily with changes in the chemical short-range order in the alloy, as opposed to long-range order.

ACKNOWLEDGMENTS

The work at Caltech was supported by the U. S. National Science Foundation under contract DMR–9816617, and the work at Argonne was supported by the U. S. Department of Energy under contract W–31–109–ENG–38.

REFERENCES

1. F. Ducastelle, Order and Phase Stability in Alloys, (North Holland, Amsterdam, 1991).
2. A. Zunger in Statics and Dynamics of Alloy Phase Transformations, P. E. A. Turchi and A. Gonis, eds., (Plenum Press, New York, 1994) p. 361.
3. J. K. Okamoto, C. C. Ahn and B. Fultz, Microbeam Analysis – 1990, edited by J. R. Michael and P. Ingram (San Francisco Press) 56 (1990).
4. A. F. Guillermet and G. Grimvall, J. Phys. Chem. Solids 53, 105 (1992).
5. L. Anthony, J. K. Okamoto and B. Fultz, Phys. Rev. Lett. 70, 1128 (1993).
6. L. Anthony, L. J. Nagel, J. K. Okamoto and B. Fultz, Phys. Rev. Lett. 73, 3034 (1994).
7. B. Fultz, L. Anthony, L. J. Nagel, R. M. Nicklow and S. Spooner, Phys. Rev. B 52, 3315 (1995).
8. P. D. Bogdanoff and B. Fultz, Philos. Mag. B 79, 753 (1999).
9. L. J. Nagel, L. Anthony, J. K. Okamoto, and B. Fultz, J. Phase Equilibria 18, 551 (1997).
10. B. Fultz, T. A. Stephens, W. Sturhahn, T. S. Toellner, and E. E. Alp, Phys. Rev. Lett. 80, 3304 (1998).
11. W. Sturhahn, T. S. Toellner, E. E. Alp, X. Zhang, M. Ando, Y. Yoda, S. Kikuta, M. Seto, C. W. Kimball, Phys. Rev. Lett. 74, 3832 (1995).
12. M. Seto, Y. Yoda, S. Kikuta, X. Zhang, and M. Ando, Phys. Rev. Lett. 74, 3828 (1995).
13. E. Gerdau, R. Rüffer, H. Winkler, W. Tolksdorf, C. P. Klages, and J. P. Hannon, Phys. Rev. Lett. 54, 835 (1985).
14. G. V. Smirnov, Hyperfine Intract. 97-8, 551 (1996).

15. M. T. Dove, Introduction to Lattice Dynamics, (Cambridge Univ. Press, 1993) Chapter 6.
16. V. F. Sears, Phys. Rev. A 7, 340 (1973).
17. M. Y. Hu, W. Sturhahn, T. S. Toellner, P. M. Hession, J. P. Sutter, and E. E. Alp, Nucl. Instrum. Meth. A 428, 551 (1999).
18. T. S. Toellner, M. Y. Hu, W. Sturhahn, K. Quast, and E. E. Alp, Appl. Phys. Lett. 71, 2122 (1997).
19. Z. Q. Gao and B. Fultz, Philos. Mag. B 67, 787 (1993).
20. G. Le Caër and J. M. Dubois, J. Phys. E 12, 1083 (1979).
21. J. K. Okamoto, Ph.D. Thesis in Applied Physics, California Institute of Technology, May 6, 1993.
22. M. M. Disko, C. C. Ahn, and B. Fultz, eds., Transmission Electron Energy Loss Spectrometry in Materials Science, TMS EMPMD Monograph Series Vol. 2 (TMS, Warrendale, 1992) ISBN Number 0-87339-180-2. Second edition in preparation for publication by J . Wiley.
23. W. Marshall and S. W. Lovesey, Theory of Thermal Neutron Scattering, (Oxford, London, 1971) p. 94.
24. L. J. Nagel, B. Fultz and J. L. Robertson, Philos. Mag. B 5, 681 (1997).
25. C. Van Dijk, Phys. Lett. A 34, 255 (1970).
26. I. M. Robertson, J. Phys.: Condens. Matter 3, 8181 (1991).
27. G. Beni and P. M. Platzman, Phys. Rev. B 14, 1514 (1976).
28. E. Kentzinger, M. C. Cadeville, V. Pierron-Bohnes, W. Petry and B. Hennion, J. Phys.: Condens. Matter 8, 5535 (1996).
29. J. D. Althoff, D. Morgan, D. de Fontaine, M. D. Asta, S. M. Foiles, et al., Phys. Rev. B 56, R5705 (1997).
30. J. D. Althoff, D. Morgan, D. de Fontaine, M. D. Asta, S. M. Foiles, et al., Comp. Mater. Sci. 10, 411 (1998).
31. R. Ravelo, J. Aguilar, M. Baskes, J. E. Angelo, B. Fultz, and B. L. Holian, Phys. Rev. B 57, 862 (1998).
32. S. J. Liu, S. Q. Duan and B. K. Ma, Phys. Rev. B 58, 9705 (1998).
33. A. Van de Walle, G. Ceder, and U. V. Waghmare, Phys. Rev. Lett. 80, 4911 (1998).

Interplay of charge, orbital and magnetic order in $Pr_{1-x}Ca_xMnO_3$

M. v. Zimmermann[1], J.P. Hill[1], Doon Gibbs[1], M. Blume[1], D. Casa[2], B. Keimer[2,3], Y. Murakami[4], Y. Tomioka[5], and Y. Tokura[6]

[1]Department of Physics, Brookhaven National Laboratory, Upton, New York 11973, USA
[2]Department of Physics, Princeton University, New Jersey 08544, USA
[3]Max-Planck-Institut für Festkörperforschung, 70569, Stuttgart, Germany.
[4]Photon Factory, Institute of Materials Structure Science, High Energy Accelerator Research Organization, Tsukuba, 305-0801, Japan
[5]Joint Research Center for Atom Technology (JRCAT), Tsukuba 305-0046, Japan
[6]Department of Applied Physics, University of Tokyo, Tokyo 113-0033, Japan and JRCAT

Abstract

We report resonant x-ray scattering studies of charge and orbital order in $Pr_{1-x}Ca_xMnO_3$ with x=0.4 and 0.5. Below the ordering temperature, T_O=245 K, the charge and orbital order intensities follow the same temperature dependence, including an increase at the antiferromagnetic ordering temperature, T_N. High resolution measurements reveal, however, that long range orbital order is never achieved. Rather, an orbital domain state is formed. Above T_O, the charge order fluctuations are more highly correlated than the orbital fluctuations. We conclude that the charge order drives the orbital order at the transition.

Introduction

Disentangling the origins of high temperature superconductivity and colossal magnetoresistance in the transition metal oxides remains at the center of current activity in condensed matter physics. An important aspect of these strongly correlated systems is that no single degree of freedom dominates their response. Rather, the ground state properties are thought to reflect a balance among several correlated processes, including orbital and charge order, magnetism, and the lattice degrees of freedom.

The perovskite manganites provide an especially illuminating example of the interplay among these interactions, since in these materials the balance may be altered, for example by doping. As a result, much work has been done to understand their magnetic ground states and lattice distortions, dating back to the seminal experiments of Wollan and Koehler [1]. Less is known about the roles of charge and orbital order, for which (until recently) there has been no quantitative experimental probe. The classic work of Goodenough [2] has nevertheless served as a guide to their ordered arrangements, as supplemented by measurements of the temperature dependence of the lattice constants, and other properties. Recently this situation has changed with the detection of charge and orbital order by x-ray resonant scattering techniques [3, 4, 5, 6, 7, 8, 9, 10]. Specifically, it has been found that the sensitivity of x-ray scattering to these structures is dramatically enhanced by tuning the incident x-ray energy to the Mn K-absorption edge. Thus, it is now possible to characterize the orbital and charge order in detail, and to study their response to changes of temperature or magnetic field.

In this paper, we report x-ray resonant scattering studies of $Pr_{1-x}Ca_xMnO_3$ with x=0.4 and 0.5. We have detected both charge and orbital order below a common phase transition temperature (T_O=245 K), and confirmed the ground state originally proposed by Goodenough for the isostructural compound $La_{0.5}Ca_{0.5}MnO_3$ [2]. Below the transition, the intensities of the charge and orbital order have the same temperature dependences, suggesting that they are linearly coupled. There is, moreover, an increase in the scattering at the Néel temperature (T_N=170 K), implying a coupling to the magnetic degrees of freedom. Intriguingly, high resolution measurements reveal that long range orbital order is never achieved at these concentrations. Rather, a domain state is formed. At temperatures above T_O, we observe critical charge and orbital scattering. Remarkably, the correlation lengths differ, with the length scale of the charge order exceeding that of the orbital order. From this we conclude that charge order drives the orbital order in these systems. This picture is supported by studies in which the phase transition is driven by an applied magnetic field.

For small x, $Pr_{1-x}Ca_xMnO_3$ has an orbitally ordered ground state at low temperature analogous to that observed in $LaMnO_3$. The electronic configuration of the Mn^{3+} ions is (t_{2g}^3, e_g^1), with the t_{2g} electrons localized. The e_g orbitals are hybridized with the oxygen p orbitals, and participate in a cooperative Jahn-Teller distortion of the MnO_6 octahedra. This leads to $(3x^2 - r^2)$-$(3y^2 - r^2)$-type orbital order of the e_g electrons in the ab-plane. For $0.3 \leq x \leq 0.7$, charge order among Mn^{3+} and Mn^{4+} ions is believed to occur in addition to the orbital order. The fraction of Mn ions in the Mn^{4+} state equals the concentration of Ca ions. Thus, by varying the Ca concentration, it is possible to alter the balance between the charge and orbital order. The proposed ground state for x =0.4 is illustrated in fig. 1b [2, 11, 12, 13]. In orthorhombic notation, for which the fundamental Bragg peaks occur at (0,2k,0) with k integer, the charge order reflections occur at (0,2k+1,0) and the orbital order reflections at $(0, k+\frac{1}{2},0)$. The magnetic structure is of the modified CE-type [11].

Experiment

The x =0.4 and 0.5 single crystals used in this study were grown by a float zone technique. A (0,1,0) surface was cut out of a cylinder of radius $\sim$3 mm and polished with diamond paste. The mosaic spreads were about 0.25 degrees for each sample. X-ray scattering experiments were carried out at the National Synchrotron Light Source on Beamlines X22B and C. X22C is equipped with a bent, toroidal focussing mirror and a Ge(111) double crystal monochromator arranged in a vertical scattering geometry. The incident energy resolution is about 5 eV. Two analyzer configurations were used. The first employed a standard Ge(111) crystal, and provided a longitudinal resolution of 4.5 $\times 10^{-4}$ Å^{-1} (HWHM). The second provided linear polarization analysis of the scattered beam via rotation of a Cu (220) crystal about the scattered beam direction [14]. The incident polarization was 95% linearly polarized in the horizontal plane (σ). Experiments in a magnetic field were performed on X22B, which supports a bent, toroidal mirror and a standard single crystal Ge(111) analyzer-monochromator combination. A 13 T superconducting magnet was utilized in a horizontal reflection geometry.

The present experiments were carried out using x-ray resonant scattering techniques. As shown in a series of recent papers [3, 4, 5, 6, 7, 8, 9, 10], the sensitivity of x-ray scattering to orbital ordering in transition metal oxides is enhanced when the incident x-ray energy is tuned near the K-absorption edge. The resonant scattering may be thought of as Templeton scattering, arising from the anisotropic charge distribution induced by orbital ordering [15,

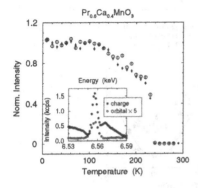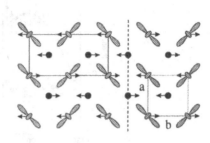

Figure 1: *left: Orbital and charge order parameters versus temperature for the x=0.4 sample. Open and closed circles: Resonant intensities of the orbital and charge order measured at the (0,1.5,0) and (0,3,0) reflections, respectively. Inset: Energy dependence of an orbital (0,2.5,0) and charge order (0,3,0) peak in the σ-π and σ-σ geometries, respectively. right: schematic of charge and orbital order, showing an orbital anti phase domain boundary (dashed line) in the ab-plane. Black points represent Mn^{4+} and open symbols Mn^{3+}. The arrows indicate the magnetic ordering. The dotted and solid lines show the charge and orbital unit cells, respectively.*

16]. In the dipole approximation, this corresponds to a $1s \rightarrow 4p$ transition at the metal site. The sensitivity to orbital ordering in manganites arises from the splitting of the Mn 4p levels as a result of the 3d ordering. However, the microscopic origin of this splitting has been controversial [3, 4, 5, 6, 7, 8, 9, 10]. Two mechanisms have been proposed. One involves the Coulomb coupling of the Mn 3d and Mn 4p states, either directly or through the O 2p levels[5], while the other involves the perturbation of the O 2p and Mn 4p wavefunctions as a result of the Jahn-Teller distortion [7, 10]. Insofar as we are aware, the experimental data obtained to date do not distinguish either possibility conclusively, and this remains an open question. Regardless of the microscopic origin, however, the resonant scattering reflects the symmetry of the orbital ordering through the redistribution of the local charge density at the Mn^{3+} sites in both cases. In terms of the Jahn-Teller distortion, specifically, the orientation of the e_g orbitals and the oxygen motion reflect the same order parameter. This is true even if the d-orbitals are not directly involved as intermediate states of the resonance. It follows that the peak positions and widths determined in the x-ray experiments measure the orbital periodicity and correlation lengths, respectively. (Supporting evidence for this is given by the consistency of our results and neutron magnetic scattering studies of $La_{1-x}Ca_xMnO_3$ [17] discussed below). It remains to interpret the x-ray peak intensities on an absolute scale, which we believe will require additional calculations and experiments. The anomalous charge order scattering originates in the small difference in K absorption energies associated with Mn^{3+} and Mn^{4+} sites as measured at difference reflections.

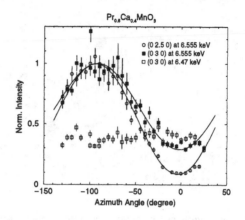

Figure 2: *Azimuthal dependence of the (0,3,0) charge order peak measured at resonance energy (filled squares) and non resonant (open squares) and the (0,2.5,0) orbital order reflection (circles) measured at resonance energy in a σ to π scattering geometry.*

Results

The inset of fig. 1 shows the energy dependence of the intensities of the charge and orbital order as the photon energy is tuned through the K absorption edge. The data were taken at the (0,3,0) and (0,2.5,0) reflections, respectively, of the $x =0.4$ sample. Each scan shows an enhancement at 6.555 keV, characteristic of dipole resonant scattering. Maximum count rates of 800 and 3000 s^{-1} were obtained for the orbital and charge order scattering, respectively, with a Ge(111) analyzer. The structure observed in the lineshape for the charge order reflects the interference of the resonant and nonresonant contributions, and will be discussed in detail elsewhere [18]. In addition to the resonance, polarization analysis further revealed that the orbital scattering is predominantly rotated (σ-π), whereas the charge scattering is, to within experimental accuracy, unrotated (σ-σ). Both the charge and orbital order intensities exhibit peculiar dependences on the azimuthal angle, which defines a rotations about the scattering vector. Figure 2 shows the peak intensity of the (0,2.5,0) orbital and the (0,3,0) charge order peak normalized on the intensity of the (0,2,0) Bragg reflection and scaled to unity at the maximum. It is observed that the orbital order reflection, measured in the σ-π geometry shows a $\sin^2(\psi)$ dependence [3, 4]. (The residual intensity at $\psi=0$ is due to leakage of the polarization analyzer.) While the non resonant scattering from charge ordering does not show any azimuthal dependence, reflecting the strain associated with the the charge order, the resonant intensity at the charge order reflection shows a $\sin^2(\psi)$ on top of the non resonant signal. These results are consistent with predictions of the resonant cross-section for orbital and charge order [3, 4, 5], and confirm the picture of charge and orbital order given for this class of compounds by Goodenough [2] (fig. 1b).

The temperature dependences of the charge and orbital order obtained at resonance for the $x =0.4$ sample are shown in fig. 1. For comparison purposes, the intensities have been scaled together at 10 K. Between 10 and 120 K, both intensities are approximately constant, but decrease by about 25% on passing through the Néel temperature ($T_N=170$ K). They

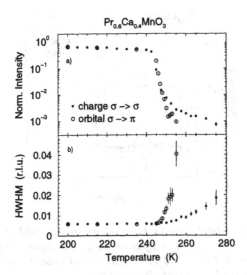

Figure 3: *a) Temperature dependence of the peak intensities of the (0,3,0) charge order peak (closed circles) and the (0,2.5,0) orbital order peak (open circles) for the x =0.4 sample. b) Temperature dependence of the half widths at half maximum (HWHM).*

drop sharply to zero at T_O=245 K. This is coincident with an orthorhombic-to-orthorhombic structural transition, as determined from measurements of the lattice constants [18]. It follows that the temperature dependences of the charge and orbital order are identical, which suggests that the corresponding order parameters are linearly coupled. It seems clear in this regard that the growth of orbital (and charge) order below 245 K enhances the antiferromagnetic correlations, and thereby promotes the magnetic phase transition. This is consistent with the results of inelastic neutron scattering studies of $(Bi,Ca)MnO_3$, in which orbital order was found to quench ferromagnetic fluctuations [19]. Qualitatively similar results have been found for x =0.5 [18].

The behavior of the charge and orbital scattering in the vicinity of the phase transition at T_O is illustrated for the x =0.4 sample in fig. 3. Longitudinal scans were taken (upon warming) of the (0,3,0) reflection in a σ-σ geometry and of the (0,2.5,0) reflection in a σ-π geometry. We observe measurable charge order fluctuations (shown on a log scale in fig. 3a) to much higher temperature above T_O than for the orbital order fluctuations. The corresponding peak widths are considerably narrower for the charge order (fig. 3b), implying that the correlation lengths of the charge order are longer than those of the orbital order at any given temperature above T_O [20].

The picture these data then present is one in which the phase transition proceeds via local charge order fluctuations which grow as the transition is approached, nucleating long range order at the transition temperature. The orbital fluctuations are induced by these charge fluctuations through the coupling discussed above, and become observable only close to the transition.

The phase transition may also be driven by applying a magnetic field, as demonstrated

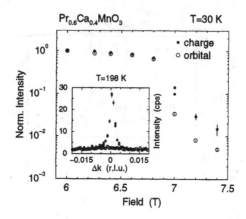

Figure 4: *Charge and orbital order as function of magnetic field at 30 K. The intensities of the two peaks have been scaled to agree at low fields. Inset: Charge and orbital order superlattice reflection at 198 K and 11 T. The orbital order is no longer observable, but scattering from charge order remains clearly visible.*

by Tomioka *et al.* [21]. It is an interesting question whether the same phenomenology of the fluctuations applies when the temperature is held fixed. We have carried out studies of the transition at two temperatures, T=30 K and 200K, with critical fields of H_O=6.9(1) T, and H_O=10.4 T, respectively. Data taken at T=30 K are illustrated in figure 4. The two order parameters exhibit identical field dependences below the transition. Above the transition, the charge order fluctuations are markedly stronger than the orbital fluctuations. Similar behavior was observed at T=200 K, i.e. charge order fluctuations were observed at fields for which orbital fluctuations were no longer observable (inset fig. 4). Thus, it appears that the transition is driven by charge order fluctutations for both temperature and field driven cases. (As a result of experimental constraints, it was only possible to measure charge and orbital order scattering at a photon energy of 8 keV in an applied magnetic field. The corresponding nonresonant intensities are sufficiently weak above T_O that it was not possible to obtain reliable values of the half-width).

Finally, we performed high q-resolution measurements of the charge and orbital order below T_O to investigate the extent of their order in detail. Remarkably, in the x=0.4 sample, we found an orbital order correlation length of $\xi_{OO} = 320 \pm 10$ Å and a charge order correlation length of $\xi_{CO} \geq 2000$ Å. This difference is even more apparent in the x=0.5 sample, as shown in fig. 5. Here longitudinal scans through the (0,2,0), (0,1,0) and (0,2.5,0) reflections are superimposed for comparison. The width of the (0,2,0) scan approximates the effective resolution. The (0,1,0) charge order reflection shows only a slight broadening (9.1(1) $\cdot 10^{-4}$ Å^{-1}), corresponding to a correlation length of $\xi_{CO} \geq 2000$ Å [22]. The orbital order, however, is substantially broadened, with a correlation length of $\xi_{OO} = 160 \pm 10$ Å [23]. It follows that the orbitals do not exhibit long range order, but instead form a domain state with randomly distributed anti-phase domain walls (see fig. 1b). In contrast, the charge order is much more highly correlated.

A possible explanation for the difference in orbital domain sizes observed in the two

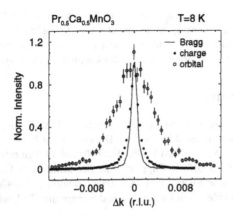

Figure 5: *Longitudinal scans of the (0 2 0) Bragg reflection, the (0 1 0) charge order peak and the (0 2.5 0) orbital order peak. The orbital order peak is significantly broadened due to the formation of orbital order anti phase domains.*

samples follows from the fact that the $x = 0.5$ sample is closer to tetragonal than the $x=0.4$ sample: $\delta(x = 0.5) = \frac{2(a-b)}{(a+b)} = 1.48 \times 10^{-3}$ compared to $\delta(x = 0.4) = 4.23 \times 10^{-3}$ at room temperature [11]. In the more tetragonal sample, the a and b domains are nearly degenerate and the energetic cost of an orbital domain wall is correspondingly reduced [24].

The presence of an orbital domain state is consistent with powder neutron diffraction studies of $La_{0.5}Ca_{0.5}MnO_3$, which also exhibits the CE magnetic structure with orbital and charge order [17]. Here, magnetic correlation lengths of $\xi_{3+} = 200 - 400$ Å and $\xi_{4+} \geq 2000$ Å were found for the respective sublattices, and anti-phase domain walls were postulated to explain the disorder. It seems likely that these domain walls are the orbital domains observed in the present experiment, which lead to magnetic disorder through the coupling mentioned above. The presence of such orbital domain boundaries breaks the magnetic coherence of the 3+ sublattice only, as long as charge order is preserved (fig. 1b). These results taken together suggest that orbital domain states may be common to these systems – at least in this range of doping.

Summary

In summary, we have used resonant x-ray scattering techniques to study charge and orbital order in doped manganites, $Pr_{1-x}Ca_xMnO_3$, with $x=0.4$ and 0.5. We have found that the transition into a charge and orbitally ordered state proceeds via charge order fluctuations, which grow as the transition is approached from above, until at an abrupt, first-order-like transition long range charge order is nucleated. At the same time, less well correlated orbital fluctuations are observed. While the correlation length of the orbital order grows as the transition is approached, long range order is not achieved, and below the charge order transition an orbital domain state is observed. The orbital correlation length appears to be concentration dependent, with a more highly correlated orbital state being observed in the $x=0.4$ sample. Phenomenologically similar behavior is observed when the phase transition is driven by a magnetic field. Below the transition, the charge and orbital order parameters exhibit identical temperature dependences, indicative of a linear coupling between these degrees of freedom. At the Néel temperature a jump in their intensities is observed demonstrating the importance of the magnetic -charge/orbital order coupling in these systems.

Acknowledgments

We acknowledge useful conversations with S. Ishihara, A.J. Millis and G.A. Sawatzky. The work at Brookhaven was supported by the U.S. Department of Energy, Division of Materials Science, under Contract No. DE-AC02-98CH10886 and at Princeton University by the N.S.F., under grant DMR-9701991. Support from the Ministry of Education, Science and Culture, Japan, by the New Energy and Industrial Technology Development Organization (NEDO) and by the Core Research for Evolutional Science and Technology (CREST) is also acknowledged.

References

[1] E.O. Wollan and W.C. Koehler, Phys. Rev. **100**, 545, (1955).

[2] J.B. Goodenough, Phys. Rev. **100**, 555, (1955).

[3] Y. Murakami, H. Kawada, H. Kawata, M. Tanaka, T. Arima, Y. Morimoto and Y. Tokura, Phys. Rev. Lett. **80**, 1932, (1998).

[4] Y. Murakami, J.P. Hill, Doon Gibbs, M. Blume, I. Koyama, M. Tanaka, H. Kawata, T. Arima, Y. Tokura, K. Hirota and Y. Endoh, Phys. Rev. Lett. **81**, 582, (1998).

[5] S. Ishihara and S. Maekawa, Phys. Rev. Lett. **80** 3799, (1998).

[6] M. Fabrizio, M. Altarelli and M. Benfatto, Phys. Rev. Lett., **80** 3400 (1998), *ibid* **81** 4030 (1998).

[7] I.S. Elfimov, V.I. Anisimov and G.A. Sawatzky Phys. Rev. Lett. **82**, 4264 (1999).

[8] Y. Endoh, K. Hirota, S. Ishihara, S. Okamoto, Y. Murakami, A. Nishizawa, T. Fukuda, H. Kimura, H. Nojiri, K. Kaneoko and S. Maekawa, Phys. Rev. Lett., **82**, 4328 (1999).

[9] L. Paolasini, C. Vettier, F. de Bergevin, D. Mannix, W. Neubeck, A. Stunault, F. Yakhou, J.M. Honig and P.A. Metcalf, Phys. Rev. Lett. **82**, 4719 (1999).

[10] M. Benfatto, Y. Joly and C. R. Natoli, Phys. Rev. Lett., **83** 636 (1999).

[11] Z. Jirák, S. Krupica, Z. Simsa, M. Dlouhá and S. Vratislav, J.of Mag. and Mag. Mat. **53**, 153, (1985).

[12] Y. Okimoto, Y. Tomioka, Y. Onose, Y. Otsuka and Y. Tokura, Phys. Rev. B **57**, R9377 (1998).

[13] The CE-type charge order structure is stable for $0.4 < x < 0.7$ in $Pr_{1-x}Ca_xMnO_3$. For $x < 0.5$, the excess electrons are believed to reside in partially occupied $3z^2 - r^2$ orbitals [11].

[14] D. Gibbs, M. Blume, D.R.Harshman and D.B. McWhan, Rev. Sci. Instrum., **60** 1655 (1988).

[15] see e.g. M. Blume in *Resonant Anomalous X-ray Scattering* Ed.s G. Materlik, C.J. Sparks, and K. Fischer, North Holland, 1991, p. 495, and A. Kirfel, *ibid* p. 231.

[16] K.D. Finkelstein, Q. Shen and S. Shastri Phys. Rev. Lett., **69**, 1612 (1992).

[17] P.G. Radaelli, D.E. Cox, M. Marezio, S.-W. Cheong, Phys. Rev. B **55** 3015 (1997).

[18] M. v. Zimmermann *et al.*, in preparation.

[19] Wei Bao, J.D. Axe. C.H. Chen and S.-W. Cheong, Phys. Rev. Lett. **78** 543, (1997).

[20] The correlation length of the charge order must be at least as long as that of the orbital order, since the unit cell of the orbital order is defined on the charge order lattice.

[21] Y. Tomioka, A. Asamitsu, H. Kawahara, Y. Morimoto and Y. Tokura, Phys. Rev. B **53**, R1689 (1996).

[22] As estimated by performing a 1-d deconvolution of the resolution and using $\xi = \frac{b}{2\pi\Delta k}$, were Δk is the fitted HWHM.

[23] Note the q-dependence of the resolution function is much too small to explanin the observed broadening.

[24] A.J. Millis, private communication.

CHARACTERIZATION OF MIXED-METAL OXIDES USING SYNCHROTRON-BASED TIME-RESOLVED X-RAY DIFFRACTION AND X-RAY ABSORPTION SPECTROSCOPY

José A. Rodriguez[*], Jonathan C. Hanson[*], Joaquín L. Brito[**], and Amitesh Maiti[***]

[*]Department of Chemistry, Brookhaven National Laboratory Upton, NY 11973, USA
[**]Centro de Química, Instituto Venezolano de Investigaciones Científicas (IVIC), Apartado 21827, Caracas 1020-A, Venezuela
[***]Molecular Simulations Inc, 9685 Scranton Road, San Diego, CA 92121, USA

ABSTRACT

Experiments are described showing the utility of synchrotron-based time-resolved x-ray diffraction (TR-XRD) and x-ray absorption near-edge spectroscopy (XANES) for characterizing the physical and chemical properties of mixed-metal oxides that contain Mo and a second transition metal (Fe, Co or Ni). TR-XRD was used to study the transformations that occur during the heating of a $FeMoO_4/Fe_2(MoO_4)_3$ mixture and the $\alpha \rightarrow \beta$ phase transitions in $CoMoO_4$ and $NiMoO_4$. The Mo L_{II}- and O K-edges in XANES are very useful for probing the local symmetry of Mo atoms in mixed-metal oxides. The results of XANES and density-functional calculations (DMol3, DFT-GGA) show large changes in the splitting of the empty Mo 4d levels when going from tetrahedral to octahedral coordinations. XANES is very useful for studying the reaction of H_2, H_2S and SO_2 with the mixed-metal oxides. Measurements at the S K-edge allow a clear identification of S, SO_2, SO_3 or SO_4 on the oxide surfaces. Changes in the oxidation state of molybdenum produce substantial shifts in the position of the Mo L_{II}- and M_{III}-edges.

INTRODUCTION

Mixed-metal oxides play a relevant role in many areas of materials science, physics, chemistry, and the electronic industry. In principle, the combination of two metals in an oxide matrix can lead to materials with novel properties and a superior performance in technological applications. In this article we illustrate the use of two synchrotron-based techniques, time-resolved x-ray diffraction (TR-XRD) and x-ray absorption near-edge spectroscopy (XANES), in the characterization of the physical and chemical properties of mixed-metal oxides. We focus on the behavior of compounds of the $MeMoO_4$ type (Me= Fe, Co or Ni), which result from adding oxides of the MeO type to MoO_3 [1] and constitute an interesting group of materials due to their structural, electronic and catalytic properties [1-5].

Investigations at Brookhaven National Laboratory has established the feasibility of conducting sub-minute, time-resolved x-ray diffraction experiments under a wide variety of temperature and pressure conditions (-190 °C < T < 900 °C; P < 45 atm) [6]. This important advance results from combining the high intensity of synchrotron radiation with new parallel data-collection devices [6]. Examples of problems studied to date with TR-XRD include the hydrothermal synthesis and thermal dehydration of zeolites, the binding of substrates and inhibitors in porous catalytic materials, and phase transformations in oxides [4,6,7]. X-ray absorption near-edge a spectroscopy has emerged as a powerful tool for studying the electronic and structural properties of solids [8]. Recently, we have shown that the results of theoretical calculations based on density-functional theory (DFT) can be very useful for explaining the XANES spectra of oxides [7].

EXPERIMENTAL

The time-resolved powder diffraction patterns of the $MeMoO_4$ compounds were collected on beamline X7B of the National Synchrotron Light Source (NSLS) at Brookhaven National Laboratory. The diffraction patterns were accumulated on a flat image plate (IP) detector and retrieved using a Fuji BAS2000 scanner. The XANES spectra of the mixed-metal oxides were collected at the NSLS on beamlines U7A (Mo M-edges and O K-edge) and X19A (Mo L-edges and S K-edge). The measurements at X19 were performed in the "fluorescence-yield mode", whereas the data at U7A were acquired in the "electron-yield mode". The theoretical spectra for the Mo L-edges [7] were calculated using DFT and commercial versions of the CASTEP [9] and DMol3 [10] codes available from Molecular Simulations Inc.

The different isomorphs of the $MeMoO_4$ compounds were prepared following the methodology described in ref. [3]. In a set of experiments, the molybdates were exposed to H_2 or SO_2 in a RXM-100 instrument from Advanced Scientific Designs. The reduction with hydrogen was carried out in a flow-reactor with a 15% H_2/85% N_2 mixture (flow rate= 50 cm³/min) and the temperature was ramped from 40 to 650 or 800 °C (heating rate = 20 °C/min). Pure oxide powders were exposed to SO_2 in a reaction cell ("batch-reactor mode") at 50 °C for 15 min with a constant SO_2 pressure of 10 Torr.

RESULTS

Phase transitions and studies using time-resolved x-ray diffraction. Under atmospheric pressure the $MeMoO_4$ compounds can adopt a low temperature α-phase in which Mo and the second metal (Fe, Co or Ni) are in an octahedral coordination, or a high temperature β-phase in which Mo is in a tetrahedral environment with the second metal remaining in octahedral coordination [1,4,7]. Using TR-XRD we monitored the α→β phase transitions in $CoMoO_4$ and $NiMoO_4$ [4,7]. Figure 1 shows that in the case of the cobalt molybdate the transformation occurs

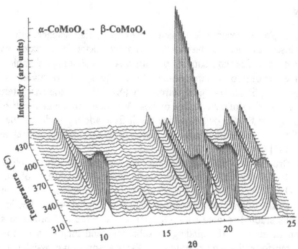

Fig 1   Time-resolved x-ray diffraction patterns for the α→β transition in
$CoMoO_4$. Heating rate= 1.6 °C/min

between 330 and 410 °C. For the nickel molybdate, the transformation takes place at much higher temperatures: 530-630 °C [4,7]. There are no intermediates in the $\alpha \rightarrow \beta$ transitions for these molybdates: the diffraction lines for α-MeMoO$_4$ disappear and simultaneously the lines for β-MeMoO$_4$ appear. First-principles density-functional calculations carried out with the CASTEP program (DFT-GGA level) show that α-CoMoO$_4$ is ~ 5 kcal/mol more stable than β-CoMoO$_4$ [7,11]. On the other hand, α-NiMoO$_4$ is ~ 9 kcal/mol more stable than β-NiMoO$_4$ [7]. For the $\alpha \rightarrow \beta$ transition in NiMoO$_4$, the DFT calculations predict an energy barrier of ~ 50 kcal/mol [7]. The results of TR-XRD indicate that the α-NiMoO$_4 \rightarrow \beta$-NiMoO$_4$ transformation follows a kinetics of first order with an *apparent* activation energy of ~ 80 kcal/mol [4,7].

In the case of the iron molybdate, the procedure followed for the synthesis of the compound [3] in some situations can yield a mixture of β-FeMoO$_4$, Fe$_2$(MoO$_4$)$_3$ and an amorphous FeMo$_x$O$_y$ phase due to the possibility of changes in the oxidation states of the second metal and molybdenum (i.e. Fe$^{2+} \rightleftharpoons$ Fe^{3+} and Mo$^{6+} \rightleftharpoons$ Mo^{5+}). The evolution of this type of system as a function of temperature can be monitored using TR-XRD. In Figure 2 diffraction lines are observed for β-FeMoO$_4$ [1,3] and Fe$_2$(MoO$_4$)$_3$ [12] at low temperatures (< 300 °C). The relative intensity of these lines indicates that the Fe$_2$(MoO$_4$)$_3$/β-FeMoO$_4$ ratio in the sample is ~ 0.2. Upon heating one sees the disappearance of the lines for Fe$_2$(MoO$_4$)$_3$ at temperatures between 350 and 430 °C. Above 430 °C, only diffraction lines for β-FeMoO$_4$ are observed but there is a continuous increase in the intensity of these lines. On the basis of the TR-XRD results, one can conclude that Fe$_2$(MoO$_4$)$_3$ and the amorphous FeMo$_x$O$_y$ phase transform into β-FeMoO$_4$ at high temperatures (350-700 °C).

In addition of being a valuable technique for studying phase transitions, we have found that TR-XRD is also very useful for studying structural changes that occur during cycles of reduction (with H$_2$)/oxidation(with O$_2$) or during the reaction of H$_2$S with the molybdates [11,13].

Studies using x-ray absorption near-edge spectroscopy. The left-side panel in Figure 3 shows Mo L$_{II}$-edge XANES spectra for MoO$_3$, the β-FeMoO$_4$/Fe$_2$(MoO$_4$)$_3$/FeMo$_x$O$_y$ mixture of Figure 2 (T= 25 °C), and β-FeMoO$_4$. These spectra involve transitions from the 2p core levels into the

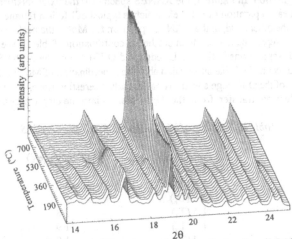

Fig 2   *Time-resolved x-ray diffraction patterns for the heating (1.8 °C/min) of a mixture of β-FeMoO$_4$, Fe$_2$(MoO$_4$)$_3$, and an amorphous FeMo$_x$O$_y$ phase.*

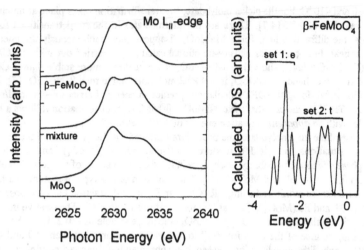

Fig 3 *Left panel: Mo L_{II}-edge XANES spectra for MoO_3, a mixture of β-FeMoO$_4$/ Fe$_2$(MoO$_4$)$_3$/FeMo$_2$O$_y$, and β-FeMoO$_4$. Right panel: Calculated DOS for the empty Mo 4d bands of β-FeMoO$_4$ (DMol3, DFT-GGA results).*

empty 4d orbitals of Mo. The right-side panel in Figure 3 displays the calculated density of states (DMol3, DFT-GGA) for the empty 4d bands of Mo in β-FeMoO$_4$. In octahedral and tetrahedral environment, the Mo 4d orbitals split in two sets of bands with t and e symmetry as a result of the ligand field generated by the oxygens around the Mo atoms. In a tetrahedral coordination the set with e symmetry is more stable than the set with t symmetry (Fig 3). The opposite is valid for an octahedral coordination [7]. The energy separation between the two sets of bands is larger in an octahedral configuration. In Figure 3, the XANES spectra of the oxide mixture and β-FeMoO$_4$ show a peak-to-peak separation of ~ 1.7 eV which is typical of Mo in a tetrahedral environment (see Table I). On the other hand, in the XANES spectrum for MoO_3, the peak-to-peak separation is ~ 3.4 eV, a value representative of Mo in octahedral coordination (Table I). We have found that the peak-to-peak separations in the Mo L_{II}-edge and O K-edge of the XANES spectra for the molybdates give direct and reliable information about the coordination of Mo in these systems (see Table I). The case of the O K-edge spectra is particularly interesting since, due to orbital mixing, one can have electron transfer from the O 1s orbitals into the empty orbitals of both metals

Table I: *Splitting of Mo L_{II}- and O K-edges in XANES*

Compound	Mo L_{II}-edge (eV)	O K-edge (eV)
MoO_3	3.2 (O$_h$)	3.4
α-NiMoO$_4$	3.1 (O$_h$)	3.2
α-CoMoO$_4$	2.8 (O$_h$)	3.2
β-FeMoO$_4$	1.6 (T$_d$)	1.8
β-CoMoO$_4$	1.5 (T$_d$)	1.9
β-NiMoO$_4$	1.7 (T$_d$)	1.9
β-MgMoO$_4$	1.6 (T$_d$)	1.9

present in a molybdate. The results of DFT calculations indicate that this transfer should occur mainly into the unoccupied Mo 4d bands [7,11], and indeed the XANES experiments show that the line-shapes in the O K-edge spectra track very well those observed in the corresponding Mo L_{II}-edge spectra [4,7,11].

XANES is very useful to study electronic perturbations induced on the molybdates by reaction with H_2, H_2S or SO_2. As an example, Figure 4 shows Mo L_{II}-edge XANES spectra recorded after partially reducing samples of β-FeMoO$_4$ by exposing them to a stream of 15%-H_2/85%-N_2. Spectra of temperature programmed reduction (TPR) indicate that the molybdate reacts only with a minor amount of H_2 at temperatures below 650 °C, and the consumption of hydrogen becomes larger at temperatures between 650 and 800 °C [3]. This is consistent with the trends seen in Figure 4. In this figure, the reaction with H_2 induces a shift of the Mo L_{II}-edge features towards lower photon energies which indicates a reduction in the oxidation state of Mo in the sample.

The MeMoO$_4$ compounds are precursors of hydrodesulfurization catalysts [3] and can be used as sorbents of SO_2. Changes in the position of the Mo M_{III}-edge in XANES have proven to be quite useful for monitoring the degree of sulfidation of Mo during the reaction of nickel molybdate with H_2S [13,14]. Measurements at the S K-edge show the formation of metal sulfides and sulfates as a result of the reaction of hydrogen sulfide with iron, cobalt and nickel molybdates [11,13,14]. Figure 5 displays a S K-edge XANES spectrum collected after dosing SO_2 to β-FeMoO$_4$ at 50 °C. There are two clear peaks that correspond to SO_3 (~ 2478 eV [14]) and SO_4 (~ 2482 eV [14]) produced by reaction of SO_2 with O atoms from the lattice of the oxide substrate. A very weak feature is seen around 2474 eV that could be attributed to a very small amount of chemisorbed SO_2. There was no dissociation of the adsorbed SO_2 and no peaks are seen for metal sulfides or pure sulfur in the XANES spectrum. Our experiments indicate that β-FeMoO$_4$ is much more reactive towards SO_2 than MoO$_3$, but less reactive than Fe$_2$O$_3$. The main product of the interaction of SO_2 with β-FeMoO$_4$ and Fe$_2$O$_3$ is SO_4. This sulfate species is stable on the oxides up to temperatures well above 300 °C.

In summary, the experiments described above show that the combination of two synchrotron-based techniques like time-resolved x-ray diffraction and x-ray absorption spectroscopy can provide detailed information about the structure, phase transitions, electronic properties and

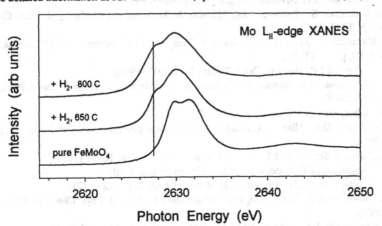

Fig 4 *Mo L$_{II}$-edge XANES spectra acquired before and after exposing β-FeMoO$_4$ to a 15%-H$_2$/85%-N$_2$ mixture at 40-650 °C or 40-800 °C*

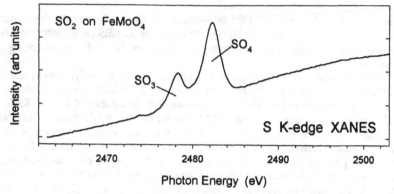

Fig 5   Adsorption of SO_2 on β-$FeMoO_4$ at 50 °C

chemical activity of mixed-metal oxides. Following this approach one can get a better understanding of the behavior of this important type of materials.

ACKNOWLEDGMENT

This work was supported by the US Department of Energy (DE-AC02-98CH10886), Basic Energy Sciences, Chemical Science Division. Travel grants from the American Chemical Society made possible visits of J.L. Brito to BNL and J.A. Rodriguez to IVIC. The facilities used at the NSLS (X7B, X19 and U7A beamlines) are supported by the Chemical and Materials Science Divisions of the US Department of Energy.

REFERENCES

1. A.W. Sleight and B.L. Chamberland, Inorg. Chem. 7, p. 1672 (1968).
2. J. Zou and G.L. Schrader, J. Catal. 161, p. 667 (1996).
3. J.L. Brito and A.L. Barbosa, J. Catal. 171, p. 467 (1997).
4. J.A. Rodriguez, S. Chaturvedi, J.C. Hanson, A. Albornoz and J.L. Brito, J. Phys. Chem. B, 102, p. 1347 (1998).
5. C. Mazzochia, C. Aboumrad, C. Diagne, E. Tempesti, J.M. Herrmann and G.Thomas, Catal. Lett. 10, p. 181 (1991).
6. P. Norby and J.C. Hanson, Catal. Today, 39, p. 301 (1998).
7. J.A. Rodriguez, J.C. Hanson, S. Chaturvedi, A. Maiti and J.L. Brito, J. Chem. Phys. 112, in press (2000).
8. J.G. Chen, Surf. Sci. Rep. 30, p. 1 (1997).
9. M.C. Payne, D.C. Allan, T.A. Arias and J.D. Johannopoulus, Rev. Mod. Phys. 64, p. 1045 (1992).
10. B. Delley, J. Schefer and T. Woike, J. Chem. Phys. 107, p. 10067 (1997).
11. J.A. Rodriguez, J.C. Hanson and J.L. Brito, to be published.
12. A.W. Sleight and L.H. Brixner, J. Solid State Chem. 7, p. 172 (1973).
13. J.A. Rodriguez, S. Chaturvedi, J.C. Hanson and J.L. Brito, J. Phys. Chem. B, 103, p. 770 (1999).
14. S. Chaturvedi, J.A. Rodriguez and J.L. Brito, Catal. Lett. 51, p. 85 (1998).

IN-SITU STUDIES OF THE PROCESSING OF SOL-GEL PRODUCED AMORPHOUS MATERIALS USING XANES, SAXS AND CURVED IMAGE PLATE XRD

DM Pickup[+], G Mountjoy, RJ Newport, ME Smith[*], GW Wallidge[*] and MA Roberts[#]
School of Physical Sciences, University of Kent at Canterbury, CT2 7NR, UK.
+ contact author: d.m.pickup@ukc.ac.uk
* Department of Physics, University of Warwick, Coventry, CV4 7AL, UK.
# Daresbury Laboratory, Warrington, WA4 4AD, UK.

ABSTRACT

Sol-gel produced mixed oxide materials have been extensively studied using conventional, *ex situ* structural techniques. Because the structure of these materials is complex and dependent on preparation conditions, there is much to be gained from *in situ* techniques: the high brightness of synchrotron X-ray sources makes it possible to probe atomic structure on a short timescale, and hence *in situ*. Here we report recent results for mixed titania- (and some zirconia-) silica gels and xerogels. Titania contents were in the range 8-18 mol%, and heat treatments up to 500 °C were applied. The results have been obtained from intrinsically rapid synchrotron X-ray experiments: *i*) time-resolved small angle scattering, using a quadrant detector, to follow the initial stages of aggregation between the sol and the gel; *ii*) the use of a curved image plate detector in diffraction, which allowed the simultaneous collection of data across a wide range of scattering at high count rate, to study heat treatments; and *iii*) X-ray absorption spectroscopy to explore the effects of ambient moisture on transition metal sites.

INTRODUCTION

Mixed titania-silica materials, $(TiO_2)_x(SiO_2)_{1-x}$, and zirconia-silica materials, $(ZrO_2)_x(SiO_2)_{1-x}$, are useful in a number of technological applications [1,2]. The sol-gel process [3] has the advantages of using liquid precursors and occurring at low temperatures. A sol of water and metal alkoxide(s) undergoes hydrolysis and condensation reactions to produce a gel, which may be dried to produce a xerogel: generally an amorphous, porous, hydrated solid. The structures of xerogels are strongly dependent on the details of preparation and heat treatment. The present study concerns acid-catalysed titania- and zirconia-silica xerogels. Our group has studied these materials using diffraction [4], IR [5, 6], NMR [5, 6] and X-ray absorption spectroscopies [5, 6, 7] (and references therein). Generally, metal atoms are homogeneously incorporated into the silica network at low concentrations, i.e. $x<0.2$, and phase separation begins to occur at higher concentrations. Small angle X-ray scattering (SAXS) has been applied to silica [8], titania-silica [9] and zirconia-silica [10] sols. Here we are particularly interested in *in situ* observations.

METHOD

Sample preparation

The $(TiO_2)_x(SiO_2)_{1-x}$ samples were prepared using a two-step hydrolysis procedure [11]. In the first step, tetraethoxyorthosilicate (TEOS) was mixed with water (and isopropanol as a mutual solvent) in equimolar ratios, with an HCl catalyst of pH=1. After 2 hrs, Ti isopropoxide, and then water, were added dropwise while stirring, such that the final ratio of alkoxide to water, R_w, equals 2. With this ratio, complete condensation occurs by utilising all of the water released during hydrolysis. The resulting gels were dried under vacuum to produce xerogels. Heat treatment consisted of heating at 5 °C/min followed by 2 hrs at constant temperature. Estimates of the sample composition were based on a combination of ^{29}Si NMR, TGA and ICP analyses; densities were measured using the Archimedes method. The $(ZrO_2)_x(SiO_2)_{1-x}$ samples were prepared in the same way, except that the solvent was propanol, and the metal alkoxide was Zr propoxide, diluted 1:5. The above method was used to prepare a $(TiO_2)_{0.18}(SiO_2)_{0.82}$ xerogel sample for the Ti XANES experiment. However, the Ti isopropoxide was slowly stirred with acetyl-acetone, acac, in a 1:2 ratio to replace some propoxide ligands with acac. This increases the homogeneity of the resulting sample [5].

Experiment and data analysis - SAXS

The SAXS experiment was carried out on station 8.2 of the SRS, Daresbury Laboratory, U.K. The angular scale of scattering is given by $Q = 4\pi\sin(\theta/2)/\lambda$. Absorption was measured using standard ion chambers; the SAXS intensity was recorded using a quadrant detector. The detector response was calibrated using a radioactive ^{55}Fe source. X-rays of wavelength $\lambda = 1.54$ Å were used with a camera length of $L_c \approx 3.5$m. The SAXS intensity, I(Q), was obtained using standard data reduction methods [12]. The liquid pre-cursors (TEOS, water, and alcohol solvent) have little contrast in electron density, and TEOS shows no SAXS features [13]. However, sol samples are liquids containing random density fluctuations due to the partial polymerisation of alkoxides. Such inhomogeneity can be modelled using the Ornstein-Zernike equation [12]

$$I(Q) \propto \frac{1}{(1+\xi^2 Q^2)} \tag{1}$$

where ξ is the "correlation length", or characteristic length-scale of polymerisation. Models of this kind have previously been applied to acid-catalysed sols [8, 9].

Experiment and data analysis – diffraction, curved image plate detector

The X-ray diffraction data were collected on station 9.1 at the Daresbury Laboratory SRS, using the curved image plate technique developed by Finney and Bushnell-Wye [14]. The wavelength was set at $\lambda = 0.487$ Å. The sample was contained in a 1 mm diameter silica capillary with axis perpendicular to the incident X-ray beam. A 400 x 148 mm image plate (Fuji Bass III) was mounted in a 185 mm arc radial to the sample axis. Heating under an ambient atmosphere was provided by a custom-made furnace fitted with Kapton windows. Data were collected for 45 minutes at each temperature. Each stored image was transferred from the image plate within 10 min. using a Molecular Dynamics scanner; calibration of the data were achieved by using a 'grid calibration' technique. Preliminary analysis of the X-ray diffraction data involved the integration of the Debye-Scherrer cone recorded by the image-plate, a correction for incident X-ray beam polarisation and background subtraction. Corrections were then made to account for absorption and for inelastic, or Compton scattering. The self scattering and Compton scattering were calculated from tables [15].

Real-space structural information was obtained from the corrected X-ray data by defining the normalised interference function, $i(Q)$; inverting this by Fourier transform yields the total differential pair correlation function $d(r)$. I_{eu} is the corrected scattered X-ray intensity, $Q = 4\pi\sin\theta/\lambda$ (where 2θ is the scattering angle) and c_i and f_i are the respective fractional concentration and atomic form factor for element i. $M(Q)$ is a Hanning modification function which is zero for $Q \geq Q_{max}$ and is required to reduce the truncation effects associated with the finite range of Q:

$$i(Q) = \frac{I_{eu} - N\sum_i c_i f_i^2}{(\sum_i c_i f_i)^2} \qquad \text{and} \qquad d(r) = \frac{2}{\pi}\int_0^\infty Qi(Q)M(Q)\sin Qr \, dQ \qquad (2)$$

Structural parameters were obtained from the X-ray data presented here by modelling the atomic correlations in Q-space using [16]:

$$i(Q)_{ij} = \frac{N_{ij}\omega_{ij}}{c_j}\frac{\sin QR_{ij}}{QR_{ij}}\exp\left[\frac{-R_{ij}^2\sigma_{ij}^2}{2}\right] \qquad \text{where} \quad \omega_{ij} = \frac{(2-\delta_{ij})c_i c_j Z_i Z_j}{[\bar{Z}]^2} \qquad (3)$$

i and j are the elements in the correlation being modelled, ω_{ij} is the X-ray weighting factor and Z_i are the atomic numbers; N_{ij} is the coordination number, R_{ij} is the distance, and σ_{ij} is the standard deviation. The Fourier transforms were carried out over the Q-range 0.7-21.1 Å^{-1}.

Experiment and data analysis - Ti K-edge XANES

The X-ray absorption near edge structure, XANES, experiments were carried out on station 8.1 of the SRS, Daresbury Laboratory, U.K. Samples were prepared by mixing with boron nitride and pressing into pellets. *In situ* measurements were obtained using a furnace with mylar windows and an N_2 atmosphere. Spectra were collected at the Ti K-edge (4966 eV) in transmission mode; a Ti foil and third ion chamber were used simultaneously to calibrate the energy of the monochromator. Energies reported here are relative to the first inflection point of the Ti foil absorption edge. The XANES spectra were processed in the usual way to obtain normalised absorbance [17]. Different Ti coordinations can be distinguished using the XANES due to the presence of a pre-edge peak corresponding to p-d mixing; this method has been firmly established through studies of Ti in reference compounds [18]. The height of the pre-edge peak increases sequentially from [6]Ti to [5]Ti to [4]Ti due to decreasing centrosymmetry as the coordination changes from octahedral to square pyramidal to tetrahedral, respectively [18]. We have recently applied this technique to titania-silica xerogels using *ex situ* Ti XANES [19].

RESULTS AND DISCUSSION

SAXS of TiO$_2$-SiO$_2$ and ZrO$_2$-SiO$_2$ sol-gels

Figure 1 shows the results for (*a*) the correlation length, ξ, and (*b*) the intensity, I(0), obtained by fitting Equation (1) to I(Q) measured *in situ* during the ageing of the sols during a

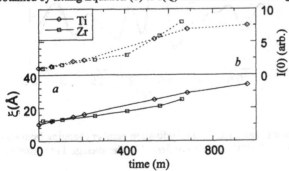

FIGURE 1: Characteristic length-scale of fluctuations, ξ, and intensities, I(0), obtained by fitting Ornstein-Zernike equation (Equation 1) to I(Q) for sol-gels with composition $(TiO_2)_{0.08}(SiO_2)_{0.92}$ ($\diamond$) and $(ZrO_2)_{0.07}(SiO_2)_{0.93}$ ($\square$).

total ageing time of ~12 hrs. The values of ξ observed here are consistent with values reported for acid-catalysed pure silica [8], titania-silica [9] and zirconia-silica [10] sol-gels with $1 < R_w <$ 5. The aggregation of polymers during gelling is expected to produce fractal polymer clusters with mass, and hence I(0), proportional to ξ^d with d<3 [9,12]. Our results follow this behaviour.

Diffraction, using curved image plate detection

Four correlations were used in the X-ray diffraction modelling procedure: a Si-O distance of ~1.6 Å, a Ti-O distance of 1.8-2.0 Å, an O-O distance of ~2.6 Å and a M-M (M=Si or Ti) distance of ~3.1 Å. The data collected from the empty SiO_2 capillary were analysed and compared with those from the literature in order to check the validity of our data collection and analysis. Figure 2 shows the real space fits to the X-ray data collected at 25 and 500 °C as examples. The structural parameters obtained *in situ* at various temperatures from the $(TiO_2)_{0.18}(SiO_2)_{0.82}$ xerogel are shown in Table 1. The errors quoted are estimates of the statistical uncertainties associated with the fitting process.

Temp. (°C)	25 (silica)	25	210	310	500
N_{Si-O}	4.3	4.4	4.4	4.3	4.2
R_{Si-O} /Å	1.61	1.62	1.63	1.62	1.62
A_{Si-O} / Å²	0.002	0.002	0.007	0.004	0.010
N_{Ti-O}		6.0	4.8	4.0	4.0
R_{Ti-O} /Å		1.97	1.98	1.82	1.82
A_{Ti-O} / Å²		0.005	0.034	0.005	0.007
N_{O-O}	5.5	5.3	4.6	4.6	5.0
R_{O-O} /Å	2.60	2.62	2.63	2.69	2.73
A_{O-O} / Å²	0.005	0.02	0.034	0.051	0.045
N_{M-M}	4.0	6.8	5.8	5.9	4.3
R_{M-M} /Å	3.04	3.09	3.08	3.12	3.09
A_{M-M} / Å²	0.01	0.045	0.039	0.051	0.024

TABLE I. *In-situ* structural parameters for $(TiO_2)_{0.18}(SiO_2)_{0.82}$ xerogel at various temperatures obtained from modelling of X-ray diffraction data. Note that the errors are ±0.02 Å in R, ±10% in N and ±20% in A.

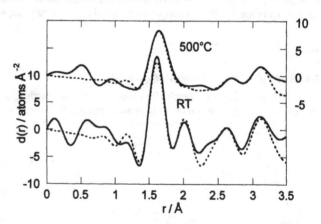

FIGURE 2. Experimental(solid line) and calculated (broken line) differential correlation functions, d(r), for the $(TiO_2)_{0.18}(SiO_2)_{.82}$ xerogel at 25 °C and (offset) heated *in-situ* to 500 °C. Note that the fit is calculated only to 3.5 Å.

The structural parameters show the average Ti-O coordination number changing from ~6 to ~4 between the temperatures of 25 and 310 °C. This is confirmed by the average Ti-O distance shortening from 1.98 Å (close to that found in TiO_2 anatase) to 1.82 Å (compares well with that found in TS-1). Looking at Debye-Waller factors, one can see that the peak assigned to Si-O at ~1.6 Å is much broader by 210 °C, illustrating that by this temperature it includes a significant contribution from the short Ti-O correlation distance. The Ti-O distance of 1.98 Å at 210 °C represents only the contribution from the 6-fold Ti.

The change in Ti local environment probably occurs by two mechanisms. One is the substitution of TiO_4 groups into the silica network with heat treatment and the other is the reversible loss of H_2O from some 6-fold Ti sites (the latter mechanism is identified using *in situ* XANES). The results show very little change in the Ti-O correlation on heating from 310 to 500 °C indicating that the Ti is stable within the silica network up to 500 °C. In pure silica the shortest O-O distance is constant up to 1036 °C. In the xerogel the O-O distance increases by ~0.1 Å with heating to 500 °C, which is probably due to Ti substituting into the silica network, since the M-O distance is greater for Ti than Si. The O-O coordination numbers for the xerogel are all close to five and reflect incomplete condensation of the network structure at these temperatures.

Ti K-edge XANES of $(TiO_2)_{0.18}(SiO_2)_{0.82}$ xerogels

The Ti K-edge XANES pre-edge peak heights and positions for a xerogel with 18 mol% TiO_2 are shown in Figure 3. This reveals a large increase in pre-edge peak height (and corresponding decrease in energy) for the *in situ* compared to *ex situ* measurements. In this case, a significant proportion of [4]Ti has been created after only 10 min. heating at 250 °C, but has largely disappeared upon cooling (*ex situ*) to 25 °C, i.e. the change is reversible. Because this occurs in a sample which had previously been heated at 250 °C for 2 hrs, it is clear that Ti coordination is affected by ambient moisture. This implies that the isolated, distorted [6]Ti precursor to [4]Ti has a coordination similar to [4]Ti plus two weak additional bonds due to coordination by water/hydroxyl groups.

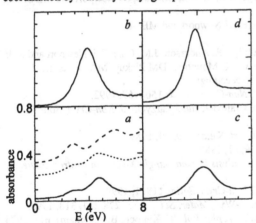

FIGURE 3: Solid lines show pre-edge peaks for a xerogel with 18 mol% TiO_2 (a,b) before heat treatment, and (c,d) after heat treatment at 250°C. The effect of ambient moisture is seen by comparing measurements made (a,c) *ex situ* and (b,d) *in situ* after 10mins at 250°C. Dotted and dashed lines show pre-edge peaks for $ZrTiO_4$ and anatase (TiO_2) with distorted and regular [6]Ti respectively.

CONCLUSIONS

In-situ SAXS of $(TiO_2)_{0.08}(SiO_2)_{0.92}$ and $(ZrO_2)_{0.07}(SiO_2)_{0.93}$ sols during ageing shows the growth of polymeric clusters of characteristic length-scale ~20 Å.

Combining the curved image-plate technique with a synchrotron radiation source is a flexible and powerful tool for collecting *in situ* high temperature X-ray diffraction data on an amorphous material on a short time scale. For the first time the calcination of an amorphous

$(TiO_2)_{0.18}(SiO_2)_{0.82}$ xerogel has been followed *in situ* with X-ray diffraction. The data were of sufficient quality to show a change in Ti coordination number from ~6 to ~4 as the sample was heated to 310 °C. Indeed, it is evident that the data could have been obtained in shorter periods of time and that the potential for significant further gains using the developing high intensity "third-generation" synchrotron sources is correspondingly higher still.

Comparison of 25 °C ambient and *in situ* Ti K-edge XANES of titania-silica xerogels unequivocally shows rapid and reversible conversion of $^{[6]}$Ti into $^{[4]}$Ti. This clearly demonstrates the distinction between $^{[4]}$Ti in the SiO_2 network which is exposed to, and hence interacts with ambient moisture, and that which is inaccessible.

ACKNOWLEDGEMENTS

The EPSRC is thanked for its support through various grants/studentships. WA Steer and BU Komanschek are thanked for their help with the data collection and analysis, and JC Dore for helpful discussions.

REFERENCES

[1] PC Shultz and HT Smyth *Amorphous materials,* ed EW Douglas and B Ellis (London: Wiley) 1972. M Itoh, H Hattori, and KJ Tanabe *J. Catalysis* **35**, 225, 1974. M Nogami *J. Non-Cryst. Solids* **69**, 415, 1985. JB Miller and EI Ko, *J. Catalysis* **159**, 58, 1996.
[2] MA Holland, DM Pickup, RJ Newport, G Mountjoy and ME Smith, *these proceedings.*
[3] CJ Brinker and GW Scherer, *Sol-gel science: the physics and chemistry of sol-gel processing* (San Diego: Academic Press) 1990.
[4] JS Rigden, JK Walters, PJ Dirken, ME Smith, G Bushnell-Wye, WS Howells and RJ Newport, *J. Phys.: Condens. Matter* **9**, 4001, 1997.
[5] DM Pickup, G Mountjoy, GW Wallidge, R Anderson, JM Cole, RJ Newport and ME Smith, *J. Mater. Chem.* **9**, 1299, 1999.
[6] DM Pickup, G Mountjoy, GW Wallidge, RJ Newport and ME Smith *Phys. Chem. Chem. Phys.* **1**, 2527, 1999.
[7] G. Mountjoy, D.M. Pickup, G.W. Wallidge, R. Anderson, J.M. Cole, R.J. Newport and M.E. Smith, *Chem. Mater.* **11**, 1253, 1999. See also G Mountjoy, DM Pickup, MA Holland, RJ Newport, GW Wallidge, ME Smith, *these proceedings.*
[8] T Kamiya, M Mikami and K Suzuki *J. Non-Cryst. Solids* **150**, 157, 1992.
[9] M Ramirez-del-Solar, L Esquivias, AF Craievich and J Zarzycki, *J. Non-Cryst. Solids* **147-148**, 206, 1992.
[10] M Nogami and K Nagasaka, *J. Non-Cryst. Solids* **109**, 79, 1989.
[11] BE Yoldas, *J. Non-Cryst. Solids* **38-39**, 81, 1980.
[12] LA Feigin and DI Svergun, *Structure analysis by small-angle X-ray and neutron scattering* (New York: Plenum Press) 1987.
[13] B Himmel, T Gerber and H Burger *J. Non-Cryst. Solids* **119**, 1, 1990.
[14] MA Roberts, JL Finney and G Bushnell-Wye, *Mater. Sci. Forum* **278-281**, 318, 1998.
[15] *International Tables for X-ray Crystallography, Vol. IV* (Kynoch, Birmingham), p71, 1974. F Hajdu, *Acta. Cryst.* A**28**, 250 , 1972.
[16] PH Gaskell, in *Material Science and Technology Volume 9: Glasses and Amorphous Materials,* edited by J Zarzycky (VCH, Weinhaim) p175, 1991.
[17] A Bianconi, *X-ray absorption: principles, applications, techniques of EXAFS, SEXAFS and XANES* ed. Koninbsberger D.C. and Prins R., Chapter 11 (Wiley, New York, 1987).
[18] F Farges, GE Brown, A Navrotsky, H Gan and JJ Rehr *Geochimica et Cosmochimica acta* **60**, 3023, 1996.

DIAMOND IN-LINE MONITORS FOR SYNCHROTRON EXPERIMENTS

P. BERGONZO, D. TROMSON, A. BRAMBILLA, C. MER, B. GUIZARD, F. FOULON
LETI (CEA - Technologies Avancées)/DEIN/SPE, CEA/Saclay, F-91191 Gif-sur-Yvette
pbergonzo@cea.fr

ABSTRACT

Diamond polycrystalline films have been synthesised using the Chemical Vapour Deposition (CVD) technique in order to fabricate new types of photo-detectors for the characterisation of X-ray light sources as encountered in synchrotron experiments. Since diamond exhibits a low absorption to low energy X-Ray photons, these devices allow beam position monitoring with very little beam attenuation at photon energies as low as 2 keV. We present here diamond based new devices for four different applications, including (*i*) semitransparent beam intensity and (*ii*) position monitors with high position resolution (< 2 μm), (*iii*) beam profile monitors with 20 μm pitch resolution, and (*iv*) ultra-fast diamond detectors (response time < 100 ps) that enable the intensity and temporal monitoring of fast X-ray pulses. These devices can be used for in-line characterisation of synchrotron beam line experiments for permanent in-situ monitoring of beam instabilities during experiments as well as for synchrotron machine diagnostics.

INTRODUCTION

Advances in experiments using synchrotron light sources have generated a demand for permanent in-line radiation hard X-ray monitors in the energy range of 1.5 keV to 25 keV. In fact, for demanding experiments such as XAFS on ultra dilute samples or polarisation dependent X-ray spectroscopy, it may become necessary to control the beam instabilities with respect to both position and time. CVD diamond has two distinct advantages over all other detector materials: its high radiation hardness allowing long term in-situ analysis and low atomic number resulting in a low X-ray absorption cross-section. This combination enables thin-film photo-detectors to be inserted permanently in synchrotron beam lines causing little intensity perturbations downstream. Diamond exhibits several remarkable properties in comparison to other semiconducting materials [1], such as high band gap, high electron hole pair mobility, short carrier lifetime, and extreme resilience to harsh environments. The recent progresses in diamond growing techniques have made the synthesis of this material readily available. Synthetic diamond layers are grown using the microwave enhanced chemical vapour deposition (CVD) technique from the dissociation of a methane and hydrogen precursor mixture [2]. The material obtained has a polycrystalline structure with a grain size of about 10 % of the layer thickness. Under optimised conditions, electronic grade diamond can be obtained [3]. We have grown diamond using optimised processes in order to obtain a material that exhibits suitable electronic properties enabling its use for detection device applications [4]. Typical growth rates are of the order of 0.5 μm/h. After growth, a series of annealing steps and chemical treatments are performed in order to improve the properties of the films [5]. According to the thickness deposited on the silicon substrate, it is possible either to keep the silicon substrate to give the thin layers mechanical resilience, or to etch it in acids and use the diamond film as a free standing layer (t ≥ 50 μm).

In the last few years, diamond has essentially been exploited for its use as a radiation detector [6-9]. We report in the present paper four diamond based devices developed for the characterisation of synchrotron light sources, namely (*i*) beam intensity and (*ii*) position monitors, (*iii*) beam profile monitors, and (*iv*) ultra-fast pulse monitors that enable the intensity and temporal monitoring of the pulse tracks distribution

Beam Intensity and Position Monitors (BPM)

The typical energy ranges of interest are in the 4 keV to 20 keV domain, but there is also a growing demand in the region below 4 keV. Conventional sensors based on semiconductor photodetectors, mostly silicon, generally involve the complete absorption of the incoming radiation [10]. To make in line devices, for measuring the beam characteristics without a significant attenuation of the X-ray flux, a low optical absorption cross section semiconductor, such as diamond, is required. The fabrication of these devices enables permanent monitoring and interactive compensation of possible beam position shifts.

From the polycrystalline diamond layer grown on silicon by the CVD technique, it is possible to back etch a hole in the silicon to obtain a thin diamond membrane supported by a silicon ring. At a typical thickness of 20 µm, a 1 cm-diameter membrane has a good mechanical resilience that enables normal handling and transportation. Membranes as thin as 2 µm could also be fabricated but they would require special care when used. In the 4-20 keV region, a 20 µm thick membrane absorbs no more than 22% of the incident beam, making this thickness ideal in this energy range. Below 4 keV, a region where most materials are opaque, diamond is unique in that a 2 µm thick membrane permits about 70 % transmission at 1.5 keV [11].

Detectors have been fabricated from 20 µm thick diamond membranes with evaporated thin contact layers on both sides. The contact thicknesses used, typically 500 Å, exhibit a negligible cross section to the X-rays. For operation above 4 keV, gold was used, while for devices at lower energies thin graphite layers could be used. The samples were biased with a typical DC polarisation of 5×10^4 V.cm^{-1}. By dividing the electrical contact in one side of the membrane into four sections closely separated by a gap, four quadrant beam position monitors were fabricated. The gap separating the electrodes can be adjusted typically from a few micrometers to a few millimetres according to the requirements of beam size, signal to noise, and sensitivity. The X-ray interaction in the diamond creates electron hole pairs that drift in the material and result in four cu--rrents that are measured in each pad. If interaction occurs in the centre of the device, the four currents are equal, hence enabling the localisation of the spot size. Using a inter-electrode distance of 500 µm, measurements at the European Radiation Synchrotron Facility (ESRF) have enabled the probing of the beam displacements on a 200 x 200µm square beam. Significant beam shifts were observed in normal operation with typically a few tens of microns per half hour. In the horizontal direction, a periodic position shift of a few micrometers was observed that corresponds to the machine global feedback period, revealing a position resolution better than 2 µm (fig. 1). A more detailed description of this device is given in [12].

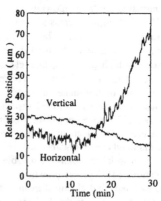

Fig. 1. Time scan measurements using the diamond BPM detector revealing typical beam position shifts. Beam energy is 4 keV, beam size is 200 x 200 µm. The use of a semi-transparent BPM allows the correction of the shifts.

One inherent drawback however with this configuration is that the BPM resolution depends on the size of the beam. Another configuration has therefore been developed, based on the use of a resistive layer in place of the four quadrant structure. A 0.1μm thick layer of boron doped amorphous silicon (a-Si:H), with a resistivity of 10^5 Ω.cm has been used (Fig 2). Here, the current induced in the diamond is detected in each of the 4 corners of the a-Si:H layer which are connected to the four channels of the electrometer. The relative high resistivity of the layer enables high discrimination between the values of the four currents, without inducing a significant drop in the diamond layer bias voltage. The induced current in each pad divided by the sum of the four currents varies then linearly as x/L where L is the distance between pads and x the distance from that pad to the position of interaction. Figure 3 illustrates this excellent linearity measured as the beam is scanned within microns or millimetres from the centre of the device. Unlike the 4 quadrant BPM, this resistive readout BPM therefore enables continuously variable signal detection with respect to the interaction position.

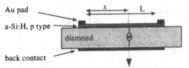

Au pad
a-Si:H, p type
diamond
back contact

Fig. 2. (above) Schematic representation of the resistive readout beam position monitor

Fig. 3. (right) Normalised currents measured on two pads of the resistive readout BPM showing the excellent linearity of the device in its central region with a 500 x 500 μm beam on a 1mm scale (top) as well as on a 200 x 200 μm beam on a 50 μm scale (bottom). Signals on the two other pads are equivalent and have been removed for clarity.

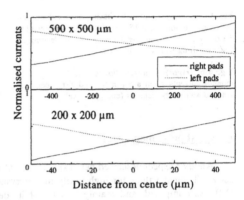

Distance from centre (μm)

Semitransparent Beam Profile Monitors

From the low X-ray cross section of diamond, other device structures have been developed in order to enable the measurement of synchrotron X-ray beam profiles with a high spatial resolution. In fact, for demanding experiments where focused X-ray beams are involved, such as high resolution X-ray imaging and spectro-microscopy and especially in the low energy region (5-10 keV), it may become necessary to verify the shape of the X-ray beam spatial intensity distribution not only during set up and alignment procedures but also while measurements are performed. This latter requirement could only be satisfied using a semitransparent beam profile monitor from a low atomic number detecting media as diamond.

A 12 x 12 mm unpolished CVD diamond layer with a 100 μm thickness grown on polished silicon was used. The silicon substrate was fully removed in order to obtain a free standing layer with one optically flat surface on the side where the silicon was removed. This flat surface enables high resolution lithography and micrometer size metallic strips were fabricated. Here we report on a device consisting of eight 50 nm thick and 15 μm wide gold lines. An inter-electrode distance of 5 μm was used, and the strip length was of 2 mm (fig. 4a). On the other rougher surface of the CVD diamond a 50 nm aluminium contact was evaporated. The device was assembled on a perforated alumina holder, and connected to a eight channel electrometer. Operation in vacuum is not necessary. Figure 4b shows the characteristic response measured when a 5 μ diameter focused beam at 5 keV is scanned across the device. Hardly any cross talk is observed even though the thickness (100 μm) is rather large with respect to the line pitch (20 μm). The combination of two profiling systems, perpendicular to each other, allows the complete beam 2D profile monitoring.

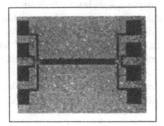

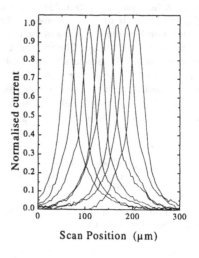

Scan Position (μm)

Fig. 4-a. (above) Photo of the beam profile monitor: in the central region, eight 15 μm width lines separated by 5 μm are fabricated on the diamond surface, leading to beam profiling with a 20 μm pitch.

Fig. 4-b. (right) Representation of the beam profile monitor currents when a 5 μm focused X-ray beam is scanned accross the detector (energy is 5 keV)

Ultra-fast pulse monitoring

Another aspect described here exploits the extremely small carrier lifetime that CVD diamond shows. Typical values for high purity mono-crystalline diamond have been reported above 1 ns [13]. As compared, and mainly because of its defective polycrystalline nature, CVD diamond carrier lifetimes can vary from 700 ps down to less than 100 ps according to the growth conditions [14]. This latter value approaches those obtained from III-V or II-VI materials (GaAs, InP, and CdTe), that have enabled the fabrication of detectors in the pico-second range, after pre-irradiation of the bulk material with fast neutrons [15-17]. However, since diamond combines fast operation capabilities with radiation and temperature hardness, it is an extremely attractive candidate for the characterisation of ultrafast VUV or X-ray photon sources where the dose levels are high, and particularly for the monitoring of electron distributions in synchrotron storage rings.

For X-ray pulse monitoring, detectors were fabricated using two electrode contact evaporation either on the same side of the diamond layer (coplanar) or on both sides in a vis-à-vis configuration (sandwich) with respect to the incident X-ray energy to monitor. The former is more suitable for low energy X-ray pulse characterisation typically below 1 keV, whereas at higher energies free standing thicker diamond layers (> 50 μm) with sandwich electrodes are commonly used. With DC bias, the response of diamond detectors to X-ray pulses reflects the change in diamond conductivity due to the generation of photo-excited carriers in the film. The resulting photo-current and its transient decay were measured using an ultra-fast sampling oscilloscope (Tektronics CSA803A).

In synchrotrons, the X-ray light is composed by a succession of ultra-short individual pulses. We have used CVD diamond detectors to monitor the temporal distribution of the individual X-ray pulse tracks. For example, at the ESRF in 2/3 filling mode, the X ray light is composed of 670 individual pulses distributed in the 2/3 of the storage ring and occurring every 2.8 ns. The pulse duration is typically 40 ps at FWHM. Also, since these detectors exhibit very strong radiation and temperature hardness, the measurements could be performed in white light (mostly 8-20 keV), conditions where despite of air cooling the detector temperature exceeded 100°C. Measurements from a 400 μm thick device in the sandwich configuration led to a response that showed a linear rise in the number of electrons in the beginning of the track (inset in fig. 5). Further, figure 5 shows pile-up effects observed in the middle region of the track, as the machine is adding consecutively two trains distributed each along one third of the ring. These features were confirmed by the machine diagnostic specifications obtained from the use of streak cameras. It demonstrates that the detector enables direct monitoring of the X-ray beam intensity and temporal distribution. The ability to separate single pulse characteristics directly gives us a simple device for machine diagnostics. It is clear that the use of a small portable detector connected directly to its ultra-fast scope can be very attractive from its simplicity, cost and ease of operation. Similar measurements have also been performed at LURE (Laboratoire pour l'Utilisation du Rayonnement Electromagnétique, Orsay, France) at lower energies (typ. 1 keV) in white light (here 300 eV - 2 keV) using a coplanar configuration of the detection devices, and gave perfect satisfaction as the pulse duration and intensity were probed.

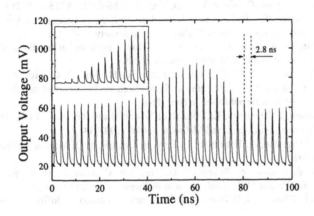

Fig. 5. Evidence of pileup in the middle of the track in the 2/3rd filling mode at ESRF. Inset shows the linear rise of the individual X-ray pulse intensity in the beginning of the track. Measurement are confirmed by machine diagnostics obtained from the use of streak cameras. This device enables direct monitoring of the X-ray beam intensity and temporal distribution. The observations are for white light : mostly 8-20 keV.

CONCLUSION

The potential of diamond films grown in our laboratory using the CVD technique for the characterisation of X-ray light sources as encountered in synchrotrons has been described. Further to the inherent properties such as the strong radiation and temperature hardness, diamond also enables the fabrication of detection devices that exploit its low atomic number for reduced X-ray cross section, and its short carrier lifetime for ultrafast pulse monitoring. New types of devices were developed for in-situ characterisation of synchrotron beam lines, from semitransparent beam position and profile monitors to beam temporal distribution probing. These CVD diamond detectors demonstrate the unique feature of enabling position and profile measurements of low energy synchrotron X-ray light sources, together with a low attenuation of X-ray photons, thus allowing real time monitoring of the beam with excellent position sensitivity. Further, very simple, and low cost devices can be fabricated in order to enable the measurement of the temporal distribution of the X-ray beam intensity distribution, with ultra-fast time responses well below 1 ns. The excellent resilience to heat and radiation has further enabled the use of these sensors in white light beams without degradation.

ACKNOWLEDGEMENTS

This study was conducted in close collaboration with people from CEA/DAM in Bruyères le Chatel : R. Wrobel, B. Brullot, and C. Rubbelynck, and partners from the European Radiation Synchrotron Facility (ESRF), Grenoble (Fr) namely J-F. Eloy, C. Gauthier, J. Goulon, O. Mathon, A. Rogalev, and V.A. Solé.

REFERENCES

[1] M. Franklin, A. Fry, K.K. Gan, S. Han, H. Kagan, S. Kanda, D. Kania, R. Kass, S.K. Kim, R. Malchow, F. Morrow, S. Olsen, W.F. Palmer, L.S. Pan, F. Sannes, S. Schnetzer, R. Stone, Y.Sugimoto, G.B. Thomson, C. White, S. Zhao, Nucl. Inst. Methods, A315, (1992) 39

[2] T. Pochet, A. Brambilla, P. Bergonzo, F. Foulon, C. Jany, A Gicquel, in Eurodiamond 96, Edited by C. Manfredotti and E. Vittone (Italian Physical Society 1996) 111

[3] C. Jany, A. Tardieu, A. Gicquel, P. Bergonzo, F. Foulon, presented at the diamond'99 meeting, Prague (1999), submitted to diamond and related materials.

[4] F. Foulon, D. Tromson, R.D. Marshall, L. Rousseau, B. Guizard, C. Mer, A. Brambilla, P. Bergonzo, presented at the Sixth International Symposium on Diamond Materials. Electrochem. Soc., Hawaï, (Oct. 1999), to be published in conf. Proc.

[5] C. Jany, F. Foulon, P. Bergonzo, R.D. Marshall, Diamond and Rel. Mat., 7 (7), (1998) 951

[6] RD 42 Collaboration, Nucl. Inst. and Methods. A 400, (1997) 69

[7] F. Foulon, P. Bergonzo, A. Brambilla, C. Jany, B. Guizard, R.D. Marshall, MRS Soc. Symp. Proc. 487 (1998) 591

[8] P. Bergonzo, F. Foulon, A. Brambilla, D. Tromson, C. Jany, S. Haan, presented at the diamond'99 meeting, Prague (1999), submitted to diamond and related materials.

[9] A. Brambilla, P. Chambaud, D Tromson, P. Bergonzo, F. Foulon, F. Joffre, submitted to IEEE, TNS

[10] Gauthier C., Goujon G., Feite S., Moguiline M., Braichovich L., Brookes N.B., Goulon J., Physica B, 208,209, (1995) 232

[11] Evaluated from www-cxro.lbl.gov/optical_constants/filter2.html

[12] P. Bergonzo, A. Brambilla, D. Tromson, R.D. Marshall, C. Jany, F. Foulon, C. Gauthier, V.A. Solé, A. Rogalev, J. Goulon, J. Synch. Rad, 6 (1999) 1

[13] E.A. Konorova, S.F. Kozlov, V.S. Vavilov, Soviet Phys. Sol-State, 8 (1966) 1

[14] F. Foulon, P. Bergonzo, C. Jany, A. Gicquel, T. Pochet, NIM-A 380 (1996) 42

[15] M. Cuzin, F. Glasser, J. Lajzerowicz, F. Maathy, J. Rustique, MRS Soc. Symp. Proc. Vol 302 (1993) 169

[16] D.R. Kania, A.E. Iverson, D.L. Smith, R.S. Wagner, R.B. Hammond, K.A. Stetlar, J. Appl. Phys., 60 (1986) 2596

[17] F. Foulon, B. Brullot, C. Rubbelynck, P. Bergonzo, T. Pochet, IEEE Trans. Nucl. Sci, , 43 (1996) 1372

Anomalous Ultra-Small-Angle X-ray Scattering from Evolving Microstructures during Tensile Creep

P. R. JEMIAN*[†], G. G. LONG[†], F. LOFAJ[†], AND S. M. WIEDERHORN[†]
* Materials Research Laboratory, University of Illinois at Urbana-Champaign, Bldg. 438D, Argonne, IL 60439, jemian@uiuc.edu
† National Institute of Standards and Technology, 100 Bureau Drive, Gaithersburg, MD 20899-8523

ABSTRACT

Ultra-small-angle X-ray scattering provides quantitative and statistically significant information on the size distribution of electron density inhomogeneities with dimensions between ≈100 Å and ≈5 μm. All sizes are sampled simultaneously with a single experiment, removing the possibility of observational bias. In a material such as commercial silicon nitride, where the inhomogeneities are due to populations of intergranular secondary phases and voids of similar dimensions, the scattering contains contributions from each individual population. A single USAXS scan cannot separate overlapping populations of scatterers due to the different contrasts of the microstructural components. Anomalous USAXS (A-USAXS) is an element-specific contrast variation method to vary the scattering contribution from one of the populations while holding that of the other populations fixed. To follow the size evolution under tensile creep of both the cavities and the Yb disilicate secondary phases, A-USAXS data was measured near the Yb L_{III} absorption edge. Creep cavity and disilicate size distributions were each determined as a function of deformation.

INTRODUCTION

Silicon nitride is a prime candidate for structural components in advanced gas turbines operating at high turbine inlet temperatures. However, creep was found to be one of the principal factors compromising its excellent high-temperature mechanical properties in components subjected to high stresses and temperature over prolonged periods of time. Intensive studies of the creep processes in silicon nitride and similar multi-phase materials have revealed that cavitation is possibly the most important mechanism resulting in creep deformation [1].

Earlier small-angle X-ray scattering (SAXS) studies [1] showed that the deformation occurs via cavity accumulation at multigrain junctions during deformation and that their growth is small [1]. The evolution of secondary phase pockets during creep has not been investigated previously because of the lack of a suitable technique. Transmission electron microscopy confirmed that cavities have dimensions similar to that of the phases in the pockets but without any statistical evidence for their distribution [1, 2]. A recent cavitation model based on the continuous accumulation of cavities suggests that the cavities grow via redistribution of the crystalline secondary phase between different junctions [3]. Thus, knowledge of the evolution of the crystalline secondary-phase particles and the development of creep cavitation during deformation is crucial for understanding the deformation process and for developing better ceramics.

EXPERIMENTAL PROCEDURE

Material and Creep Testing

The commercial grade of gas-pressure sintered silicon nitride (designated SN88 [4, 5]) that was studied consists of β-Si_3N_4 grains, voids, and secondary phases containing Yb and a small amount of Y. The major crystalline secondary phase after heat treatment is ytterbium disilicate, $Yb_2Si_2O_7$. Minor phases include residual $Yb_4Si_2N_2O_7$, $Y_5Si_3NO_{12}$, residual SiO_2 glass, and porosity.

Five tensile creep tests were carried out at 1400° C under 150 MPa load. The details of the testing procedure are described elsewhere [1]. After different periods of time and strain, the

131

tests were intentionally interrupted and the specimens were cooled under load. The samples for A-USAXS were cut from each specimen from the undeformed grip area and from the gage area parallel to the stress axis. The samples were ground and hand-polished to thicknesses of ≈180 μm and ≈100 μm, respectively.

Instrument

To calibrate the X-ray energy (to within 0.5 eV) and determine the Yb anomalous dispersion coefficients, an X-ray absorption spectrum about the Yb L_{III} edge was measured from one of the samples. The absorption edge energy ($\Delta E=0$) was taken as the first inflection point of the spectrum while the dispersion corrections were determined [6] from the optical theorem (to obtain f") and a Kramers-Kronig transformation (to obtain f'). The measured anomalous dispersion coefficients were then used to calculate the energy-dependent scattering contrast of the Yb disilicate scatterers in the silicon nitride matrix, as shown in Fig. 1.

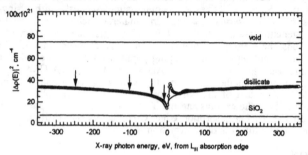

Fig. 1 X-ray scattering contrast of scatterers in SN-88 silicon nitride near the Yb L_{III} edge. Data points use data from absorption spectra, solid line uses anomalous dispersion corrections to the atomic scattering factor by the calculation of Cromer and Liberman [7]. Estimated errors are smaller than the symbols. The arrows indicate the energies used to measure the A-USAXS.

The USAXS instrument [8] at UNICAT beam line 33ID at the Advanced Photon Source was used to measure the USAXS of each sample. USAXS at each of four energies (-250 eV, -100 eV, -40 eV, and -10 eV from the edge) was measured in the vicinity of the Yb L_{III} absorption edge to vary the scattering contrast of the Yb disilicate, $Yb_2Si_2O_7$, with respect to that of the voids. A blank (no sample) was measured at each energy, as well. The slit length of the USAXS instrument was ≈0.03 $Å^{-1}$. The USAXS data were scaled to units of absolute cross-section per unit volume per steradian by a primary method [9] and desmeared by an iterative method [10].

Anomalous Ultra Small-Angle X-ray Scattering Analysis (A-USAXS)

Processing of the scattering data, taking into account the contrast variation, is conducted piecewise [11]. First, each scan is interpreted in terms of an energy-dependent size distribution. Thus we have four different distributions, one for each energy measured. Each distribution so obtained is the sum of all size distributions of scatterers in the sample, each weighted by their energy-dependent scattering contrast. Second, the four energy-dependent size distributions are combined with the energy dependence of the scattering contrasts to produce the anomalous scattering results. By fitting a straight line at each size bin in the energy-dependent size distributions vs. the energy-dependent scattering contrasts of the Yb disilicate, the slope reveals the size distribution of the Yb disilicate while the intercept indicates the size distribution of voids. Assuming that the deformation during tensile creep is due to creep cavities in the gage section, a subtraction of the void size distribution within the gage section sample from that within the grip section sample yields the distribution of cavities due to creep. This subtraction will also remove from consideration any residual non-creep porosity found in both sections.

The small-angle scattering can be considered to derive from the presence in the sample of a number of non-interacting scatterers k, each with a size distribution, $N_k(D)$, and X-ray energy-dependent scattering contrast, $|\Delta\rho_k(E)|^2$. The scattered intensity (in absolute units) is

$$I(Q) = \frac{d\Sigma}{d\Omega}(Q,E) = \sum_k |\Delta\rho_k(E)|^2 \int_0^\infty N_k(D) V^2(D) |F(Q,D)|^2 \, dD, \qquad (1)$$

where $N_k(D)dD$ is the number of scatterers of type k per unit volume of sample with diameter between D and $D+dD$, and $V(D)$ and $F(Q,D)$ are the volume and scattering form factor of the scatter. The magnitude of the scattering vector,

$$Q = (4\pi/\lambda)\sin\theta, \qquad (2)$$

where λ is the wavelength of the X-rays, and 2θ is the angle through which the scattering occurs. We assume that the different scattering types have similar morphologies so that they may be described by a common scattering form factor and volume term. For spherical scatterers,

$$F(Q,D) = \frac{3}{(QD/2)^3}[\sin(QD/2) - (QD/2)\cos(QD/2)], \qquad (3)$$

At small-angles, the scattering contrast, $|\Delta\rho_k(E)|^2$, is due to the difference in scattering length densities,

$$\Delta\rho_k(E) = r_e(E)\sum_k \Delta c_z[Z + f'_z(E) + if''_z(E)], \qquad (4)$$

where f' and f'' are the real and imaginary parts of the anomalous dispersion corrections to the atomic scattering factor of the element with atomic number Z and r_e is the classical electron radius. $\Delta\rho_k(E)$ is due to the difference in atomic concentration, Δc_z, of element Z between the scatterer of type k and the average sample material.

For simplicity, make the two definitions

$$\varphi(D,E) = \sum_k |\Delta\rho_k(E)|^2 N_k(D) V(D) = \sum_k |\Delta\rho_k(E)|^2 f_k(D), \qquad (5)$$

noting the volume fraction of scatterers of type k, $V_{V,k} = \int_0^\infty f_k(D)$, and

$$G(Q,D) = V(D)|F(Q,D)|^2, \qquad (6)$$

so that the scattered intensity can be rewritten,

$$I(Q,E) = \int_0^\infty \varphi(D,E) G(Q,D) \, dD. \qquad (7)$$

The solution method to obtain $\varphi(D,E)$ from $I(Q,E)$ maximizes the smoothness of $\varphi(D,E)$ with the additional constraint that the $I(Q,E)$ calculated from $\varphi(D,E)$ fits the data within the experimental errors.

X-ray diffraction and microscopy has shown that the majority secondary phases are Yb disilicate and voids. The scattering contrast of the Yb disilicate (with respect to the matrix) changes dramatically near the L_{III} absorption edge while the contrasts of the voids, SiO_2, and $Y_5Si_3NO_{12}$ are nearly constant. Thus, we consider the scattering contributions from the other phases ($Yb_4Si_2N_2O_7$, $Y_5Si_3NO_{12}$, and SiO_2) as negligible and make the approximation that $\varphi(D,E)$ is the sum of Yb disilicate and voids,

$$\varphi(D,E) = |\Delta\rho_{disilicate}(E)|^2 f_{disilicate}(D) + |\Delta\rho_{void}|^2 f_{void}(D). \qquad (8)$$

That is, the volume fraction size distributions of Yb disilicate, $f_{disilicate}(D)$, and voids, $f_{void}(D)$, are determined from the slope and intercept, respectively, of a linear fit of $\varphi(D,E)$ vs. $|\Delta\rho_{disilicate}(E)|^2$ at each diameter in the distribution.

The size distributions of cavities formed by tensile creep are determined from the difference in void size distributions of the samples from the gage and grip sections of the tensile creep coupon,

$$f_{creep\,cavities}(D) = f_{void,gage}(D) - f_{void,grip}(D). \tag{9}$$

RESULTS AND DISCUSSION

All samples measured show a strong change in the scattered intensity, such as in Fig. 2a, (sample test 50 h, grip section) as a function of X-ray energy near the L_{III} edge. Statistical errors are much smaller than the symbol size in this figure. The energy-dependence of the scattering intensity, shown in the inset, is consistent with the change in scattering contrast of Yb disilicate as shown in Fig. 1. By restricting our attention at this stage of the analysis to the data range $0.0002 < Q$, Å^{-1} < 0.01 for which the signal-to-noise ratio is excellent, the effects of underlying instrumental profile at low Q and background at high Q are negligible (less than 1 part in 10^3). All data sets show a change in the power law behavior of $d\Sigma/d\Omega(Q)$ in the vicinity of $Q \approx 0.001$ Å^{-1}. The presence of such a transition is indicative of the presence of a well-defined distribution of scatterers with a peak diameter of a few thousand Å. The scattering data $Q > 0.001$ Å^{-1} shows Porod law behavior ($d\Sigma/d\Omega \sim Q^{-4}$). If there are scatterers in the sample with diameters smaller than a few thousand Å, their product of V_v and $|\Delta\rho(E)|^2$ is not sufficient to cause a change in the $d\Sigma/d\Omega(Q)$ power law slope over the measured Q range.

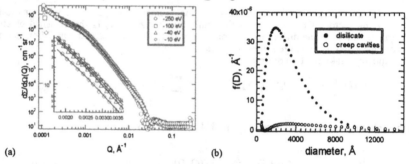

(a)   Q, Å^{-1} (b)   diameter, Å

Fig. 2. Anomalous USAXS (a) measured at four X-ray energies near the Yb L_{III} absorption edge in SN-88 tensile creep sample, grip section, tested for 50 h. The inset shows details for part of the curve. The measured data points are plotted as open symbols. Also shown as lines are the fitted intensities calculated from the size distribution analyses. Size distributions of Yb disilicate and voids derived using the above method are shown in (b). The distribution of Yb disilicate is shown with solid symbols, that of the creep cavities is shown with open symbols.

Fig. 2b compares the volume fraction size distribution of Yb disilicate in the grip section, using (8), to the volume fraction size distribution of creep cavities, using (9), for a sample tested 50 h. While the distributions appear to have a log-normal shape, no such assumption was imposed by the analysis. At this stage, we impose a lower limit of 500 Å on size distributions derived from the scattering data. For $D < 500$ Å, the regression error, from (8), becomes larger than the values of $f(D)$. The peak diameters of the two distributions are comparable while the amount of disilicate is much greater than the creep cavities. The volume fraction, V_v, of Yb disilicate is much greater (by 5 to 8 times for all samples similarly) than that of the creep cavities.

Fig. 3 shows the size distribution results for Yb disilicate (as derived from the grips of the samples) and the creep cavities (as derived from the difference between the porosity results for each grip-gage pair). Note the difference of a factor of 7 in the $f(D)$ scale between the two plots. An interpretation of the detail in the size distributions requires an examination of the propagation and amplification of noise through the analysis, which is beyond the scope of this paper. At this stage, the analysis indicates that the mean size of Yb disilicate does not change appreciably over these test conditions while the volume fraction increases with test time, as shown in Fig. 3a. The Yb disilicate distributions are very broad with mean sizes in the range of 2000-3000 Å and Yb disilicate volume fractions of 8%-12%. The creep cavity volume fractions are also broad and are 0.4%-3.1%. In Fig. 4, we show volume fraction of creep cavities vs. strain. The strain is related to the tensile test samples as shown in Table 1.

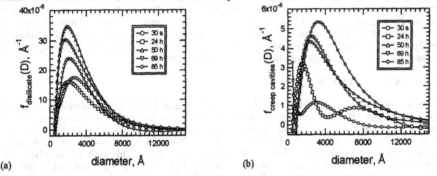

(a)

(b)

Fig. 3. Size distributions of Yb disilicate (a) and tensile creep cavities (b) in SN-88 tested at 1400 C and 150 MPa. The legend indicates the creep test time for each sample.

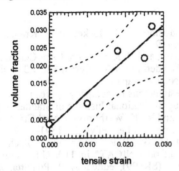

Fig. 4. A-USAXS determination of creep cavity volume fraction vs. strain. The solid line is a least squares fit line and the dashed lines indicate the 95% confidence level on the least squares fit.

The volume fraction of creep cavities is proportional to the measured tensile strain of the creep coupons as shown in Table 1, confirming that the population of creep cavities falls within the size range measured by these experiments. The volume fraction and mean diameter of creep cavities increases linearly with tensile strain and the proportionality constant is 1. This demonstrates excellent agreement with earlier density change results [1].

Table 1. Comparison of ASAXS results with tensile creep strain.

test time	30 s	24 h	50 h	69 h	85 h
tensile test remarks	broken	interrupted	interrupted	broken and slow-cooled	broken
strain	~0	1%	1.8%	2.5%	2.7%
V_v (ASAXS)	0.4%	0.9%	2.4%	2.2%	3.1%

CONCLUSIONS

A-USAXS data were analyzed to determine simultaneously the evolutions of the size distributions of Yb disilicate and creep cavities as a function of tensile creep sample of SN-88 silicon nitride. These distributions were all of comparable dimensions in the size range examined from 500 Å up to 1 μm. The volume fraction of Yb disilicate was found to be about 5-8 times greater than that of creep cavities and nearly the identical size. A linear relationship between V_v measured from A-USAXS and tensile strain agrees well with density change data and confirms that cavitation is the main creep mechanism. A-USAXS obtained statistically significant measurements of the evolution of Yb-rich secondary phase pockets during creep in the presence of a creep cavity population of similar dimensions.

ACKNOWLEDGMENTS

The assistance of J. Ilavsky and A.J. Allen with the A-USAXS measurements is greatly appreciated. The UNICAT facility at the Advanced Photon Source (APS) is supported by the Univ of Illinois at Urbana-Champaign, Materials Research Laboratory (U.S. DoE, the State of Illinois-IBHE-HECA and the NSF), the Oak Ridge National Laboratory (U.S. DoE under contract with Lockheed Martin Energy Research), the National Institute of Standards and Technology (U.S. Department of Commerce) and UOP LLC. The APS is supported by the U.S. DoE, BES, OER under contract No. W-31-109-ENG-38. The support for one of the authors (F.L) provided by the Fulbright Commission is acknowledged.

REFERENCES

1. W. E. Luecke, S. M. Wiederhorn, B. J. Hockey, R. F. Krause, Jr., and G. G. Long, J Am Ceram Soc, 78, 2085-2096 (1995)
2. F. Lofaj, A. Okada, H. Usami, and H. Kawamoto, J Am Ceram Soc, 82, 1009-1019 (1999)
3. W. E. Luecke, and S. M. Wiederhorn, J Am Ceram Soc, 82, 2769-2778 (1999)
4. NGK Insulator Co., Ltd., Nagoya, Japan
5. The use of commercial designations or company names is for identification only and does not indicate endorsement by the National Institute of Standards and Technology.
6. J. J. Hoyt, D. de Fontaine, and W. K. Warburton, J Appl Cryst, 17, 344-351 (1984)
7. D. T. Cromer, and D. Liberman, J Chem Phys, 53, 1891-1898 (1970)
8. G. G. Long, A. J. Allen, J. Ilavsky, P. R. Jemian, and P. Zschack, "The Ultra-Small-Angle X-ray Scattering Instrument on UNICAT" in 11th U.S. National Synchrotron Radiation Instrumentation Conference (SRI'99), edited by P. Pianetta, and H. Winick New York: American Institute of Physics, 1999), pp. in press.
9. G. G. Long, P. R. Jemian, J. R. Weertman, D. R. Black, H. E. Burdette, and R. Spal, J Appl Cryst, 24, 30-37 (1991)
10. J. A. Lake, Acta Cryst, 23, 191-194 (1967)
11. P. R. Jemian, J. R. Weertman, G. G. Long, and R. D. Spal, Acta Metall Mater, 39, 2477-2487 (1991)

Synchrotron Small Angle X-ray Scattering Study of Melt Crystallized Polymers

Georgi Georgiev[1], Patrick Shuanghua Dai[1], Elizabeth Oyebode[1], Peggy Cebe[1,*], and Malcolm Capel[2]
[1]Dept. of Physics and Astronomy, Tufts University, Science and Technology Center, Medford, MA 02155
[2]Biology Dept., Brookhaven National Laboratory, Upton, NY 11973
*To whom correspondence should be addressed

ABSTRACT

In this paper we report a synchrotron small angle X-ray scattering (SAXS) study of development of structure in semicrystalline Poly(Ether Ether Ketone), (PEEK) and an 80/20 blend with amorphous Poly(Ether Imide) (PEEK/PEI). Samples were treated to dual stage melt crystallization scheme involving initial isothermal crystallization at T_1 followed by a second isothermal period at T_2 ($T_1 < T_2$). Intensity of small angle scattering was measured in real-time. Structural parameters characterizing the lamellar thickness, l_c, long period, L, and SAXS invariant were deduced from the one-dimensional electron density correlation function assuming an ideal, two-phase structural model.

INTRODUCTION

High intensity synchrotron sources of X-radiation give polymer researchers a valuable tool to study development of structure in real-time. Structural information is obtained about formation, annealing, and melting of crystals. The assignment of structural parameters derived from scattering intensity gives us information about the size and relative perfection of lamellar crystals, and the dependence upon melt processing history.

Poly(Ether Ether Ketone), PEEK, an engineering thermoplastic polymer, has a very high glass transition temperature (145°C) and crystal melting point (337°C), which make it important in high temperature applications. Poly(Ether Imide), (PEI), is a noncrystallizable material, which has much higher glass transition temperature (above 210°C). Blends of PEEK with PEI are known to be miscible in the melt state, but undergo phase separation upon cooling as the PEEK component crystallizes. Addition of PEI changes the PEEK crystallization kinetics, degree of crystallinity and thermal properties [1-5], such as expansion coefficient and stability. Prior SAXS and thermal studies of this polymer have been used to address melting behavior and changes in structural parameters in PEEK [6-9] and PEEK/PEI blends [1-3,5]. Here, we report a study of PEEK and an 80/20 PEEK/PEI blend using small angle X-ray scattering (SAXS) to study structural changes during, and subsequent to, a two-stage melt crystallization treatment. This particular blend was chosen because PEI can remain between the PEEK lamellae when the composition of PEI is at or below 0.20 [3]. The two-stage melt treatment scheme involves annealing/crystallization at T_1 followed by annealing/crystallization at T_2, where $T_1 < T_2$. Our results show that both inter-lamellar spacing (long period) and crystal thickness increase during heating between the two stages of isothermal treatment as a result of melting of a small population of crystals. During the final heating after the second stage, the long period increases continuously during melting while the crystal thickness first increases, as thinner crystals melt, and then begins to decrease, probably as a result of surface melting.

EXPERIMENTAL DETAILS

PEEK pellets were obtained from ICI Americas. Compression molding at 400°C followed by quenching into ice water yielded amorphous films as indicated by absence of birefringence. PEEK/PEI blend films of composition ratio 80/20 were kindly provided by Dr. B. Hsiao. SAXS experiments were performed at the Brookhaven National Synchrotron Light Source (NSLS) with samples encapsulated in Kapton™ tape, and heated in a Mettler FP80 hot stage.

Mat. Res. Soc. Symp. Proc. Vol. 590 © 2000 Materials Research Society

Dual stage crystallization was used to study the relative perfection of the crystals formed from the melt. Figure 1 shows the treatment schemes for dual stage melt crystallization. The samples were heated to 375°C for three minutes to erase crystals from the previous thermal history, before cooling to 280°C at 20 °C/min. After holding 10 min., samples were heated to the second stage temperature of 295°C at 5°C/min. After a variable holding time at 295°C, the samples were finally heated at 5°C/min to 360°C.

The SAXS system at NSLS was equipped with a histogramming two-dimensional position sensitive detector. The sample to detector distance was 172.7 cm and the X-ray wavelength was 1.54Å. SAXS data were taken continuously during the isothermal periods, heating/cooling between stages, and final heating to 360°C. Each SAXS scan was collected for 30 sec. Circular integration of the isotropic scattered intensity, I, was used to increase the signal to noise ratio. The following corrections were made to the SAXS raw intensity: background subtraction, sample absorption, changes in incident beam intensity, and thermal density fluctuation correction from Is^4 vs. s^4 plot ($s = 2\sin\theta/\lambda$ where θ is the half-scattering angle). Structural parameters were determined from the one dimensional electron density correlation function [10], K(z), obtained by discrete Fourier analysis of the Lorentz corrected intensity. K(z) was determined from:

$$K(z) = \sum_{j=1}^{N} (4\pi \, I_{corr}) \, s^2 \omega_N^{(j-1)(z-1)} \tag{1}$$

where

$$\omega_N = e^{-2\pi i/N} \tag{2}$$

is the $N\underline{th}$ root of unity [11]. In equation 1, z is the direction normal to the lamellar stacks; N, is the number of actual data points; and Icorr is the intensity corrected for background and thermal density fluctuations. Linear extrapolation from the beam stop region to s=0 was used in the summation. The long period, L, scattering invariant, Q, and crystal thickness, lc, are determined from K(z) according to the method of Strobl

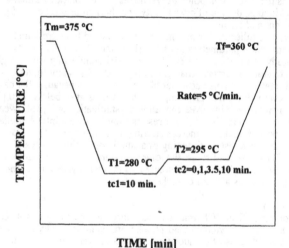

Figure 1. *Thermal treatment scheme for PEEK and the PEEK/PEI blend.*

and Schneider [10]. The amorphous layer thickness, l_a, is determined by $l_a=L-l_c$. We select l_c so that $l_c < l_a$, in agreement with results of others on PEEK [1] and with our own prior work in poly(phenylene sulfide) [12]. The opposite assignment ($l_c > l_a$) has been taken by other researchers [13]. While either choice is consistent with Babinet's reciprocity theorem, results presented in this paper support our assignment $l_c < l_a$.

DISCUSSION

In Figure 2, intensity after the first isothermal crystallization stage is compared for PEEK homopolymer (upper curve) and 80/20 PEEK/PEI blend (lower curve). The intensity scattered from the blend is weaker than from the homopolymer. The integrated areas under these two peaks are different by about a factor of five. The SAXS invariant, Q, which is the total integrated area underneath the Lorentz corrected scattering curve is given by:

$$Q = \chi_s \, \chi_c \, (1-\chi_c) \, (\rho_c - \rho_a)^2 = \int 4\pi I_{corr} s^2 \, ds \qquad (3)$$

where χ_s is the spherulite filling fraction, and χ_c is the crystallinity of the lamellar stacks, and $\rho_c - \rho_a$ is the difference between the electron densities in the crystal and amorphous phases. At the completion of crystallization, spherulites completely fill the available volume and $\chi_s = 1$ for both the homopolymer and the blend. Thermal analysis results show that the overall degrees of crystallinity in the blend and homopolymer samples are nearly the same, and this is confirmed by nearly identical lamellar stack crystallinities found from SAXS analysis. Thus, differences in χ_c cannot account for the difference in scattered intensity.

The reduction in intensity in the blend is a result of reduction in the electron density difference (i.e., scattering contrast) between the crystalline phase and the amorphous phase, $\rho_c - \rho_a$. Q is strongly affected by changes in the electron density difference. For PEEK homopolymer at room temperature, the density of the crystalline unit cell is 1.400 g/cm^3 [8]. Amorphous PEEK has a room temperature density of ρ_a(PEEK) = 1.263 g/cm^3 [8] while PEI has a density of ρ_a(PEI) = 1.28 g/cm^3 [14]. Mixing amorphous PEI

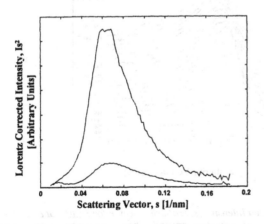

Figure 2. Lorentz corrected intensity vs. scattering vector, after holding for ten minutes at 280°C. Upper curve, PEEK homopolymer; lower curve, 80/20 PEEK/PEI blend.

with amorphous PEEK results in lowering of ($\rho_c - \rho_a$) for interlamellar placement of the PEI. Also, effects of thermal expansion will tend to reduce the amorphous phase density in the homopolymer, improving its scattering contrast.

Figure 3 shows real-time Lorentz corrected intensity, Is^2, vs. scattering vector for PEEK homopolymer. The curves from top to bottom represent different temperatures and times as indicated in the legend on the right hand side of the figure. Small but systematic variation in intensity is seen during the isothermal periods at 280°C and 295°C. As the holding time at 280°C increases from 0 to 10min, a weak shoulder grows up on the high-s side of the Bragg peak. After 10 min. at 280°C the intensity in the shoulder is as large as intensity in the main Bragg peak. When the temperature first increases from 280°C to 295°C, the weak shoulder diminishes, and increases once again as the holding time at 295°C increases. Results of differential scanning calorimetry (DSC) have been presented before [9] and are not shown for the sake of brevity. Both PEEK and 80/20 PEEK/PEI show dual endotherms, the lower of which increases in area as the time at 280°C increases. This lower, minor endotherm occurs at about 287°C and represents the melting of a small population of imperfect, secondary crystals. During heating from 280°C to 295°C, the reduction in the high s shoulder corresponds to the melting of these least perfect crystals formed initially at 280°C, and melting just above this temperature. When the non-isothermal heating from 295°C begins, and the temperature reaches 310°C, overall intensity begins to decrease. The decrease becomes substantial at 330°C, as melting begins and the major endotherm in the DSC scan is reached. The Bragg peak shifts to lower scattering vector, and eventually when the sample is fully melted at 360°C, no Bragg peak can be detected.

Figure 4a,b shows the structural parameters, L and l_c, vs. time for PEEK homopolymer. In Figure 4a, the three lower L curves have been displaced downward by ~10Å for clarity. The samples have 10min at 280°C, followed by: 0min. at 295°C (open circles), 1min. at 295°C (crosses), 3.5min. at 295°C (stars), and 10min. at 295°C(solid line). The isothermal period at 280°C is marked with a horizontal bar. An arrow for each curve marks the end of the isothermal period at 295°C. The sample treated with no

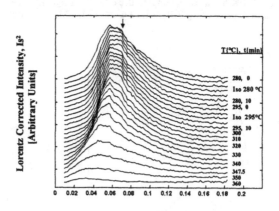

Figure 3. *Lorentz corrected intensity vs. scattering vector for PEEK, at a sequence of temperatures and times during two-stage crystallization. Temperatures and times are given in the plot legend. The appearance of a shoulder at high s (indicated by an arrow) shows the development of a tiny population of less perfect crystals, which melts at lower temperatures than the main population of crystals.*

second stage treatment (open circles) actually had only 9 min. at 280°C (due to human error), and the first two data points shown are representative of non-isothermal cooling to 280°C, followed by 9 min. isothermal period. After the isothermal period at 280°C, this sample was directly heated to 360°C to melt it. The other three samples experience a heating ramp between the 280°C and 295°C stages, followed by the variable isothermal time period at 295°C, and finally a second heating ramp to melt the sample.

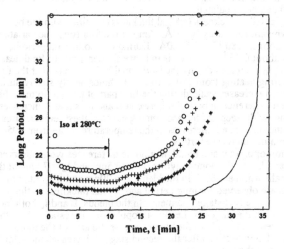

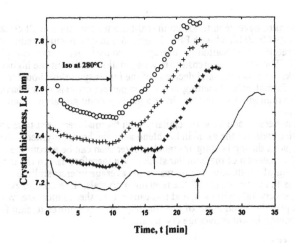

Figure 4. *a.) Long period, L, and, b.) crystal thickness, l_c, vs. time during thermal treatment. Each sample had 10 min. at 280°C, followed by annealing at 295°C for: 0min. (circles), 1min. (crosses), 3.5min. (stars) and 10min.(line). The end of the second isothermal period is marked with a vertical arrow.*

Long period decreases during the isothermal holding time at 280°C as crystals form. During the heating ramp from 280°C to 295°C, L increases by about 10Å, a value that is too large to be the result of thermal expansion effects. The increase in L is due to the melting of the small population of imperfect crystals formed at 280°C. During holding at 295°C, L decreases very slightly, probably as a result of crystals becoming more perfect upon annealing. During subsequent heating to melt the sample, L increases in roughly two stages of differing slope. The change in slope to steeply rising occurs at about 335°C, just at the beginning of the major endotherm. When the temperature increases to 350°C no long period can be calculated.

Figure 4b shows lamellar thickness (crystal thickness) vs. time with the bottom three curves displaced downward for clarity by ~2Å. Among the four runs, the variability in l_c during initial equilibration at 280°C is 0.5-1.0Å. During the isothermal periods, and heating, variability is about 0.25Å. The trends in l_c are similar, prior to final heating, to those seen for L. Viz., l_c decreases during the isothermal holding period at the first stage, increases (by ~1Å) during heating from 280°C to 295°C, holds nearly steady during annealing at 295°C, then increases again during the first part of heating to melt the sample. The important difference in behavior between L and l_c is that L increases throughout the melting, with steepest increase coming at the highest temperatures. The crystal thickness, on the other hand, increases in the temperature range from 295-340°C, then flattens as temperature increases further.

Comparing the homopolymer and blend, negligible differences were observed in the absolute values of L and l_c. In the homopolymer, L = 20.5nm and l_c = 7.5nm at the end of the first stage crystallization at 280C. In the blend, L = 20nm and l_c = 7.05nm. Exactly similar trends are observed for long period and crystal thickness in the two materials as a function of the two-stage treatment. In the blends, L and l_c both increase during the heating between the stages as the minor population of crystals melts, though the effect is not expressed as strongly in the blend because the area of the minor endotherm is so small. Upon heating after the second stage, L increases monotonically, while l_c first increases and then decreases.

CONCLUSIONS

Changes in structure have been followed in real-time for two-stage melt crystallization of PEEK and its 80/20 blend with PEI. Scattered intensity is much reduced in the blend as a result of reduction of scattering contrast caused by larger amorphous phase density in the blend. The scattered intensity shows a small shoulder on the high-s side of the main Bragg peak. The increase and decrease of the intensity in this shoulder correlates well with formation and melting of the least perfect crystals which have grown in a constrained environment, and therefore have a melting point just above their formation temperature.

Structural parameters, L and l_c, vary regularly with thermal treatment time and temperature, and the trends are the same in the blend as in the homopolymer. The increase in L and l_c seen during heating are much greater than can be accounted for on the basis of thermal expansion of the lamellar stacks. We therefore assign the increase in L and l_c to the melting of crystals, which increases the average intercrystalline separation, as well as the average size of the remaining crystals. Eventually, even the most perfect (thickest) crystals will melt, and this causes L further to increase, while l_c levels off. We interpret the leveling off of l_c at the highest temperatures to signify surface melting, resulting in lower average crystal size.

ACKNOWLEDGMENTS

This work was supported by the National Aeronautics and Space Administration, Grant NAG8-1167.

REFERENCES

1. A. Jonas, T. Russell, D. Yoon. *Macromolecules*, **28**, 8491 (1995).
2. B. Hsiao, B. Sauer. *J. Polym. Sci., Polym. Phys. Ed.*, **31**, 901 (1993).
3. A. Jonas, D. Ivanov, D. Yoon, *Macromolecules*, **31**, 5352 (1998).
4. B. Hsiao, R. Verma, B. Sayer, *J. Macromol. Sci.*, Phys., **B37(3)**, 365 (1998).
5. H. Chen, R. Porter, *J. Polym. Sci. B:Polymer Physics*, **31**, 1845 (1993).
6. B. Hsiao, K. Gardner, D. Wu, B. Chu. *Polymer*, **34**, 3996 (1993).
7. K. Kruger, H. Zachmann. *Macromolecules*, **26**, 5202 (1993).
8. D. J. Blundell, B. N. Osborn. *Polymer*, **24**, 953 (1983).
9. G. Georgiev, P. S. Dai, E. Oyebode, P. Cebe, M. Capel, Proceedings of the American Chemical Society Division of Polymeric Materials: Science and Engineering, **78**, 215 (1998).
10. G. R. Strobl, M. Schneider. *J. Polym. Sci., Polym. Phys. Ed.*, **18**, 1343 (1980).
11. Matlab Reference Guide, *The Mathworks Inc.*, Natick, 1996.
12. S. X. Lu, P. Cebe, M. Capel. *Macromolecules*, **30(20)**, 6243 (1997).
13. B. Hsiao, B. Sauer, R. Verma, H. Zachmann, S. Seifert, B. Chu, P. Harney. *Macromolecules*, **28**, 6931 (1995).
14. Material Data Sheet for General Electric ULTEM (PEI), Boldeker Plastics, (1999).

MAXIMUM ENTROPY METHOD CHARGE DENSITY DISTRIBUTIONS OF NOVEL THERMOELECTRIC CLATHRATES

B. IVERSEN,* A. BENTIEN,* A. PALMQVIST,** D. BRYAN,*** S. LATTURNER,*** G. D. STUCKY,*** N. BLAKE,*** H. METIU,*** G. S. NOLAS,**** D. COX*****

* Dept. of Chemistry, University of Aarhus, Århus, Denmark
** Dept. of Applied Surface Chemistry, Chalmers University of Technology, Göteborg, Sweden
*** Dept. of Chemistry, University of California, Santa Barbara, CA, USA
**** R&D Division, Marlow Industries Inc, Dallas, Texas, USA
***** NSLS, Brookhaven National Laboratory, Upton, NY, USA

ABSTRACT

Recently materials with promising thermoelectric properties were discovered among the clathrates. Transport data has indicated that these materials have some of the characteristics of a good thermoelectric, namely a low thermal conductivity and a high electrical conductivity. Based on synchrotron powder and conventional single crystal X-ray diffraction data we have determined the charge density distribution in $Sr_8Ga_{16}Ge_{30}$ using the Maximum Entropy Method. The MEM density shows clear evidence of guest atom rattling, and this contributes to the reduction of the thermal conductivity. Analysis of the charge distribution reveals that $Sr_8Ga_{16}Ge_{30}$ contains mixed valence alkaline earth guest atoms. The Sr atoms in the small cavities are, as expected, doubly positively charged, whereas the Sr atoms in the large cavities appear negatively charged. The MEM density furthermore suggests that the Ga and Ge atoms may not be randomly disordered on the framework sites as found in the conventional least-squares refinements.

INTRODUCTION

Thermoelectricity is a unique phenomenon of energy conversion suitable for power generation and cooling applications [1]. Recently thermoelectric materials have recieved a lot of attention spurred by theoretical and experimental developments as well as by growth in commercial markets for thermoelectric technology. New materials related to the skutterudite and the clathrate structures have been reported to have great promise for exceptionally high performance [2].

Understanding chemical and physical properties of molecular systems requires knowledge of their charge distributions [3]. Experimentally, electron density distributions (EDDs) can be reconstructed from accurate X-ray diffraction data through a series of elaborate data reduction and data analysis steps [4]. The most widely used method entails least-squares optimization of models containing atom-centered aspherical density functions [5]. The aspherical modelling schemes have the great virtue of providing a deconvolution of the thermal motion from the EDD thereby allowing a direct comparison with results from quantum mechanical calculations [6]. However, the result will inherently depend on the explicit nature of the model used in the least-squares analysis.

In recent years a new method, the maximum entropy method (MEM), has been introduced in charge density reconstruction. When X-ray data are used, the MEM yields the EDD [7], whereas neutron diffraction data allows the direct space nuclear density distribution (NDD) to be determined [8]. From limited numbers of X-ray diffraction data, EDDs have been reconstructed by the MEM in a number of systems, and maps that qualitatively reveal bonding features have been obtained [9]. The virtue of the MEM is that it is much less model dependent than the least-squares method. Thus an initial least-squares refined model is used only to phase the data, to bring them on an absolute scale, and to correct for systematic errors such as extinction. In the subsequent density reconstruction minimal bias is introduced with regard to aspherical features. This allows estimation of bonding features in the case of X-ray data, or anharmonic motion in the case of neutron data. It also provides a direct assessment of disorder, which is often extremely difficult to model with atom-centered least-squares models. In this paper we apply the MEM to the reconstruction of the EDD of a germanium based type I clathrate with composition $Sr_8Ga_{16}Ge_{30}$.

EXPERIMENTAL

Single crystal X-ray diffraction. Measurements were done at room temperature on a Bruker SMART CCD diffractometer (MoKα) using 0.3° ω-scans. To facilitate proper absorption correction a nearly spherically shaped crystal was prepared in a racetrack crystal grinder. Data were collected with very large redundancy to optimize intensity corrections. Data collection, integration and data reduction was carried out with the SMART software. A summary of experimental and crystallographic details is given in Table 1.

Least-squares refinements were made with SHELXTL [10] in the centro symmetric space group Pm-3n. This space group corresponds to a framework having disorder between the Ga and Ge atoms. Refinements were also carried out in space groups corresponding to ordered framework structures, but in all cases the refinement residuals were worse. Similar results were reported by Chakoumakos et al [11] based on single crystal neutron diffraction data on $Sr_8Ga_{16}Ge_{30}$, whereas theoretical calculations by Blake et al [12] suggest an ordered structure to be of lower energy. In all refinements anisotropic thermal parameters and an isotropic extinction parameter were employed. The SHELXTL extinction model indicates that the data are severely affected by extinction in the low order data. Thus around 30 reflections have more than 10 % extinction. For the most affected reflections the SHELXTL model greatly overcorrects the data making them unsuitable for MEM analysis. It was therefore decided to construct an extinction corrected data set in which reflections with more than 10 % extinction were replaced by reflections from synchrotron powder diffraction measurements.

Synchrotron radiation powder diffraction data. Room temperature high resolution synchrotron X-ray powder diffraction data were collected at beam line X7A at the National Synchrotron Light Source at Brookhaven National Laboratory. This beam line, which is positioned at a bending magnet provided a beam with λ = 0.6996(5) Å from a channel-cut double crystal Ge(111) monochromator. The diffracted intensities were detected with a position sensitive detector, and counting times were increased with 2θ to allow for the fall-off in scattered intensity at high angles. The data were corrected for dead time and also normalized with respect to the incident beam intensity measured by an ion chamber placed just before the sample assembly. The wavelength and zero point offsets were determined by calibration with a Si standard. The sample was contained in a thin 0.2 mm capillary to minimize absorption effects. Absorption effects were experimentally estimated by scanning with a very narrow 50 μm vertical beam over the sample volume. Subsequently the data were corrected for absorption using the measured absorption factors before further analysis. Full pattern Rietveld analysis was carried out with the program GSAS in the cubic space group Pm-3n. On the basis of the refined Rietveld model, structure factors were extracted on an absolute scale from the powder patterns. Only the structure factors affected by more than 10 % extinction in the single crystal data were used for MEM calculations. Further experimental details are given in Table 1.

MEM calculations. Absorption, extinction and anomalous dispersion corrected observed structure factors on an absolute scale were used for MEM calculations. The basic premise of the method is that it includes an entropy term besides the χ^2 function [13]. In the approximation introduced by Collins [7a] and implemented in the MEED program [14], the MEM EDD is calculated on a grid using an iterative procedure. In the present study calculations were performed using both uniform and non-uniform prior densities. The non-uniform prior was calculated by a MEM optimization on *calculated* structure factors corresponding to the SHELXTL model. The non-uniform prior therefore corresponds to the EDD of an assembly of neutral, spherical atoms having anisotropic harmonic thermal motion. This means that the observed data are used to estimate the effects of disorder, chemical bonding and charge transfer, i.e. the reorganization of charge taking place upon formation of the real molecular system. In all MEM calculations a 128 x 128 x 128 pixel grid was used and iterations were stopped at $\chi^2=1$. Estimated standard uncertainties on the observed structure factors were obtained from the data averaging procedure. The MEM density using a non-uniform prior closely resembles the density obtained with a uniform prior. However, it has been shown that MEM densities calculated with non-uniform priors are of higher quality than with uniform priors [15], and in the rest of this paper we therefore only discuss results obtained with the non-uniform prior. The R_F factor for the MEM density is 0.0310 compared to 0.0588 for the SHELXTL refinement. This shows that significant redistribution of density has taken place during the MEM optimization.

	Single crystal	Powder
Radiation	MoKα	synchrotron
λ (Å)	0.7101	0.6996(5)
a (Å)	10.7830(1)	10.73521(3)
Space group	Pm-3n	Pm-3n
μ_i (mm^{-1}) /	34.5	0.43
$(\mu^*R)_{measured}$		
Crystal/capillary	0.225	0.2
diameter (mm)		
Pattern range (°)		4 - 74
Step size (°)		0.01
N_{obs}	43019	7122
N_{unique}	703	733
R_{int}	0.0428	
N_{par}	19	33
R_P		0.0157
R_{wP}		0.0173
R_F / R_F^2	0.0588	0.0629
GoF	1.89	1.39

Table 1. Experimental and crystallographic details for Sr$_8$Ga$_{16}$Ge$_{30}$

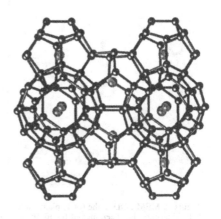

Figure 1. Structure of clathrate type- I. The two cage types are present in a 2:6 ratio in the unit cell.

DISCUSSION

The structure consists of two types of cages, the 20-atom dodecahedron and the 24-atom tetrakaidecahedron, which are present in a 2:6 ratio in the unit cell, Figure 1. The cage framework is built up from germanium and gallium atoms, and the alkaline earth atoms reside inside the cages as weekly bound guests. A number of studies have shown that the guest atoms exhibit extreme motion inside the oversized cavities [16]. This motion is contributing to the effective reduction of the thermal conductivity of the clathrate materials, which is important for obtaining a high thermoelectric figure of merit (ZT) [17]. However, ZT will not improve if the "rattling" behaviour of the guest atoms also lowers the electrical conductivity. Due to the composition of known clathrates there is a general belief that the guest atoms donate valence electrons to the framework atoms. It is therefore likely that considerable coupling exists between the framework and the guest atoms, and thus *a priori* the clathrates are unlikely candidates as good thermoelectric materials. Nevertheless several clathrate compositions have been reported with promising ZTs.[2] Theoretical calculations surprisingly show, that contrary to common belief, the Sr guest atoms in Sr$_8$Ga$_{16}$Ge$_{30}$ seem to be close to neutral, and that the electrical conduction takes place through the framework [12]. The low degree of charge transfer between the Sr atoms and the framework was recently confirmed by Palmqvist et al using X-ray spectroscopy [18].

In Figure 2 the MEM EDD in the (001) plane through z = 0 is shown. For comparsion the density corresponding to the SHELXTL refinement (i.e. the non-uniform prior density) is shown in Figure 3. The most interesting feature of the MEM EDD is the abnormally large smearing of the Sr(2) density at (1/4,1/2,0). This feature must be due to very large thermal motion of the Sr(2) guest atom, which is convoluted on the static density to give the thermally smeared MEM density. If MEM analysis is carried out on neutron data one obtains NDDs, which can directly be interpreted as a measure of the shape and extent of the nuclear potential because atoms are point scatterers of neutrons. In this way force constants between the atoms may be determined [19]. Such analysis is much more involved if based on X-ray data since any deconvolution must be based on assumptions about the MEM EDD, which stems from the convoluted effect of the static EDD and the NDD. Nevertheless, comparison of the shape and extent of the EDD around the atomic positions is a measure of the relative thermal movement of the atoms. As can be seen in Figure 2 all other atoms except Sr(2) have almost spherical densities of quite localized character. Based on chemical knowledge we expect the covalent bonding between the framework and Sr(2) to be very weak (the distance is greater than 3.5 Å), i.e. we expect the EDD to have limited aspherical bonding components. Not only is the density of Sr(2) very smeared, it is also far from elliptical, as seen when comparing Figure 2 and 3. In a

147

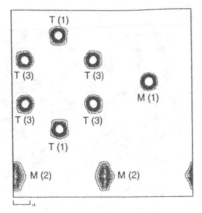

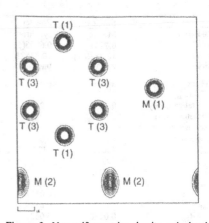

Figure 2. MEM EDD in the (001) plane through z = 0. The origin has been moved by 30 pixels in x and 50 pixels in y. The density is plotted on a linear scale with 5 e/Å³ contour intervals and truncated at 100 e/Å³.

Figure 3. Non-uniform prior density calculated from model structure factors based on a free, spherical, neutral atom procrystal model. Contours as for 2.

single crystal neutron diffraction study of $Sr_8Ga_{16}Ge_{30}$ Chakoumakos et al [11] found that the Sr(2) site is best modelled with four disordered sites. Neutron Fourier maps do not contain the convoluted effects from the EDD and are therefore better in revealing disorder. The shape of the Sr(2) MEM density is suggestive of a disordered site, but it should be noted that non-elliptical shapes also can be due to anharmonic motion besides disorder. However, it is unlikely that the shape of Sr(2) is entirely due to anharmonicity, which tends to produce "superelliptical" NDDs rather than the star shape seen in Figure 2 [18].

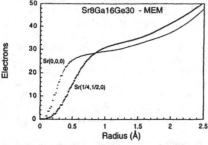

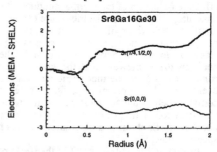

Figure 4. Spherical integration of the MEM EDD around the Sr guest atom sites. The Sr atom at (0.25,0.5,0) is situated in the large cage, the Sr atom at (0,0,0) is in the small cage

Figure 5. Spherical integration of the difference between the MEM EDD and the non-uniform prior density

Theoretical calculations by Blake et al [12] suggest that the Sr atoms in $Sr_8Ga_{16}Ge_{30}$ are close to neutral and thus have limited electronic interaction with the framework. The MEM EDD allows an experimental validation of this suggestion. It is, however, very difficult to unambiguously define an atomic charge. As discussed by for example Coppens [6] a large number of atomic charge definitions are in common use. Only the definition based on the quantum theory of atoms in molecules (QTAM) has a rigorous quantum mechanical foundation [3]. In the case of numerical densities on a grid, very accurate densities are needed to apply the QTAM definition with success. We are therefore forced to apply simpler and more approximate criteria to examine the atomic charges. However, since we expect the guest atoms to be fairly localized due to the lack of directional bonding, we can integrate the MEM EDD over spherical regions. Figure 4 shows the integrated electron counts for the guest atoms as a function of spherical integration radius. The diffuse nature of the Sr(2) atom at (0.25,0.5,0) in the large cavity is clearly seen in the plot. Thus Sr(2) contains much less electrons than Sr(1) at small radii. Yet at R ≈ 0.85 Å, Sr(2) becomes more electron rich than Sr(1). The problem with the spherical integration plots is that the exact atomic radii are unknown. To get a better estimate of the atomic charges we therefore compare the MEM EDD to the non-uniform prior density. This density was constructed from an assembly of free spherical, neutral atoms, and the difference between the MEM EDD and the non-uniform prior therefore provides an indication of the charge flow upon formation of the solid. In Figure 5 the difference electron counts are plotted as a function of spherical integration radii. The difference between the two Sr atoms is striking. While the Sr(1) atoms in the small cavity are close to a +2 charge, the Sr(2) atoms in the large cavity are negatively charged ! Thus upon formation of the crystal, electron density is moved into the atomic volume of Sr(2), but removed from the volume of Sr(1). This shows that $Sr_8Ga_{16}Ge_{30}$ contains alkaline earth atom inclusions of mixed valence. Mixed valence systems are well known among transition metal complexes, but very unusual for alkali or alkaline earth metals. The MEM EDD suggest that the electronic interactions between the guests and the framework, and between the guest atoms themselves, may be very complex. It seems that comparative EDD studies of different clathrate systems could be useful for understanding these interactions better.

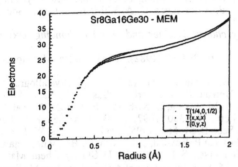

Figure 6. Spherical integration of the MEM EDD around the framework sites.

In Figure 6 we show spherical integration results for the three different framework sites. In the non-uniform prior density (i.e. in the conventional least-squares refinements) the three sites were identical. This was the case even though the Ga and Ge occupancies were varied freely in the least-squares analysis. As seen in Figure 6, the MEM EDD suggests that the framework sites are not identical, and thus the Ga and Ge atoms may not be randomly distributed. It may be that the Ga atoms have a preference for occupying sites of the small cavity in order to receive the electrons donated by the Sr(1) atom. However, since the difference in the framework sites can be due to both charge redistribution and specific Ga/Ge siting, it is not straightforward to interpret the MEM EDD result. We are currently examining this aspect in more detail.

CONCLUSION.

MEM analysis of X-ray diffraction data reveals that the EDD of the Sr guest atom in the large cavity of $Sr_8Ga_{16}Ge_{30}$ is very diffuse. The shape of the MEM EDD of this atom indicates that the diffuseness is due to both disorder and large atomic vibration. Comparison of the MEM EDD with a procrystal EDD consisting of free, spherical, neutral atoms reveal that $Sr_8Ga_{16}Ge_{30}$ contains mixed valence alkaline earth inclusions. Thus while the Sr atoms in the small cavity, as expected, are doubly positively charged, the Sr atoms in the large cavity are negatively charged. The nature of the guest-host interactions thus appear very complex. Since the guest-host interactions have profound effects on the thermoelectric properties of materials, it is of interest to carry out comparative MEM EDD studies of thermoelectric clathrates to obtain a better understanding of the interactions.

ACKNOWLEDGEMENTS

We gratefully acknowledge the beam time granted at the National Synchrotron Light Source. The X7a beam line is supported by the U.S. Department of Energy, Division of Materials Science, under contract No. DE-AC02-98CH10886. BBI thanks the Carlsberg Foundation and the DANSYNC center under the Danish Research Councils for financial support. AECP greatly appreciates support from the Swedish Foundation for International Cooperation in Research and Higher Education (STINT) and from the Swedish Technical Research Council.

REFERENCES

[1]. F. J. DiSalvo, Science, 285, 703-706 (1999)

[2]. G. S. Nolas, J. L. Cohn, G. A. Slack, S. B. Schjuman, Appl. Phys. Lett. 73, 178 (1998)

[3]. R. F. W. Bader, *Atoms in molecules. A quantum theory*. Oxford University Press, 1990.

[4]. (a) B. B. Iversen, F. K. Larsen, B. N. Figgis, P. A. Reynolds, Acta Crystallogr. Sect B, 53, 923 (1996). (b) B. B. Iversen, F. K. Larsen, A. A. Pinkerton, A. Martin, A. Darovsky, P. A. Reynolds, Acta Crystallogr. Sect B, 55, 363 (1999).

[5]. (a) R. F. Stewart, Acta Crystallogr. Sect A, 32, 565 (1976). (b) F. L. Hirshfeld, Isr. J. Chem., 16, 226 (1977). (c) N. K. Hansen, P. Coppens, Acta Crystallogr. Sect A, 34, 909 (1978). (d) B. N. Figgis, P. A. Reynolds, G. A. Williams, J. Chem. Soc. Dalton Trans., 2239 (1980).

[6]. P. Coppens, *X-ray charge densities and chemical bonding*, Oxford University Press, 1997.

[7]. (a) D. M. Collins, Nature, 298, 49 (1982). (b) M. Sakata, M. Sato, Acta Crystallogr. Sect A, 46, 263 (1990).

[8]. M. Sakata, T. Uno, M. Takata, C. J. Howard, J. Appl. Crystallogr., 26, 159 (1993)

[9]. (a) M. Takata, B. Umeda, E. Nishibori, M. Sakata, Y. Saito, M. Ohno, H. Shinohara, Nature 377, 46 (1995). b) M. Takata, E. Nishibori, B. Umeda, M. Sakata, E. Yamamoto, H. Shinohara, Phys. Rev. Lett., 78, 3330 (1997). (c) B. B. Iversen, F. K. Larsen, M. Souhassou, M. Takata, Acta Cryst. Sect. B, 51, 580 (1995). (d) R. Y. Vries, W. J. Briels, D. Feil, D. Phys. Rev. Lett, 7, 1719 (1996). (e) P. Roversi, J. J. Irwin, G. Bricogne, *Acta Crystallogr*. Sect. A., 54, 971 (1998). (f) B. B. Iversen, S. Latturner, G. D. Stucky, Chem. Mat., 11, 2912 (1999)

[10]. SHELXTL, version 5; Siemens Industrial Automation, Inc; Madison, WI 1994, USA

[11]. B. C. Chakoumakos, B. C. Sales, D. G. Mandrus, G. S. Nolas, J. Alloys and Compunds, in press (1999)

[12]. N. P. Blake, L. Møllnitz, G. Kresse, H. Metiu, J. Chem. Phys, 111, 3133 (1999)

[13]. J. Skilling, *Maximum Entropy and Bayesian Methods*, edited by J. Skilling, Kluwer Academic Publishers: Dordrecht, 1989, pp. 46.

[14]. S. Kumazawa, Y. Kubota, M. Takata, M. Sakata, Y. Ishibashi, J. Appl. Crystallogr., 26, 453 (1993).

[15]. B. B. Iversen, J. L. Jensen, J. Danielsen, Acta Cryst. Sect. A, 53, 376 (1997).

[16]. (a) B. C. Chakoumakos, B. C. Sales, D. Mandrus, V. keppens, Acta Cryst. Sect. B, 55, 341 (1999). (b) B. C. Sales, B. C. Chakoumakos, D. Madrus, J. W. Sharp, J. Solid State. Chem., 146, 528 (1999)

[17] The thermoelectric figure of merit is defined as $ZT = TS^2\sigma/\kappa$, where S is the Seebeck coeffient, σ the electrical conductivity and κ the thermal conductivity.

[18] A. E. C. Palmqvist, B. B. Iversen, L. R. Fuhrenlid, G. S. Nolas, D. Bryan, S. Latturner, G. D. Stucky, in *Applications of Synchrotron Radiation Techniques to Materials Science V*, edited by S. R. Stock, D. L. Perry, S. M. Mini, MRS 1999 Fall Meeting Symposium Proceedings, submitted

[19]. B. B. Iversen, S. K. Nielsen, F. K. Larsen, Philos. Mag. Sect. A, 72, 1357 (1995).

High real-space resolution structure of materials by high-energy x-ray diffraction

V. PETKOV[1], S. J. L. BILLINGE[1], J. HEISING[2], M. G. KANATZIDIS[2], S. D. SHASTRI[3], and S. KYCIA[4]

[1]Department of Physics and Astronomy and Center for Fundamental Materials Research, Michigan State University, East Lansing, MI 48823
[2]Department of Chemistry and Center for Fundamental Materials Research, Michigan State University, East Lansing, MI 48823
[3]Advanced Photon Source, Argonne National Laboratory, Argonne, IL 60439
[4]Cornell High Energy Synchrotron Source, Cornell University, Ithaca, NY 14853

ABSTRACT

Results of high-energy synchrotron radiation experiments are presented demonstrating the advantages of the high-resolution atomic Pair Distribution Function technique in determining the structure of materials with intrinsic disorder.

INTRODUCTION

It is well known that physical properties and technological characteristics of materials are, to a great extent, predetermined by the atomic-scale structure. Also, most of technologically important materials are not mono but polycrystalline in their nature. For that reason much effort has been exercised to develop techniques for determining the structure of polycrystalline materials. Great progress has been made in the field by employing the so-called Rietveld technique [1]. Essentially it is a least-squares refinement of crystal structure parameters, specimen characteristics, diffraction optics and instrumental factors carried out until the best possible agreement between the observed and calculated powder diffraction patterns is obtained. Naturally, Rietveld refinement relies mainly on the sharply defined Bragg peaks in the diffraction pattern [1,2]. Nowadays the structures of polycrystalline materials of variable structural complexity, ranging from disperse catalysts to ceramic high-Tc semiconductors and even simple organic macromolecules, are being almost routinely refined by the Rietveld technique [2]. However, increasingly many new interesting materials contain significant disorder on an atomic scale. Often this disorder has a direct effect on the properties which make the material technologically and/or scientifically important. It is clearly necessary to have a technique which can characterize not only the average, long-range structure but the deviation from it, i.e. the local disorder as well. Information about the local structural disorder is, however, contained in the diffuse scattering which is of low intensity and is usually widely spread in reciprocal space. A fruitful experimental approach which can handle both Bragg peaks and the diffuse component is the so-called atomic Pair Distribution Function (PDF) technique. With the PDF technique both Bragg intensities and the diffuse component(s) of the total diffraction spectrum are treated simultaneously and then Fourier transformed to yield the atomic PDF which is thus a representation of both the long-range (average) and short-range (local) atomic structure

151

of the material. Since the PDF is obtained with no assumption of periodicity, glassy materials as well as polycrystals exhibiting a different degree of local disorder can be characterized employing the same approach [3]. To apply the technique usefully the PDF has to be of high resolution. To do this, accurate total scattering intensity has to be measured over a wide range of diffraction vectors, Q. It implies collecting data with x-rays of high incident energies, i.e. the use of synchrotron sources of x-rays. In the present paper selected examples of such high-resolution PDF studies are presented.

FUNDAMENTALS OF THE *PDF* TECHNIQUE

The atomic PDF, G(r), is defined as follows:

$$G(r) = 4\pi r[\rho(r) - \rho_o], \qquad (1)$$

where $\rho(r)$ and ρ_o are the local and average atomic number densities, respectively and r is the radial distance. $G(r)$ is a measure of the probability of finding an atom at a distance r from a reference atom and so describes the atomic arrangement, i.e. structure, of materials. It is the sine Fourier transform of the experimentally observable total structure function, $S(Q)$, i.e.

$$G(r) = (2/\pi) \int_{Q=o}^{Q_{max}} Q[S(Q) - 1]\sin(Qr)dQ, \qquad (2)$$

where Q is the magnitude of the wave vector. The structure function is related to only the coherent part of the total diffraction spectrum of the material as follows:

$$S(Q) = 1 + \left[I^{coh}(Q) - \sum c_i|f_i(Q)|^2 \right]/|\sum c_i f_i(Q)|^2, \qquad (3)$$

where $I^{coh.}(Q)$ is the coherent scattering intensity per atom in electron units and c_i and f_i are the atomic concentration and scattering factor for the atomic species of type i, respectively [4]. The following important details of the PDF technique are to be noted: $G(r)$ is barely influenced by diffraction optics and experimental factors since these are accounted for in the step of extracting the coherent intensities from the raw diffraction data. In the present studies it was done with the help of the program RAD [5]. As *Eq.* 2 implies, the total, not only Bragg diffracted, intensities contribute to $G(r)$. Also, by accessing high values of Q, experimental $G(r)$s of high real-space resolution can be obtained and, hence, quite fine structural features revealed. The latest point is well demonstrated by the present studies where wave vectors as high as 45 Å^{-1} were achieved with the use of intense synchrotron sources of radiation. All these features make the PDF technique a natural approach when the real atomic, both long and short-range order, structure of materials is needed.

RESULTS

A. Local atomic structure of $In_{1-x}Ga_xAs$ semiconductor alloys

Ternary semiconductor alloys, in particular $In_{1-x}Ga_xAs$, have technological importance because they allow useful properties, such as band-gaps, to be varied continuously between the two end

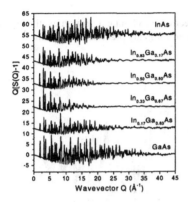 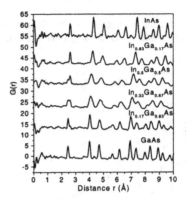

Figure 1. Experimental structure factors for $In_{1-x}Ga_xAs$ alloys.

Figure 2. Reduced atomic PDFs for for $In_{1-x}Ga_xAs$ alloys.

points by varying the composition, x. On average, $In_{1-x}Ga_xAs$ alloys are of the zinc-blende type structure where metal ($In;Ga$) and As atoms occupy two interpenetrating fcc lattices. Due to the considerably different bond lengths present, $L_{In-As} = 2.61$ Å and $L_{Ga-As}=2.437$ Å, the zinc-blende lattice of the alloy is, however, locally distorted. The real, i.e., distorted, structure is a prerequisite to any accurate band-structure and phonon dispersion calculations and for that reason we have investigated it by the PDF technique. We carried out diffraction experiments at the A2 24 pole wiggler beamline at CHESS using x-rays of energy 60 keV. More experimental details can be found in refs. [6,7]. Experimental reduced structure factors and the corresponding atomic PDFs $G(r)$ are shown in Figs. 1 and 2, respectively. Significant Bragg scattering (well-defined peaks) are seen with the end-members, $InAs$ and $GaAs$, up to 40 Å^{-1}. This implies that the samples have long-range order and there is little positional disorder. The Bragg peaks disappear at 20 Å^{-1}- 25 Å^{-1} in the $In_{1-x}Ga_xAs$ samples. Clearly the alloys are still long-range ordered but they also have significant local positional disorder giving rise to a pronounced diffuse scattering seen at higher Q-values. From the second neighbour onwards this disorder results in broad atomic-pair distributions as can be seen in Fig. 2. The nearest-neighbour peak is the only peak which remains sharp. In the alloy samples it is clearly split into a doublet with low and higher-r components corresponding to $Ga-As$ and $In-As$ bonds, respectively. A simple structure model based on the 8-atom cubic unit cell of $(In;Ga)As$ has been fit to the experimental PDFs and the way the underlying zinc-blende structure of the alloys distorts locally to accommodate the bond-length mismatch has been quantified. It has been found that both metal ($In;Ga$) and As atoms are statically displaced from their positions in the ideal lattice. Extra positional disorder, manifested by enlarged temperature factors, has been found on both metal and As sites as well. Both the static displacement and the positional disorder have been found to peak at a composition $x = 0.5$ [6,7].

B. Average atomic structure of "restacked" WS₂

Due to a unique combination of valuable structural, electronic and optical properties, the layered dichalcogenides, such as WS_2, have been studied and used for many practical applications

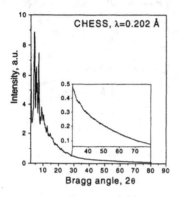

Figure 3. X-ray diffraction spectrum of "restacked" WS₂.

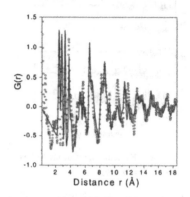

Figure 4. Experimental (symbols) and model (solid line) atomic PDFs of "restacked" WS₂.

[8 and refs. therein]. The chemistry of these materials is particularly fascinating since the individual (WS_2) layers can be exfoliated and kept apart in water for days. The material can be "restacked" by precipitation, evaporation or centrifugation and "guest species" can be encapsulated between the layers. There has been a lot of confusion about the structure of "restacked" WS_2. The structure has been proposed to be trigonal TiS_2 type [9,10]. Recent electron diffraction study, however, showed that the structure (two-dimensional xy-plane) of the single (WS_2) layers is similar to that of the orthorhombic WTe_2 [11]. In this structure W atoms within a single layer form zig-zag parallel chains via metal-metal bonds. Since no structural information was obtained in the z-direction, and in order to show that the single crystalline specimens probed by the electron diffraction were representative of the bulk, we undertook a PDF study with the use of x-rays of energy 61 keV. The experiments were carried out at the A2 beamline, CHESS. The raw X-ray diffraction spectrum obtained is shown in Fig. 3 and the resulting PDF - in Fig. 4. As can be seen in Fig. 3 the diffraction spectrum of "restacked" WS_2 contains a pronounced diffuse scattering component and only a few Bragg peaks which renders the data analysis by ordinary techniques, like the Rietveld refinement, almost impossible. The atomic PDF, however, is rich in structure-related features and lends itself to structure determination. It has been found that the experimental PDF can be fit well with a structure based on a monoclinic unit cell with parameters a=3.2545(5) Å, b=5.7092 Å, c = 12.3783(5) Å; β=87.74° which can be viewed as a distorted derivative of the unit cell of WTe_2. The result agrees with the electron diffraction study [11] suggesting that the layered structure of untreated WS_2 does undergo a considerable distortion down to a monoclinic symmetry when the material is subjected to the chemical processing described above. A more detailed account of the present study will be reported elsewhere [12].

C. Atomic ordering in $Ca_{x/2}Al_xSi_{1-x}O_2$ glasses

Calcium aluminosilocate glasses are among the most frequently man-made glasses. That is why they are subjected to extensive studies. It is generally accepted that these glasses are built of $Si-O$ and $Al-O$ polyhedral units linked together by common oxygen atoms. The so linked units form a continuous random network with Ca ions occupying large irregular cavities in it. Oxygens linking two polyhedral units from the network are called "bridging" while those connecting one

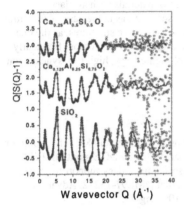

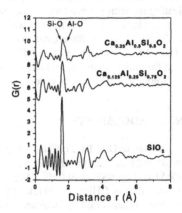

Figure 5. Experimental structure factors for calcium aluminosilicate glasses (dots) together with the optimum smooth line.

Figure 6. Atomic PDFs, G(r), of calcium aluminosilicate glasses obtained by Fourier transforming the smoothed data of Fig. 5.

Al or Si with Ca cation - "non-bridging" (NBO). Although NBO are not an integral part of the polyhedral network they play an important role in determining the thermodynamic properties of the glasses. We undertook a PDF study to investigate the nature of $Si-O$ and $Al-O$ polyhedral units and the distribution of NBOs on them. The experiments were carried out at 1-BM beamline at the Advanced Photon Source, Argonne with the use of x-rays of energy 65 keV [13]. The experimental structure factors are shown in Fig. 5 and the corresponding PDFs - in Fig. 6. Due to the fact that the glassy samples investigated exhibited weak diffuse scattering, statistics of the experimental data turned out to be somewhat poor despite the powerful synchrotron source employed. It necessitated extra smoothing of the data, as shown in Fig. 5. The smoothed $S(Q)$ data show oscillating behaviour, i.e. structure relevant features, up to the maximum Q-value studied (40 Å^{-1}). The corresponding high-resolution $G(r)s$ have a sharp first peak which is obviously composed of two components in Ca and Al containing glasses. The first component is positioned at 1.61 Å and reflects the presence of well-defined SiO_4 units. The higher-r component is positioned at approximately 1.75 Å which is the $Al-O$ distance usually found in AlO_4 tetrahedral units. Thus the present study provides a strong experimental evidence supporting the model picture viewing calcium aluminosilicate glasses as a network of linked SiO_4 and AlO_4 tetrahedra. New PDF experiments aimed at improving the statistical accuracy of the data will be carried out soon and the fascinating atomic ordering of $Ca_{x/2}Al_xSi_{1-x}O_2$ glasses, including the distribution of NBOs on the individual SiO_4 and AlO_4 tetrahedra, revealed in more detail.

CONCLUSIONS

The combination of intense high-energy sources of x-ray radiation, such as synchrotrons, and the atomic Pair Distribution Function technique offers new opportunities for exploring the structure of materials with higher real-space resolution. It can be fruitfully applied for investigating the atomic arrangement in completely disordered materials such as glasses, revealing fine local deviations from an well known average structure, as in the case of $In_{1-x}Ga_xAs$ semiconductor alloys considered, and even for determining of unknown structures, as the one of "restacked" WS_2. Such high-resolution PDF studies are envisaged to be much more frequently employed for materials structure studies with the number of high-energy, high flux synchrotron sources rapidly increasing worldwide.

ACKNOWLEDGEMENTS

Thanks are due to Drs. Th. Proffen, D. Haeffner, J.C. Lang, B. Himmel and Mr. I-K. Jeong for discussions and help with the experiments. The work was supported by DOE grant DE FG02 97ER45651 and by NSF grant CHE 99-03706. S.J.L.B. also acknowledges support from Alfred P. Sloan Foundation. CHESS is operated by NSF through grant DMR97-13242. Advanced Photon Source is supported by DOE under contract W-31-109-Eng-38.

REFERENCES

1. H.M. Rietveld, J. Appl. Cryst. **2**, 65 (1969).
2. R. A. Young, in *The Rietveld Method* edited by R.A. Young (Oxford University Press, 1995).
3. T. Egami, in *Local structure from Diffraction* edited by S.J.L. Billinge and M.F. Thorpe (New York, Plenum, 1998), p. 1.
4. Y. Waseda, *The structure of non-crystalline materials*, (New York, McGraw Hill, 1980).
5. V. Petkov, J. Appl. Cryst. **22**, 387 (1989).
6. V. Petkov, I-K. Jeong, J. Chung, M.F. Thorpe, S. Kycia and S.J.L. Billinge, Phys. Rev. Lett. **83**, 4089 (1999).
7. V. Petkov, I-K. Jeong, F.M-Jacobs, Th. Proffen, J.S.L. Billinge and W. Dmowski, J. Appl. Phys., submitted.
8. R. R. Chianelli, A.F. Ruppert, and M. Jose-Yacaman, A. Vazquez-Zavala, Catal. Today **23**, 269 (1995).
9. D. Yang and R.F. Frindt, J. Phys. Chem. Solids **57**, 1113 (1996).
10. H.-L. Tsai, J. Heising, J.L. Schindler, C.R. Kannewurf, M.G. Kanatzidis, Chem. Mater. **9**, 879 (1997).
11. J. Heising and M. G. Kanatzidis, J. Am.Chem. Soc. **121**, 638 (1999).
12. V. Petkov, S.J.L. Billinge, J. Heising and M. G. Kanatzidis, in preparation
13. V. Petkov, I-K. Jeong, M. Gutmann, P.F . Peterson, S.J.L. Billinge, S. Shastri and B. Himmel, Advanced Photon Source User Activity Report - www.aps.anl.gov/xfd/communicator/user2000/sri-cat.html

SYNCHROTRON X-RAY STUDY OF ELASTIC PHASE STRAINS IN THE BULK OF AN EXTERNALLY LOADED Cu/Mo COMPOSITE

A. WANNER*, D.C. DUNAND**
* Institut für Metallkunde, Universität Stuttgart, Seestr. 71, D-70174 Stuttgart, Germany.
** Dept. of Materials Science and Engineering, Northwestern University, Evanston, IL 60208, U.S.A.

ABSTRACT

High-energy, high-flux x-rays from a third-generation synchrotron source were used to measure average elastic strains in the bulk of 1.5 mm thick composites consisting of a copper matrix reinforced with 7.5 vol.% molybdenum particles. From the evolution of lattice strains in both phases during uniaxial tensile deformation, the internal load transfer between phases and reinforcement damage were characterized during elastic and plastic deformation of the composite. The graininess of the diffraction rings, which is related to the Bragg peak broadening, was quantified as a function of applied stress and related to plastic deformation in the matrix.

INTRODUCTION

The load-bearing capacity of a composite is dictated by the load transfer occurring from the soft matrix to the rigid reinforcements [1]. Experimental measurement of load partitioning between the individual phases of composites can give a wealth of information on the micromechanical evolution of composites during deformation. The classical method to characterize this load transfer is to measure the lattice strains individually for both phases of the composite. The standard techniques applied for such measurements in the bulk of crystalline solids are based on the diffraction of thermal neutrons [2]. However, these techniques require long measurement times or a large diffraction volume because of the small neutron flux available at existing neutron sources and the inherently weak interaction of neutrons with crystalline solids. The availability of high-energy photons from 3rd-generation synchrotron research facilities offers alternative approaches to internal strain diffraction measurements which do not have these restrictions and are therefore a complementary tool to neutron diffraction. We have applied and further developed a synchrotron radiation transmission technique recently described by Withers and coworkers [3,4] to study the internal load transfer and damage in a two-phase Cu-Mo alloy during plastic deformation.

EXPERIMENTAL

We investigated a copper matrix composite with nominal content of 7.5 vol.% molybdenum particles fabricated by powder metallurgy [5]. The starting materials were 99.9% pure, spherical copper powder less than 44 µm in size and 99.8% pure molybdenum powder with a reported size range of 10-44 µm. The molybdenum powder particulates exhibit a complex morphology as they are partially sintered aggregates of fine, micron-size molybdenum particles. The powder blends were packed in steel cans and densified by hot isostatic pressing (HIP) at 900°C and 100 MPa for 125 minutes. Below the eutectic temperature of 1083.4°C, Cu and Mo exhibit very little mutual solubility [6]. Therefore, the Cu/Mo composite is expected to be consist of the two pure, untextured phases. Fig. 1 is a representative backscatter SEM micrograph of a particulate embedded in the copper matrix, showing that consolidation was effective and that the particulates exhibit a two-phase, interpenetrating microstructure. A likely explanation is that the initially porous molybdenum powder particulates were completely filled with copper forced into the fine porosity during solid-state consolidation. Quantitative metallographic examination of optical micrographs yielded a volume fraction of reinforcement particulates of 16%, i.e. roughly twice the nominal volume fraction of Mo phase, which is consistent with the idea that the particulates contain a significant amount of Cu phase. The grain size in the copper matrix surrounding the particulates as determined by the linear intersect method on etched cross sections (Fig. 2) was 8.8 µm.

Mat. Res. Soc. Symp. Proc. Vol. 590 © 2000 Materials Research Society

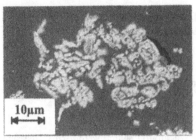

Fig. 1: Typical two-phase microstructure reinforcement particulate embedded in Cu matrix (backscatter SEM; dark phase is Cu, light phase Mo)

Fig. 2: Light-optical micrograph of etched cross-section, showing the grain structure of the Cu matrix. The Mo particles are etched away.

Dogbone-shaped, flat tensile specimens with gauges 8 mm long, 3.0 mm wide and 1.5 mm thick were fabricated by electric discharge machining and subsequent mechanical polishing. Uniaxial tensile tests were performed at room temperature using a miniature, screw-driven tensile device. These tests were performed *in-situ* at a bending magnet beamline of the Advanced Photon Source (Argonne Natl. Lab.). The tensile strain was increased stepwise and held constant during x-ray exposure times. The general experimental setup for the type of high-energy x-ray transmission experiments performed in the present study has recently been described by Daymond and Withers [3]. The specimen was irradiated with a monochromatic, parallel beam of high-energy x-rays. Complete, ring-shaped diffraction patterns were recorded with a two-dimensional detector, as shown in Figures 3 a and b. Unlike Daymond and Withers, we attached a calibration powder to the specimen, as described in more detail later. The tensile specimen was irradiated by a beam of 65 keV x-rays (with wavelength $\lambda \approx 0.19$ A) with approximate dimension 0.2 mm × 0.2 mm and positioned approximately in the center of the gauge section. Hence, the path length of the incident beam inside the specimen was 1.5 mm and the diffracting volume, over which the average strain results were measured, was about 0.06 mm³. Based on the metallographic findings, it is estimated that this volume contained roughly 10^5 Cu matrix grains and more than 100 polycrystalline reinforcement particulates.

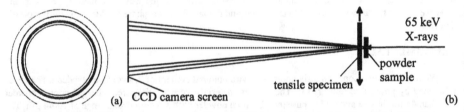

Fig. 3: (a) Typical diffraction pattern recorded by the CCD camera showing the five diffraction rings. (b) Schematic of the experimental arrangement used for the high-energy x-ray transmission experiments.

The diffraction geometry is similar to that present in a transmission electron microscope in the sense that the wavelength of the high-energy x-rays is considerably smaller than the typical spacing of low-index lattice planes of highly symmetric crystals. Hence, in diffracting condition, these planes are oriented almost parallel to the incident beam. This allows for recording the strains in the plane perpendicular to the incident beam, and this plane contains the loading direction as well as a transverse direction of the tensile specimen. In the present study we have not attempted to measure any initial lattice strains but restricted our analysis to the additional strains which develop in the course of deformation under the applied tensile stress.

The ring-like diffraction patterns produced on a plane normal to the incident beam were recorded using a CCD camera providing intensity readings over a circular screen 132 mm in diameter consisting of an orthogonal array of square, 64μm × 64μm, pixels. The exposure time was always 600 seconds. The distance L between the CCD screen and the specimen was set to 510 mm, which allowed for simultaneous recordings of the Mo (110), Cu (111), Cu (200), and Mo (200) rings from the specimen as well as the Fe (110) ring from an iron powder sample inserted into the x-ray beam for error correction purposes. This iron powder was filled in a small polyethylene container fixed directly to the specimen gauge section using rubber bands.

For small Bragg angles, the lattice spacing d is related to λ and L and to the diameter D of the diffraction ring on the camera screen as:

$$d \approx \frac{2\lambda L}{D} \tag{1}$$

Hence, the lattice strain ε with respect to a reference state denoted by superscript 0 is directly correlated to the measured ring diameters

$$\varepsilon = \frac{d - d^0}{d^0} \approx \frac{D^0 - D}{D} \approx \frac{D^0 - D}{D^0} \tag{2}$$

This simple equation, which has been widely used to evaluate the kind of diffraction patterns produced by the present technique [3,4], is based on the tacit assumption that the product λL in Eq. 1 is unchanged for the two independent diameter measurements D and D_0. However, if λ or L change during a series of measurements, misleading results are obtained. We avoided this issue by taking into account an additional diffraction ring produced by the unstrained powder sample attached to the specimen. In the following, the quantities related to this calibration substance and the specimen are denoted by subscripts C and S, respectively. If, as in our experiments, the spacing $\Delta L = L_S - L_C$ remains unchanged during the measurements and is also much smaller than the specimen-to-camera distance L_S, the following expression can be derived based on Equations 1 and 2:

$$\varepsilon_S = \frac{d_S - d_S^0}{d_S^0} \approx 1 - \frac{D_S}{D_S^0} \cdot \frac{D_C^0}{D_C} \tag{3}$$

This expression for ε_S is robust against small changes of λ and against cooperative movement of specimen and calibration substance along the beam axis during the experiment.

The simplest approach to evaluate the diffraction patterns would be to merely measure the horizontal and vertical diameters of the diffraction rings. This approach is successful if the rings from the specimen are extremely smooth due to a very small grain size and a favorable texture in the specimen, as was the case in Ref. [3]. In the course of the present study, it however became apparent that the diffraction rings exhibited a noticeable graininess, giving rise to very poor strain resolution if the evaluation was based on narrow strips extracted from diffraction patterns. We therefore developed an algorithm which takes the whole diffraction rings into account, thus reducing the effect of graininess on the strain results. This algorithm consists of the following steps:

(i) Identification of the approximate center of the pattern, (ii) polar re-binning of pixel intensity data into radius/angle (R/φ) with respect to the pattern center in φ-steps of 0.225°, (iii) Gaussian peak fitting in R at each of the 1,600 angular positions, (iv) addition of the Gaussian center positions of diametrically opposite peaks to obtain the ring diameters $D(\varphi)$, (v) determination of the vertical and horizontal ring diameters by fitting an ellipse to the $D(\varphi)$ data set, (vi) calculation of the lattice strains in axial and transverse directions for both phases of the composite according to Equation 3.

Step (v) was accomplished by plotting $D(\varphi)$ versus $sin^2\varphi$ and fitting a straight line to the data points. The semi-axes are then found at $sin^2\varphi = 1$ ($\varphi = 90°$ and 270°; loading direction) and $sin^2\varphi = 0$ ($\varphi = 0°$ and 180°; transverse direction), as illustrated in Fig. 4 for the Mo (110) ring obtained at an applied stress of 100 MPa. Every angular direction has the same statistical weight, independent of the signal intensity. This is helpful for the evaluation of grainy diffraction rings, because the information from the weak-intensity gaps between high-intensity speckles forming the ring is fully taken into account.

The availability of Gaussian peak fit data obtained at a large number of different φ-positions can be used to characterize the graininess itself, which, as described in the following, gives additional insight into the deformation processes occurring within the material. In the present study we have assessed the graininess by calculating the average A_W and the standard deviation S_W of the 1600 Gaussian peak widths determined at different positions around the ring (step *(iii)* of the algorithm). The graininess of the diffraction rings results from the incident x-ray beam irradiating only a limited number of grains, only a small fraction of which are oriented in a diffracting condition. This fraction depends on the broadness of the reciprocal lattice points of the grains, which is governed by their real-space size and lattice defect structure. The broader these reciprocal lattice points are, the higher is the fraction of grains that contribute to the diffraction ring and thus the more diffuse are the speckles produced by the individual grains. Hence, for a given number and orientation distribution of grains in the gauge volume, the diffraction rings become less grainy if the lattice defect density is increased. Changes in graininess observed upon mechanical loading of the specimen can then be attributed to changes in the defect density of the phase that produces the diffraction ring.

RESULTS

In Fig. 5 the engineering stress-strain curve of the composite as measured in an *in-situ* uniaxial tensile test is shown. The Mo (110), Cu (111) and Cu (200) diffraction results from the same test are shown in Figs. 6 a-d, where the applied stress is plotted against the volume-averaged lattice strains measured in each phase for the loading and transverse directions. No data is presented for the Mo (200) diffraction ring because it was too grainy to yield consistent results. It is apparent that the strain varies between phases, as expected from their different mechanical properties, and also between crystallographic directions for the Cu phase, due to the elasto-plastic anisotropy of the individual grains. All curves show three regions with different slopes during mechanical loading, indicative of changes in the load sharing between matrix and reinforcement. The change of slope observed in the high-stress regime is very pronounced, leading to a full reversal of the lattice strain evolution in the Mo phase, i.e. the lattice strain decreases with increasing applied stress.

Diffraction ring graininess was found to evolve for the copper phase in the course of the *in-situ* experiments, while it was essentially constant for the molybdenum phase. For all Cu and Mo rings evaluated, the average ring width A_W always lay in the range of 130-150 μm initially and did not change by more than 10% as the specimen was deformed. However, the ring width standard deviation S_W was more sensitive to the visible changes in ring graininess. The S_W values of the Cu (111) and Mo (110) rings are plotted as a function of applied stress in Figure 7. A significant reduction in graininess of the Cu (111) ring was only observed during plastic deformation, but not during subsequent elastic unloading to zero applied stress, while the graininess of the Mo (110) ring did not change significantly throughout the whole tensile tests.

DISCUSSION

Three regimes of load-sharing can be identified in the curves of applied stress vs. x-ray lattice strain (Figs. 6 a-d). At low applied stresses (regime I), load sharing is constant and both phases are expected to behave linear-elastically. At intermediate stresses (regime II), the stress dependence of lattice strain with increasing applied stress decreases in the matrix but increases in the reinforcements. Thus the proportion of load borne by the matrix decreases, while the reinforcement carries proportionally more load. This type of behavior has been observed in many composites when plastic deformation occurs in the matrix, thus increasing the mismatch with, and load transfer to, the elastic reinforcement [1]. At even higher applied stresses (regime III), i.e. above about 100 MPa, this trend is reversed, indicating that the reinforcements are getting less efficiently loaded. This behavior could be attributed to relaxation of the matrix, plastic deformation of the particulates, or damage processes such as particulate fracture and particulate/matrix detachment. As discussed below, we believe that the dominant process is the disintegration of the particulates consisting of fine, sintered Mo particles infiltrated by copper. We estimate the average stress in the reinforcement phase from the Mo (110) lattice strains measured in longitudinal direction (Fig. 6d) by assuming that the longitudinal stress is independent of the grain orientation and by neglecting the effects of transverse stresses due to grain interaction within the

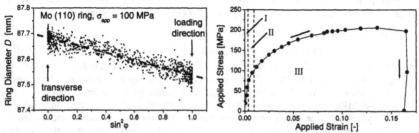

Fig. 4: Typical $D(\varphi)$ vs. $\sin^2\varphi$ diagram   Fig. 5: Stress-strain curve of the composite

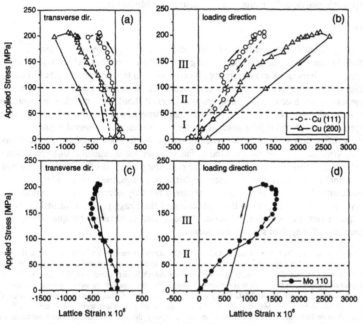

Figure 6: Applied stress vs. lattice strain. (a) Cu (111) and Cu (200) transverse direction, (b) Cu (111) and Cu (200) loading direction, (c) Mo (110), transverse direction, (d) Mo (110) loading direction.

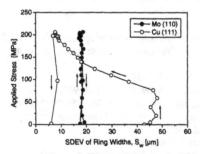

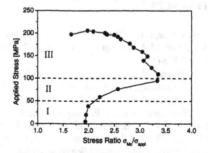

Fig. 7: Evolution of diffraction ring graininess during loading and unloading of the composite.

Fig. 8: Evolution of the reinforcement stress ratio σ_{Mo}/σ_{ap} with increasing applied stress.

reinforcement phase or from interactions between the two phases of the composite. With these simplifying assumptions the average stress in the molybdenum phase, σ_{Mo}, is estimated as:

$$\sigma_{Mo} \approx E_{<110>,Mo} \cdot \varepsilon_{<110>,Mo} \qquad (4)$$

where $E_{<110>,Mo}$ = 319 GPa [7,8] is the Young's modulus of a Mo single crystal uniaxially loaded in the 110-direction and $\varepsilon_{<110>,Mo}$ is the lattice strain measured in that direction (Fig. 6d). The evolution of the ratio σ_{Mo}/σ_{app} is shown in as a function of the applied stress σ_{appl} in Fig. 8. The average stress present in the molybdenum phase at which first signs of damage are apparent (transition from regime II to regime III) is roughly 340 MPa, which corresponds to about half of the yield stress reported for commercially pure Mo polycrystals [9]. Due to the stress concentrations existing at necks connecting the small Mo particles agglomerated into a single particulate, disintegration of the particulates is likely to occur as a result of local plastic deformation and fracture at these necks. The graininess results (Fig.7) imply, however, that large-scale plastic deformation did not occur in the Mo phase.

CONCLUSIONS

The present study shows that synchrotron x-ray diffraction experiments can provide valuable information on bulk lattice strain and load partitioning in Cu-Mo composites, where both phases have relatively high atomic numbers and x-ray absorption. As compared to the older, more established neutron diffraction technique, the measurement times were about one order of magnitude shorter and the diffracting volume was about three orders of magnitude smaller. It should be noted that the measurement time would be reduced even further if the same experiment was performed at an insertion device-beamline providing an even brighter beam of photons with even shorter wavelength.

While the small diffraction volume of the synchrotron technique has a number of positive aspects (small quantities of specimen material are required to fabricate tensile specimens, high-resolution strain mapping is possible etc.), the technique is limited by the resulting graininess of the diffraction rings. The present study shows that by fitting ellipses to complete diffraction rings, the level of tolerable graininess can be raised significantly. Moreover, as the ring graininess is not only governed by the number of grains in the diffraction volume but also related to the broadening of the Bragg peaks, information about peak broadening (and thus about the evolution of lattice defects) can be obtained by quantifying the ring graininess as a function of the applied stresses and strains.

Acknowledgements: We thank the late Professor Jerome B. Cohen (Northwestern University) for introducing us to the technique of synchrotron x-ray diffraction, for lending us his miniature tensile device, and for many helpful discussions. A.W. gratefully acknowledges the Max Kade Foundation (New York, NY) as well as the Eshbach Society (Evanston, IL) for supporting his stay at Northwestern University during which this work was performed. The diffraction experiments were performed at the DuPont-Northwestern-Dow Collaborative Access Team (DND-CAT) Synchrotron Research Center located at the Advanced Photon Source (Argonne National Laboratory, IL), whose staff we acknowledge for their excellent technical support. DND-CAT is supported by the E.I. DuPont de Nemours & Co., The Dow Chemical Company, the U.S. National Science Foundation through grant DMR-9304725, and the State of Illinois through grant IBHE HECA NWU 96. Use of the APS was supported by the U.S. Department of Energy under contract no. W-31-102-Eng-38.

REFERENCES

1 T.W. Clyne, P.J. Withers, *An Introduction to Metal Matrix Composites*, Cambridge Univ. Press, 1993.
2 A.J. Allen, M.A.M. Bourke, S. Dawes, M.T. Hutchings, P.J. Withers (1992), *Acta metall. mater.* 40, 2361-2373 (1992).
3 M.R. Daymond and P.J. Withers, *Scripta Materialia* 35, 1229-1234 (1996).
4 A.M. Korsunsky, K.E. Wells and P.J. Withers, *Scripta Materialia* 39, 1705-1712 (1998).
5 M.R. Daymond, C. Lund, M.A.M. Bourke, D.C. Dunand, *Met. Mat. Trans. A* 30, 2989-2997 (1999).
6 P.R. Subramanian and D.E. Laughlin, The system Cu-Mo, in: *Binary Alloy Phase Diagrams*, 2nd Edition, Vol. 2, T.B. Massalski (Ed.), ASM International, 1990.
7 J.F. Nye, *Physical Properties of Crystals*, Oxford University Press, 1985.
8 T.H. Courtney, *Mechanical Behavior of Materials*, McGraw-Hill Publishing Company, 1990.
9 H.E. Boyer, *Atlas of Stress-Strain Curves*, ASM International, 1987.

Surfaces

X-RAY SURFACE SCATTERING STUDIES OF MOLECULAR ORDERING AT LIQUID-LIQUID INTERFACES

MARK L. SCHLOSSMAN, MING LI, DRAGOSLAV M. MITRINOVIC, ALEKSEY M. TIKHONOV
Department of Physics, University of Illinois, Chicago, IL 60607, schloss@uic.edu

ABSTRACT

We present our recent progress in using synchrotron x-ray surface scattering techniques to study several different aspects of ordering at liquid-liquid interfaces. (1) We present measurements of the interfacial width at the water-alkane interface for a series of different chain length alkanes. The width for the shortest chain length studied is in agreement with capillary wave theory. However, significant deviations occur for longer chain lengths, indicating the presence of molecular ordering at the interface. (2) Under appropriate conditions, a surfactant monolayer forms at the interface between water and a hexane solution of a fluorinated surfactant. Reflectivity measurements that probe the electron density profile normal to the interface provide information about the surfactant ordering. This monolayer undergoes a solid to gas transition as a function of temperature. Diffuse scattering near the transition reveals the presence of islands. (3) Equilibrium interfaces between two aqueous phases containing PEG (polyethylene glycol) and potassium phosphate salts can be studied. We present studies of coherent capillary wave fluctuations between two interfaces of a thin film of this biphase system. We also demonstrate that biological macromolecules can be trapped and studied at this aqueous-aqueous interface.

INTRODUCTION

An outstanding problem in the area of interfacial phenomena is the determination of structure at liquid-liquid interfaces. These interfaces play an important role in many chemical and biological systems in addition to being interesting model systems for study of the statistical physics of interfaces and membranes. Surfactant molecules naturally arrange themselves at liquid interfaces and, in the case of aqueous-organic interfaces, bring together materials on the microscopic scale that may be otherwise immiscible. Biological membranes which exist at aqueous-aqueous interfaces play a critically important role in many cell processes. In spite of their importance, there remain fundamental questions in the study of liquid-liquid interfaces that have been barely addressed. There is currently a poor understanding of electron and molecular density profiles both normal to the interface and within the interface.

Few experimental techniques are capable of directly probing the molecular order or condensed matter states at a single liquid-liquid interface, non-linear optical studies being a recently developed, notable exception that provides information about molecular conformations [1-3]. Recently, x-ray and neutron surface scattering have been applied to the study of liquid-liquid interfaces [4-8]. Here, we discuss briefly several investigations that probe fundamental aspects of molecular ordering at liquid-liquid interfaces. In addition, we also discuss a system of potential use for probing biological processes at aqueous-aqueous interfaces.

EXPERIMENTS AND DISCUSSION

Experimental Methods

These x-ray surface scattering measurements were conducted at beamline X19C at the National Synchrotron Light Source (Brookhaven National Laboratory, USA) with a liquid surface spectrometer and measurement techniques described in detail elsewhere [9]. For x-ray reflectivity, the incident angle α is equal to the scattered angle β (see Fig. 1). The reflected x-ray intensity is normalized to the incident intensity and measured as a function of the wave vector transfer normal to the plane of the interface, $Q_z = (4\pi/\lambda)\sin(\alpha)$, where $\lambda = 0.0825\pm0.0002$ nm is the x-ray wavelength for these measurements. For off-specular diffuse scattering the incident and scattered angles are different.

For measurements at the water-alkane interface, the liquids were contained in vapor-tight, temperature-controlled, polycarbonate or stainless steel sample cells with mylar x-ray windows. The sample cells had x-ray path lengths through the upper phase of either 25, 50, or 76 mm. The liquids were first stirred and allowed to reach thermal equilibrium. To reduce most of the curvature from the meniscus formed at the cell windows, the windows are angled 25° from the vertical (see Fig. 1). However, fine tuning of the interfacial flatness is required. This is accomplished by pinning the meniscus to the cell windows (by roughening the windows), then rotating the entire sample. This twists the interface to yield a very flat region suitable for x-ray scattering. Alternatively, the liquid volume can be adjusted to flatten the interface.

Measurements at the aqueous-aqueous interface were from liquids placed in a circular teflon trough of diameter 75 mm. This trough was in a temperature controlled, closed container.

High purity water was produced from a Barnstead NanoPure system. High purity alkanes were purchased and then further purified by filtration through basic alumina. Their purity was confirmed by interfacial tension measurements. Other chemicals were used as received from the manufacturer.

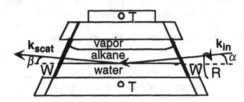

Fig. 1 Cross-sectional view of sample cell for water-alkane measurements; W - mylar windows; T - thermistor to measure temperature; R - rotation about the horizontal used to fine tune the sample flatness. The kinematics of surface x-ray reflectivity is also indicated: k_{in} is the incoming x-ray wave vector, k_{scat} is the scattered wave vector, α is the angle of incidence and reflection.

Water-Alkane Interfaces

Of fundamental importance in the study of liquid-liquid interfaces is the interface between two pure liquids such as water and alkane. This can be used as a model system to understand the interactions of alkyl chains with aqueous phases. A number of molecular dynamics simulations have focused upon the issues of the interfacial width and orientational order close to the interface. Molecular dynamics simulations by Buuren et al. found that the interfacial width depends sensitively upon the intermolecular potentials [10]. Here, we show that x-ray reflectivity can be used to measure the interfacial width between water and a series of alkanes.

X-ray reflectivity measurements as a function of the reflection angle from a bulk interface can be analyzed to yield an electron density profile along the normal to the interface (and averaged over the plane of the interface) with a resolution of a fraction of a nanometer [11]. Figure 2 illustrates the x-ray reflectivity, $R(Q_z)$, measured from the water-hexane interface at 32.00 °C [7]. The reflectivity varies with the angle of reflection, α, expressed in terms of the wave vector transfer normal to the interface, $Q_z = (4\pi / \lambda)\sin\alpha$. Also shown is the Fresnel reflectivity, $R_F(Q_z)$, predicted for an ideal, smooth and flat interface. Both of these curves have a small region of constant reflectivity (equal to 1) below a critical wave vector transfer, $Q_c = 0.123$ nm^{-1}, corresponding to a region of total reflection. Both curves then drop off rapidly and smoothly with increasing wave vector transfer.

Although the qualitative features of the ideal and measured curves are similar, the reduction in intensity of the measured reflectivity at larger wave vector transfer is the result of x-rays being scattered by interfacial roughness due to thermally induced capillary wave fluctuations. The classical capillary wave model for the fluctuations [12] corresponds to an error function profile for the electron density averaged over the plane of the interface, $\langle \rho(z) \rangle$. For this interfacial profile the distorted wave Born approximation expresses the reflected intensity as [13],

$$R(Q_z) \cong \left| \frac{Q_z - Q_z^T}{Q_z + Q_z^T} \right|^2 \exp(-Q_z Q_z^T \sigma^2), \qquad (1)$$

where $Q_z^T \equiv \sqrt{Q_z^2 - Q_c^2}$ is the z-component of wave vector transfer with respect to the lower phase. A fit of the data to Eq.(1) using a single fitting parameter, σ, is illustrated by the solid line in Fig. 2 and yields the interfacial width $\sigma = 3.3 \pm 0.25$ Å.

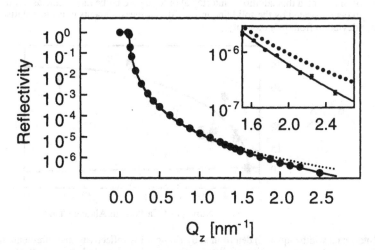

Fig. 2 X-ray reflectivity from the water-hexane interface as a function of wave vector transfer normal to the interface. ●, measurements at 32.00 °C; solid line through the soild circles, fit described in the text; dotted line, Fresnel reflectivity for ideal interface. The inset is an expanded view of the high Q_z region. Error bars on the data are similar to the symbol size in the inset.

In addition to the capillary wave contribution just discussed the interfacial width may also contain a contribution from an intrinsic profile described, for example, by van der Waals theories [14]. In the spirit of a hybrid model of the interface that describes this intrinsic profile roughened by capillary waves, the total interfacial width, σ_{total}, can be represented as a combination of an intrinsic profile width, σ_o, and the capillary wave contribution [15-17],

$$\sigma_{total}^2 = \sigma_o^2 + \frac{k_B T}{2\pi\gamma} \int_{q_{min}}^{q_{max}} \frac{q\, dq}{q^2 + \xi_{\parallel}^{-2}} \equiv \sigma_o^2 + \sigma_{cap}^2, \qquad (2)$$

where $k_B T$ is Boltzmann's constant times the temperature, γ is the interfacial tension, the correlation length, $\xi_{\parallel}$, is given by $\xi_{\parallel}^2 = \gamma / \Delta\rho_m g$ and determines the exponential decay of the interfacial correlations given by the height-height correlation function of interfacial motion [18], $\Delta\rho_m$ is the mass density difference of the two phases, and g is the gravitational acceleration. The wave vector, q, represents the in-plane capillary waves. The limit q_{min} is determined by the instrumental resolution that sets the largest in-plane capillary wavelength that the measurement probes. The limit, q_{max}, is determined by the cutoff for the smallest wavelength capillary waves that the interface can support. Direct calculation of σ_{cap} using the literature value of the interfacial tension (51.4 mN/m at $T = 22$ °C) [19] and $q_{max} = 2\pi/0.5$ nm^{-1} yields $\sigma_{cap} = 3.29$ Å, in agreement with our measurement of $\sigma = 3.3$ Å. This indicates that the intrinsic profile width, σ_o, is small for this interface, $\sigma_o < 1.5$Å.

We have extended these measurements to interfaces between water and longer chain alkanes. The x-ray reflectivity is similar in form to the measurements in Fig. 2 and can be fit well with the expression in Eq. (1). The results for the interfacial width squared, σ^2, are shown in Fig. 3. For alkanes longer than hexane there is a significant deviation of the measured interfacial width from the prediction of capillary wave theory. This indicates the presence of molecular ordering at the interface that can be characterized by an intrinsic profile width, σ_o, as in Eq. (2). Although it may have been expected that additional interfacial ordering would be monotonic as a function of chain length (as illustrated by the solid line in Fig. 3), our measurements indicate a statistically significant deviation from monotonicity.

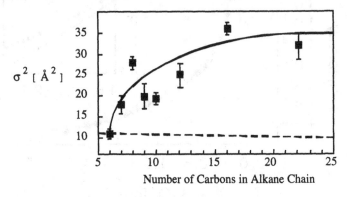

Fig. 3 Interfacial width (squared) determined by fitting x-ray reflectivity measurements from the water-alkane interface as a function of alkane carbon number. The dashed line indicates, approximately, the prediction for the interfacial width from capillary wave theory. The solid line is drawn as a guide to the eye.

Soluble Fluorinated Surfactants at the Water-Hexane Interface

The few available thermodynamic or spectroscopic measurements indicate that soluble and insoluble amphiphilic monolayers at water-oil interfaces are more loosely packed than the corresponding monolayers at the water-vapor interface [20-22]. In particular, the aliphatic tail groups of the amphiphiles exhibit greater chain disorder, possibly because the oil phase acts as a solvent for the chains at the interface [22]. Therefore, the chains are further from their neighbors than in a close packed solid. This reduces the van der Waals attractive forces between the alkyl chains, but allows for a higher conformational entropy of these flexible chains. Monolayers formed from surfactants soluble in the oil phase are expected to be disordered and in a liquid or gas phase at liquid-liquid interfaces [22].

To explore the ordering in a monolayer of soluble surfactants at the liquid-liquid interface we studied monolayers of $F(CF_2)_{10}(CH_2)_2OH$ (denoted here as $FC_{12}OH$) self-assembled from a solution (2×10^{-3} mol/kg) in hexane onto the solution-water interface [23, 24]. Figure 4 illustrates x-ray reflectivity measurements at T=32.00 (±0.03)°C from the monolayer at the water–(hexane solution) interface [8]. Also shown are measurements from the pure water-hexane interface at the same temperature and measurements from the water-(hexane solution) interface at T=48.00°C. These curves have a small region of constant reflectivity (nearly equal to 1) below a critical wave vector transfer, $Q_c = 0.0123$ Å^{-1}, corresponding to a region of total reflection. The enhanced reflectivity at higher Q_z from the water–(hexane solution) interface at T=32.00°C is due to a constructive interference of x-rays reflected from the top of the fluorinated monolayer with x-rays reflected from the bottom of the monolayer (inset, Fig. 4). The reflectivity from the water-(hexane solution) interface at T=48.00 °C is nearly identical to the measurements from the pure water-hexane interface and indicates that most of the $FC_{12}OH$ molecules have desorbed from the interface at this temperature.

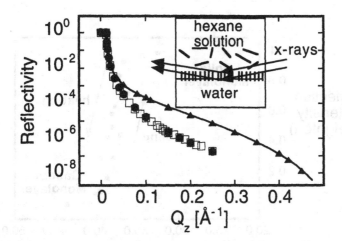

Fig. 4 X-ray reflectivity from the water–(hexane solution of $F(CF_2)_{10}(CH_2)_2OH$) interface at T=32°C, triangles; T=48°C, solid circles; pure water-hexane interface at T=32°C, open squares. Inset is a cartoon of the solid surfactant monolayer that forms at T=32°C.

These data can be analyzed by using a general expression, derived from the first Born approximation for x-ray scattering, that relates the reflectivity to the electron density gradient normal to the interface, $d\langle\rho(z)\rangle/dz$ (averaged over the interfacial plane), and written as [11]

$$\frac{R(Q_z)}{R_F(Q_z)} \cong \left| \frac{1}{\Delta\rho_{bulk}} \int dz \frac{d\langle\rho(z)\rangle}{dz} \exp(iQ_z z) \right|^2 ,$$ (3)

where $\Delta\rho_{e,bulk} = \rho_{e,bulk,lower} - \rho_{e,bulk,upper}$ and $R_F(Q_z)$ is the Fresnel reflectivity predicted for an ideal, smooth and flat interface [25]. The layer of $FC_{12}OH$ is modeled simply as a thin slab of higher electron density sandwiched between two bulk liquids. The interfaces of the top and bottom of this slab are roughened by thermal capillary waves, characterized by the roughness parameter, σ. The reflectivity calculated from Eq.(3) using the slab model for the electron density is fit to the data to yield values for the three fitting parameters: the slab thickness $L = 1.24 \pm 0.03$ nm, $\sigma = 0.36 \pm 0.02$ nm, and the slab electron density $\rho_f = 1.90 \pm 0.04$ (see Fig. 4) [8].

Our measurement of the electron density of the monolayer at T=32.00°C, $\rho_f = 1.90 \pm 0.04$, corresponds to a mass density of 2.19 ± 0.05 g/cm³. This agrees with the density of bulk solid fluoroalkane phases (e.g., for n-$C_{20}F_{42}$) which have a density of either 2.23 g/cm³ for the monoclinic crystal phase or 2.16 g/cm³ for the rhombohedral rotator solid phase [26]. The error bars on our measurement of ρ_f do not allow us to distinguish between these two different solid phases. However, these measurements exclude the possibility that the $FC_{12}OH$ monolayer at the water-hexane interface is in a liquid monolayer phase (bulk liquid fluoroalkanes have a mass density of approximately 1.7 g/cm³) [26, 27].

A similar analysis of the data from the water–(hexane solution) interface at T=48.00°C indicates that a conservative upper limit to the surface coverage of $FC_{12}OH$ is approximately 1.5% [8]. The transition from the solid $FC_{12}OH$ monolayer to this gaseous monolayer at higher temperature can be studied by measuring reflectivity as a function of temperature. Figure 5 indicates the first heating-cooling cycle through this transition.

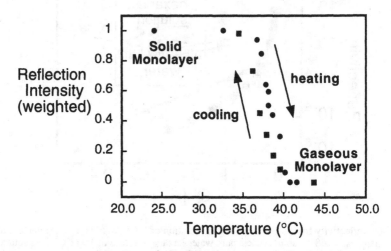

Fig. 5 Reflection intensity at $Q_z = 0.1$ Å^{-1} weighted so that a value of 1 indicates the presence of the solid monolayer phase while the value of 0 indicates the gaseous monolayer. This temperature scan reveals hysteresis through the solid–gas monolayer transition of $F(CF_2)_{10}(CH_2)_2OH$ at the water–(hexane solution) interface.

We believe that the transition region represents an interface with coexisting solid and gas monolayer phases. This is consistent with enhanced off-specular diffuse scattering that we measured in this transition region. Off-specular diffuse scattering for the heating cycle is illustrated in Fig. 6. This scattering can be described by the distorted wave Born approximation as,

$$I(\mathbf{Q}) \propto |T(\alpha)|^2 |T(\beta)|^2 |F(\mathbf{Q})|^2 |S(\mathbf{Q})|^2 \qquad (4)$$

where α is the incident angle; β is the scattering angle; the wave vector transfer, $\mathbf{Q} = \mathbf{k}_{scat} - \mathbf{k}_{in}$, is the difference between the incoming and scattered wave vectors; $F(\mathbf{Q})$ is the form factor for a particle; $S(\mathbf{Q})$ is the structure factor for the assembly of particles; $T(\alpha)$ or $T(\beta)$ are the Fresnel transmission coefficients [28, 29]. The tall, narrow peaks in the center of the scans in Fig. 6 occur when the incident angle, α, is equal to the scattering angle, β, and correspond to the specular reflectivity. In Fig. 6 a small peak appears at small β when $\beta = \theta_c$ (the angle for total reflection) and is known as a surface enhancement or Yoneda peak [30, 31].

Here, we are interested in the excess diffuse scattering in the "shoulders" immediately adjacent to the tall specular peaks in Fig. 6. The shape of this scattering is determined primarily by the form and structure factors. The shoulders in the measurement at 39.6°C illustrate essentially the shape of the form factor and reveal the presence of islands (or disks) of the solid phase on the order of a few micrometers in diameter. The sequence of lower temperatures (39.1°C, 38.9°C, and 38.3°C) indicate the presence of additional diffuse scattering due to the assembly of islands. The small peaks that appear in the shoulders of the specular peak indicate that the islands are spatially correlated. As the system is heated from 38.3°C to 39.6°C these small

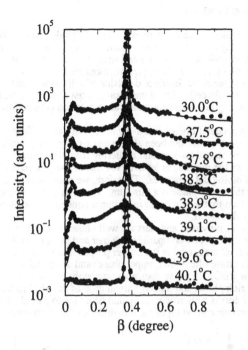

Fig. 6 Off-specular diffuse scattering for temperatures scanned through the range of the solid to gas monolayer transition of $F(CF_2)_{10}(CH_2)_2OH$ at the water–(hexane solution) interface. At 30°C the monolayer is in the low temperature solid phase; at 40.1°C it is in the high temperature gas phase. Additional diffuse scattering in the range 37.5°C < T < 40.1°C indicates the presence of islands of the solid phase. The curves are offset for clarity.

peaks move closer to the specular. This shows that the islands are spaced progressively further apart as the monolayer undergoes a transition from a solid to a gas phase. Over a narrow range in temperature of less than 0.15 °C (not shown in Fig. 6) the islands disappear.

An important feature of studies of the liquid-liquid interface is the ability to characterize the role of the upper phase on the structural order within, for example, monolayers at the interface. To gain some insight into the role of the upper-phase hexane on the ordering of the low temperature $FC_{12}OH$ solid phase monolayers, we studied a similar transition in $FC_{12}OH$ monolayers supported on the water-vapor interface. The $FC_{12}OH$ monolayer was spread in the crystalline island phase at an inverse density of 0.4 nm^2/molecule. Since the crystalline unit cell size is 0.29 nm^2/molecule there is excess area at the interface that allows the monolayer to undergo a transition to a liquid or gaseous disordered phase. X-ray grazing incidence diffraction measurements indicate that an order to disorder phase transition occurs between 58.8 °C and 62.09 °C, slightly more than 20 °C higher than the solid-gas transition in the $FC_{12}OH$ monolayer at the water–(hexane solution) interface. Therefore, the presence of the hexane solvent acts to disorder the monolayer at a lower temperature.

Aqueous-Aqueous Interfaces [32]

One of the most tantalizing areas for investigation of liquid-liquid interfaces is biological phenomena. Although biological membranes exist at liquid-liquid interfaces, it has not been possible previously to study molecular ordering and structure on the sub-nanometer scale within a membrane at the liquid-liquid interface. Biological membranes are complex bilayer structures of lipids and proteins. They separate the inside and outside of cells as well as the inside and outside of intracellular organelles within cells. These membranes exist at aqueous-aqueous interfaces and play a critical role in mediating biological communication between cells and transport of material into and out of the cell.

Although we have successfully applied x-ray reflectivity and diffuse scattering to the study of the liquid-liquid interface, we have not been able to use x-ray grazing incidence diffraction because of the large absorption and background scattering from the upper liquid phase. This diffraction technique allows the in-plane order at an interface to be probed and is likely to be very important in the study of biological structures. Therefore, we chose to develop an aqueous-aqueous interface where the upper phase is a thin film. This was also necessary because of the much larger x-ray absorption that would occur in an upper aqueous phase as compared to the upper alkane phases previously discussed.

Aqueous solutions of polymers and salts are well known to separate into two equilibrium phases. We chose to use biologically compatible polymer and salts, polyethylene glycol (PEG), and potassium phosphates. Once mixed in the proper proportions the resulting solution phase separates into a PEG-rich phase (the lighter, upper phase) and a salt-rich phase. Under the appropriate conditions the PEG-rich phase does not completely wet the salt-rich phase. We extracted the lower, salt-rich phase, and placed a small drop of the PEG-rich phase on its surface. This PEG-rich phase then forms a microscopically thin film (due to partial wetting of the interface) in equilibrium with a reservoir consisting of the remainder of the small drop (see Fig. 7).

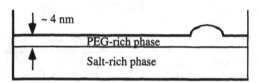

Fig. 7 Cartoon of a thin film of a PEG-rich phase in equilibrium with a PEG-rich reservoir drop and a bulk salt-rich phase.

Figure 8 shows the x-ray reflectivity (normalized to the Fresnel reflectivity) from this thin film system consisting of PEG (3400MW):K_2HPO_4:H_2O (13.2:28.9:57.9 wt %) at 35°C. The oscillations indicate the presence of the thin film of PEG-rich phase. The solid line in Fig. 8 is the result of a single slab model analysis of the data using Eq. (3). The analysis reveals that the film is 42 Å thick with an interfacial width between the PEG-rich phase and the vapor of $\sigma_{PEG\text{-}vapor} = 3.2$ Å and an interfacial width between the salt-rich and PEG-rich phases of $\sigma_{salt\text{-}PEG} = 8.3$ Å.

A combination of long and short-range forces is responsible for determining the equilibrium thickness of the thin PEG-rich film. As discussed in the wetting literature, the effect of these forces can be summarized in an interface potential that represents the potential energy between the PEG-vapor and salt-PEG interfaces as a function of the thickness of the film. Both of these interfaces fluctuate with capillary waves. If the fluctuations are conformal (in-phase, see Fig. 9a) then the equilibrium thickness is maintained locally throughout the film. If the fluctuations are non-conformal (or out of phase, see Fig. 9b) then the film thickness varies from point to point throughout the film. From the viewpoint of a local free energy, these variations in thickness will require a free energy penalty. In this local viewpoint we can determine the free energy cost to lowest order by expanding about the minimum of the interface potential. This results in a local

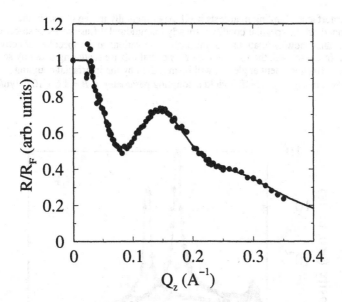

Fig. 8 X-ray reflectivity from a thin partially wet film of a PEG-rich phase on top of a salt-rich subphase. The solid line is a fit described in the text.

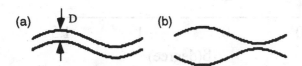

Fig. 9 (a) Conformal fluctuations of two interfaces that maintain the local equilibrium film thickness D throughout the entire film. (b) Non-conformal (out of phase) fluctuations in which the film thickness varies throughout the film.

modification of the capillary wave Hamiltonian expressed as

$$H = \int_{q_{min}}^{q_{max}} d^2s \left\{ \sum_{i=1,2} \left[\frac{1}{2}\gamma_i \left(|\nabla \zeta_i|^2 + \left(\frac{\zeta_i}{\xi_{||,cw}}\right)^2 \right) \right] + B(\zeta_1 - \zeta_2)^2 \right\}, \tag{5}$$

where i labels the two interfaces, $\zeta_i(x,y)$ represents the local height of each interface above a reference plane whose coordinates are $s = (x,y)$, and B represents the curvature of the interfacial potential near its minimum (we have assumed that the lowest order term in the interface coupling is quadratic). The Hamiltonian in Eq. (5) without the coupling term is the standard capillary wave Hamiltonian for two independent interfaces [12, 18]. This modified Hamiltonian is still Gaussian and can be analyzed by standard statistical mechanical methods to yield the ensemble averages of the correlation functions between the amplitudes of the capillary waves in reciprocal space. These, in turn, are used to predict the structure factor for off-specular diffuse scattering from the coupled capillary wave fluctuations of the two interfaces.

173

The interfacial coupling manifests itself experimentally in two ways. In the first, scans in Q_z slightly tuned off the specular condition (nearly longitudinal diffuse scans or so-called background scans) show oscillations that mimic the oscillations in the specular reflectivity curve (not shown). In the second, the form of the off-specular diffuse scattering taken by scanning the exit angle, β, at fixed incident angle, α, will be modified by the interfacial coupling. These latter scans are shown in Fig. 10. The fits yield a coupling parameter of $B = 8.6 \times 10^{-6}$ dyn/cmÅ^2.

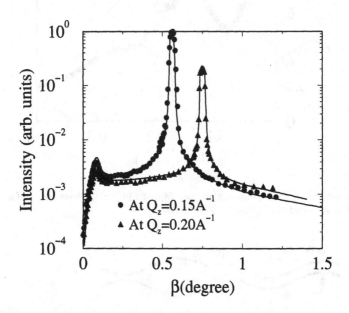

Fig. 10 Off-specular nearly transverse diffuse scattering from the capillary wave fluctuations of a thin PEG-rich film on top of a salt-rich subphase. The specular reflectivity for this thin film was shown in Fig. 8. The lines are fits to a model that contains coupled capillary wave fluctuations between the two interfaces.

To test the usefulness of this thin film system for the study of biological macromolecules or membranes the protein Ferritin (from horse spleen, purchased from Aldrich, 100 mg/ml ferritin in a NaCl 0.15 M solution) was adsorbed to the aqueous-aqueous interface between the thin PEG-rich film and the salt-rich subphase (see Fig. 11). To reduce the salt concentration to make the solution more favorable to the protein, we used a solution of PEG (8000MW):K_2HPO_4: KH_2PO_4:H_2O (5:9:6:80 wt %) at 32°C. Analysis of the x-ray reflectivity data from the thin film without the protein (shown in the bottom of Fig. 12) reveals that the film is 51 Å thick with an interfacial width between the PEG-rich phase and the vapor of $\sigma_{PEG\text{-}vapor} = 3.2$ Å and an

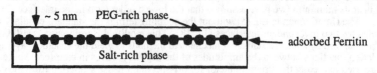

Fig. 11 Cartoon of adsorbed protein Ferritin at aqueous-aqueous interface.

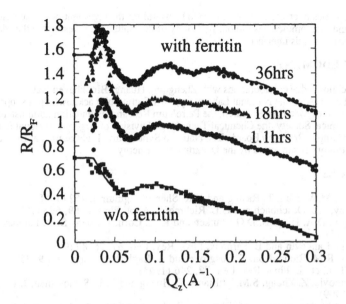

Fig. 12 X-ray reflectivity measurements show the evolution in time of the ferritin layer at the liquid-liquid interface. The lowest curve is the reflectivity before the ferritin is added. The solid line in the upper curve is a fit to a model of ferritin sandwiched between the PEG-rich and salt-rich phases. The curves are offset for clarity.

interfacial width between the salt-rich and PEG-rich phases of $\sigma_{\text{salt-PEG}} = 15.9$ Å. The protein is added to the interface by injecting 0.8 µl of a ferritin solution into the salt-rich subphase (the solution was prepared by taking 20 µl of the as purchased ferritin in NaCl and diluting that in 5 ml of a 16 wt% PEG solution). Immediately after adding the protein solution a new peak is present in the reflectivity that indicates additional electron density at the interface (Fig. 12). Adsorption of ferritin at the interface continues until, after 36 hours, a well developed structure is present in the reflectivity. Our slab model analysis is consistent with the adsorption of ferritin at the liquid-liquid interface.

CONCLUSION

We have used x-ray reflectivity and diffuse scattering to study the molecular ordering at a variety of liquid-liquid interfaces. We have addressed the fundamental issue of ordering at pure water-alkane interfaces and demonstrated significant deviations from the predictions of capillary wave theory. This indicates the presence of molecular ordering of the alkanes at the interface. These effects are also currently unexplained by molecular dynamics simulations. We have also addressed the problem of characterizing the order of an equilibrium surfactant monolayer at the water-hexane interface. X-ray reflectivity was used to determine the thickness and electron density of the monolayer. This revealed that the monolayer is a solid at room temperature and undergoes a transition to a gas phase at higher temperatures. Off-specular diffuse scattering revealed that the solid monolayer breaks up into islands of solid phase in coexistence with the gas phase. As the temperature is increased all the solid islands are converted into a gas phase.

We also demonstrated that microscopically thin films can be formed with equilibrium aqueous biphase solutions. The thickness of the films is on the order of 4 - 5 nm. Capillary waves on the top and bottom surfaces of these films are correlated. The lower phase of this film is an aqueous-aqueous interface that may prove to be useful for the investigation of biological membranes and macromolecules. In our first study of this nature we investigated the adsorption of the protein ferritin to this interface.

ACKNOWLEDGMENTS

We acknowledge collaborations with Zhengqing Huang (Brookhaven National Laboratory) and David Chaiko (Argonne National Laboratory). MLS gratefully acknowledges support from the NSF Division of Materials Research, the Petroleum Research Foundation administered by the American Chemical Society, the Chemical Technology Division of Argonne National Laboratory, and the UIC Campus Research Board. The National Synchrotron Light Source at Brookhaven National Laboratory is supported by the Department of Energy.

REFERENCES

1. S.G. Grubb, M.W. Kim, T. Rasing and Y.R. Shen, Langmuir 4, 452 (1988).
2. J.C. Conboy, J.L. Daschbach and G.L. Richmond, Appl. Phys. A 59, 623 (1994).
3. R.R. Naujok, D.A. Higgins, D.G. Hanken and R.M. Corn, J. Chem. Soc., Faraday Trans. 91, 1411 (1995).
4. L.T. Lee, D. Langevin and B. Farnoux, Phys. Rev. Lett. 67, 2678 (1991).
5. J.S. Phipps, R.M. Richardson, T. Cosgrove and A. Eaglesham, Langmuir 9, 3530 (1993).
6. B.R. McClain, et al., Phys. Rev. Lett. 72, 246 (1994).
7. D.M. Mitrinovic, Z. Zhang, S.M. Williams, Z. Huang and M.L. Schlossman, J. Phys. Chem. 103, 1779 (1999).
8. Z. Zhang, D.M. Mitrinovic, S.M. Williams, Z. Huang and M.L. Schlossman, J. Chem. Phys. 110, 7421 (1999).
9. M.L. Schlossman, et al., Rev. Sci. Instrum. 68, 4372 (1997).
10. A.R.v. Buuren, S.-J. Marrink and H.J.C. Berendsen, J. Phys. Chem. 97, 9206 (1993).
11. P.S. Pershan, Far. Disc. Chem. Soc. 89, 231 (1990).
12. F.P. Buff, R.A. Lovett and F.H. Stillinger, Phys. Rev. Lett. 15, 621 (1965).
13. S.K. Sinha, E.B. Sirota, S. Garoff and H.B. Stanley, Phys. Rev. B 38, 2297 (1988).
14. J.S. Rowlinson and B. Widom, Molecular Theory of Capillarity (Clarendon Press, Oxford, 1982).
15. J.D. Weeks, J. Chem. Phys. 67, 3106 (1977).
16. A. Braslau, et al., Phys. Rev. Lett. 54, 114 (1985).
17. M.L. Schlossman, in Encyclopedia of Applied Physics (ed. Trigg, G.L.) (VCH Publishers, New York, 1997) pp. 311-336.
18. M.P. Gelfand and M.E. Fisher, Physica A 166, 1 (1990).
19. A. Goebel and K. Lunkenheimer, Langmuir 13, 369 (1997).
20. B.Y. Yue, C.M. Jackson, J.A.G. Taylor, J. Mingins and B.A. Pethica, J. Chem. Soc. Faraday I 72, 2685 (1976).
21. M.C. Messmer, J.C. Conboy and G.L. Richmond, J. Am. Chem. Soc. 117, 8039 (1995).
22. G.L. Richmond, Analytical Chemistry 69, 536A (1997).
23. Y. Hayami, A. Uemura, N. Ikeda, M. Aratono and K. Motomura, J. Coll. Int. Sci. 172, 142 (1995).
24. T. Takiue, et al., J. Phys. Chem. 100, 20122 (1996).
25. M. Born and E. Wolf, Principles of Optics (Pergamon Press, Oxford, 1980).
26. H. Schwickert, G. Strobl and M. Kimmig, J. Chem. Phys. 95, 2800 (1991).
27. H.W. Starkweather, Macromolecules 19, 1131 (1986).
28. R.S. Becker, J.A. Golovchenko and J.R. Patel, Phys. Rev. Lett. 50, 153 (1983).
29. S. Dietrich and H. Wagner, Z. Phys. B 56, 207 (1984).
30. Y. Yoneda, Phys. Rev. 131, 2010 (1963).
31. M.L. Schlossman and P.S. Pershan, in Light Scattering by Liquid Surfaces and Complementary Techniques (ed. Langevin, D.) (Marcel Dekker Inc., New York, 1992) pp. 365-403.
32. In collaboration with David Chaiko (Argonne National Laboratory).

THE INCORPORATION OF LUNG SURFACTANT SPECIFIC PROTEIN SP-B INTO LIPID MONOLAYERS AT THE AIR-FLUID INTERFACE: A GRAZING INCIDENCE X-RAY DIFFRACTION STUDY

K.Y.C. LEE*, J. MAJEWSKI**, T.L. KUHL§, P.B. HOWES¶, K. KJAER¶, M.M. LIPP§, A.J. WARING#, J.A. ZASADZINSKI§, G.S. SMITH**
*Department of Chemistry, The University of Chicago, Chicago, IL 60637 kayeelee@rainbow.uchicago.edu
**Manuel Lujan Jr. Neutron Scattering Center, Los Alamos National Laboratory, Los Alamos, NM 87545
§Department of Chemical Engineering, University of California, Santa Barbara, CA 93106-5080
¶Condensed Matter Physics and Chemistry Department, Risø National Laboratory, DK-4000, Roskilde, Denmark
#Department of Pediatrics, Martin Luther King Jr./Drew Medical Center and UCLA, Los Angeles, CA 90095

ABSTRACT

Grazing incidence x-ray diffraction (GIXD) measurements were performed to determine the effects of SP-B_{1-25}, the N-terminus peptide of lung surfactant specific protein SP-B, on the structure of palmitic acid (PA) monolayers. In-plane diffraction shows that the peptide fluidizes a portion of the monolayer, but does not affect the packing of the residual ordered phase. This implies that the peptide resides in the disordered phase, and that the ordered phase is essentially pure lipid. The quantitative insights afforded by this study lead to a better understanding of the lipid/protein interactions found in lung surfactant systems.

INTRODUCTION

Lung surfactant (LS), a complex mixture of lipids and proteins, lines the alveoli, and is responsible for the proper functioning of the lung [1]. LS works both by lowering the surface tension inside the lungs to reduce the work of breathing, and by stabilizing the alveoli through varying the surface tension as a function of alveolar volume. To accomplish this, the LS mixture must adsorb rapidly to the air-fluid interface of the alveoli after being secreted. Once at the interface, it must form a monolayer that can both achieve low surface tensions upon compression and vary the surface tension as a function of the alveolar radius. Insufficient levels of surfactant, due to either immaturity in premature infants, and disease or trauma in adults, can result in respiratory distress syndrome (RDS), a potentially lethal disease in both populations.

LS consists primarily of the saturated dipalmitoylphosphatidylcholine (DPPC), unsaturated phosphatidylcholines, along with significant amounts of unsaturated and anionic phospholipids such as phosphatidylglycerols (PGs), lesser amounts of anionic lipids such as palmitic acid (PA), and neutral components like cholesterol. LS also contains four lung surfactant-specific proteins, known as SP-A, -B, -C, and -D. Of the four, SP-B and SP-C are small, amphipathic proteins believed to be important for the surface activity of LS. Although the complete roles of SP-B and SP-C are not yet fully understood, they are known to be essential for the proper functioning of LS in vivo and in replacement surfactants for the treatment of RDS. SP-B in particular has been shown to greatly increase the activity of LS lipids both in vitro and in vivo. It has also been demonstrated that simple peptide sequences based on the amino terminus of SP-B possess the full activity of the native protein [2-5].

177

Pure DPPC can form monolayers that attain surface tensions near zero values on compression; in the context of LS, this feature of DPPC definitely helps reduce the work of breathing. However, DPPC adsorbs and respreads slowly as a monolayer under physiological conditions, rendering it by itself not an ideal LS candidate [1, 6-8]. The unsaturated and anionic lipids present in natural LS and added to many replacement surfactants are believed to enhance the adsorption and respreading of DPPC [6-8]. PA, for example, is one of the three compounds added to exogenous surfactant in Survanta (8.5 w/w; Ross Laboratory, Columbus, Ohio) and Surfactant TA (8.5% w/w; Tokyo Tanabe) used to treat premature infants with neonatal RDS. PA by itself collapses at relatively low surface pressures, however, the addition of full length SP-B or its amino-terminus peptide, SP-B$_{1-25}$, to monolayers of PA results in much higher monolayer collapse pressures (lower surface tension values) than those of either PA or SP-B alone. This suggests that a synergistic effect between PA and SP-B may result in the retention of both the lipid and the protein in the primarily DPPC monolayer upon compression [9].

To determine how the SP-B$_{1-25}$ protein incorporates itself into the PA monolayer, we have carried out a series of GIXD experiments on both pure PA and mixed PA/SP-B$_{1-25}$ monolayers at the air-water interface at different surface pressures. Our findings give us the first quantitative information on the effect of SP-B$_{1-25}$ on the packing of the PA monolayer, and helps to pinpoint the location of the peptide in the lipid matrix.

EXPERIMENT

Palmitic acid, PA, (Sigma Chemical Co., St. Louis, MO; > 99% pure) was used as obtained. All subphases were prepared using 18.2 MΩ•cm Millipore water obtained from a Milli-Q UV Plus system (Millipore Corp.). Stock spreading solutions were made with either pure chloroform (for PA) or 4:1 vol/vol chloroform-methanol (for SP-B$_{1-25}$). Spreading solution were made by mixing aliquots of the stock solutions to obtain the desired concentration.

The synthetic peptide based on the first 25 amino acids of the NH$_2$-terminal sequence of SP-B, SP-B$_{1-25}$, shown in Fig. 1, was synthesized by the solid state method using a Fmoc strategy with an Applied Biosystems 431A peptide synthesizer. Details of the synthesis have been published elsewhere [10].

$$1 \qquad\qquad + \quad ++ \qquad\qquad +25$$

NH2–FPIPLPYCWLCRALIKRIQAMIPKG-COOH

Figure 1. Amino acid sequence of the N-terminal truncated model peptide SP-$_{B1-25}$, with the positively charged residues indicated with a plus sign.

All synchrotron x-ray measurements were performed with the liquid surface diffractometer [11, 12] at the BW1 (undulator) beam line [13] at HASYLAB, DESY (Hamburg, Germany). A thermostated Langmuir trough, equipped with a Wilhelmy balance for measuring the surface pressure (π), and a barrier for surface pressure control, was mounted on the diffractometer. In a typical experiment, a monolayer was first spread using a microsyringe at the desired temperature. At least 30 minutes was allowed for complete solvent evaporation before the two-dimensional film was compressed to the desired surface pressure. The film was then held at this surface pressure throughout the experiment. The trough was enclosed in a sealed, helium-filled canister where the oxygen level was constantly monitored.

The synchrotron x-ray beam was monochromated to a wavelength of $\lambda \sim 1.303$ Å by Laue reflection from a Be (200) monochromator crystal. By tilting the reflecting planes out of the vertical plane, the monochromatic beam could be deflected down to yield a glancing angle with the horizontal liquid surface. The x-ray beam was adjusted to strike the surface at an incident angle of $\alpha_i = 0.11° = 0.85\ \alpha_c$, where α_c is the critical angle for total external reflection. At this angle the incident wave is totally reflected, while the refracted wave becomes evanescent, traveling along the liquid surface. Such a configuration maximizes surface sensitivity [14]. The dimensions of the incoming x-ray beam footprint on the liquid surface were approximately 5 mm X 50 mm or 1 mm X 50 mm.

In three-dimensional (3D) crystals, Bragg diffraction takes place only when the scattering vector Q coincides with $\{h,k,l\}$ points of the reciprocal 3D lattice, giving rise to Bragg spots (h, k, l are the Miller indices). In our two-dimensional (2D) systems and at surface pressures of interest, the PA monolayers are a mosaic of 2D crystals with random orientation about the direction normal to the subphase, and can therefore be described as 2D powders. Due to the lack of a vertical crystalline repeat, there is no restriction on the scattering vector component Q_z along the direction normal to the crystal: Bragg scattering from a 2D crystal extends as continuous Bragg rods through the reciprocal space [11, 15, 16]. The scattered intensity is measured by scanning over a range of horizontal scattering vectors $Q_{xy} \sim (4\pi/\lambda)\sin(2\theta_{xy}/2)$, where $2\theta_{xy}$ is the angle between the incident and diffracted beam. The Bragg peaks, resolved in the Q_{xy}-direction, are obtained by integrating the scattered intensity over all the channels along the Q_z-direction in the position sensitive detector (PSD). Conversely, the Bragg rod profiles are obtained by integrating, after background subtraction, for each PSD channel, the scattered intensity across a Bragg peak. The angular positions of the Bragg peaks allow the determination of the spacings d = $2\pi/Q_{xy}$ for the 2D lattice. From the line shapes of the peaks, it is possible to determine the 2D crystalline coherence length, L (the average distance in the direction of the reciprocal lattice vector Q_{xy} over which crystallinity extends). The intensity distribution along the Bragg rod can be analyzed to determine the direction and magnitude of the molecular tilt.

RESULTS

Experiments were carried out at 16 °C on a pure water subphase. On pure water, 16 °C is well below the triple point of a PA monolayer [10, 17]. The triple point denotes the unique temperature and surface pressure in the pure component phase diagram where the condensed (LC), liquid-expanded (LE) and gaseous (G) phase all coexist in equilibrium. As the LE phase does not exist below the triple point temperature, a LC film is guaranteed at low non-zero surface pressures. These conditions thus made it possible to probe any fluidizing effect of the peptide on the PA monolayer. GIXD measurements were made on pure PA as well as mixed PA/20 wt% SP-B$_{1-25}$ monolayers, with the films held at different surface pressures. In all cases, the pure PA experiments gave a "baseline" for comparison with the mixed monolayer.

The diffraction pattern obtained for pure PA monolayers at $\pi = 15$ mN/m and $\pi = 30$ mN/m are shown in Fig. 2a. At $\pi = 15$ mN/m, two Bragg peaks are observed at $Q_{xy} = 1.45$ Å^{-1} and $Q_{xy} = 1.50$ Å^{-1}, the corresponding coherence lengths are 160 Å and 450 Å, respectively. The observation of precisely two Bragg peaks in the 2D powder pattern is indicative of a rectangular unit cell [18]. Previous work on fatty acid monolayers has also shown that in the rectangular S, $L2$, $L2'$ phases these monolayers yield two in-plane reflections: $\{1,1\}_{rect}$ and $\{0,2\}_{rect}$ [19-22]. The integrated intensity of the Bragg peak at $Q_{xy} = 1.45$ Å^{-1} peak is roughly twice that of the Q_{xy} = 1.50 Å^{-1} peak. This higher intensity results from coincident $\{1,1\}_{rect}$ and $\{1,-1\}_{rect}$ reflections, and thus leads to the assignment of the $\{1,1\}_{rect}$ reflection to the $Q_{xy} = 1.45$ Å^{-1} peak and the $\{0,2\}_{rect}$ reflection to the $Q_{xy} = 1.50$ Å^{-1} peak. The calculated d-spacings, $d_{11} = 4.33$ Å ($d_{xy} =$

$2\pi/Q_{xy}$) and $d_{02} = 4.19$ Å, give rise to a rectangular unit cell with axes $|a| = 5.06$ Å $(1/|a|^2 = (1/d_{11}^2 - 1/(2d_{02})^2)$ and $|b| = 8.38$ Å, and an area per chain, A_{15}, of 21.2 Å² (two molecules per unit cell of area = 42.4 Å²).

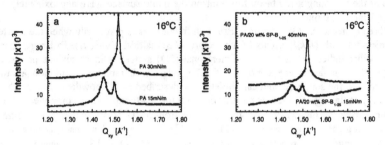

Figure 2. Bragg peaks from GIXD on pure water at 16 °C of (a) a pure PA film at 15 mN/m and 30 mN/m; (b) a mixed PA/20 wt% SP-B$_{1-25}$ film at 15 mN/m and 40 mN/m. For clarity, the high pressure data have been offset vertically.

The $\{0,2\}_{\text{rect}}$ Bragg rod has its maximum in the intensity profile at $Q_z = 0$ Å^{-1} (Bragg rod profiles not shown), which indicates that the molecular axis lies in a plane perpendicular to the b axis. We analyzed the $\{1,1\}_{\text{rect}}$ and $\{0,2\}_{\text{rect}}$ Bragg rod intensity profiles (data not shown) by approximating the PA molecule by a cylinder with constant electron density [11]. Our results show that the molecule has a tilt of ~ 21° towards the nearest neighbor (along unit vector a) direction, and the effective thickness (thickness projected on the surface normal) of the monolayer is 20.5 Å.

When the system is compressed to 30 mN/m, the two Bragg peaks at 15 mN/m collapse into one (see Fig. 2a), indicating a transition to a hexagonal lattice. The scattering vector for the single Bragg peak, indexed as $\{1,0\}_{\text{hex}}$ is $Q_{xy} = 1.52$ Å^{-1}, with an average coherence length of 470 Å, which is a significant increase over the coherence length of the $\{1,1\}_{\text{rect}}$ reflections at 15 mN/m. The hexagonal lattice has a d-spacing of 4.13 Å and the unit cell dimension (an intermolecular distance) $a_H = d/\cos 30° = 4.77$ Å. The interfacial area per chain, A_{30}, is reduced to 19.7 Å². The position of the intensity maximum of the Bragg rod ($Q_z \sim 0$ Å^{-1}; data not shown) implies ~ 0° tilt of the molecules at 30 mN/m. Using the areas obtained for surface pressures of 15 mN/m (A_{15}) and 30 mN/m (A_{30}) we find once again the tilt angle of the chain at 15 mN/m, t_{15} = arccos (A_{30}/A_{15}) ~ 21°. This agrees well with that obtained from fits to the $\{1,1\}_{\text{rect}}$ and $\{0,2\}_{\text{rect}}$ Bragg rods above. The quantitative information shows that an increase in surface pressure primarily causes a decrease in the area per PA molecule due to a decrease in molecular tilt, while maintaining the molecular packing in the plane normal to the molecules.

The Bragg peaks obtained for the lipid/peptide mixed monolayers at 15 mN/m and 40 mN/m are shown in Fig. 2b. At 15 mN/m, two Bragg peaks resulting from $\{1,1\}_{\text{rect}}$ and $\{0,2\}_{\text{rect}}$ reflections are observed at $Q_{11} = 1.45$ Å^{-1} and $Q_{02} = 1.50$ Å^{-1}, with coherence lengths $L_{11} = 130$ Å and $L_{02} = 425$ Å. The d-spacings have values $d_{11} = 4.33$ Å and $d_{02} = 4.20$ Å, indicating a rectangular unit cell with axes $|a| = 5.05$ Å and $|b| = 8.40$ Å with an interfacial area per molecule of 21.2 Å². Similar to the pure PA case, the $\{0,2\}_{\text{rect}}$ Bragg rod has its maximum intensity at $Q_z \sim 0$

Å^{-1}, and the profile of the $\{1,1\}_{rect}$ reflection gives a tilt angle of 20.4° towards the nearest neighbor (along unit vector a) direction. Comparing these results with those for pure PA, it is apparent that the presence of SP-B$_{1-25}$ does not affect the lipid packing of the condensed phase. The unit cell, the coherence lengths, and the molecular tilt found in the condensed phase of PA are all preserved in this mixed system. This suggests that the peptide is completely excluded from the condensed region of the film, consistent with fluorescence imaging [10, 17]. As a PA monolayer at 16 °C is well below the triple point and should be in a pure LC phase at 15 mN/m, our results indicate that incorporating the peptide into the monolayer creates a disordered phase whose structure cannot be probed by the GIXD techniques.

Corroborative evidence of this disordered phase can be found by the drop (by a factor of *ca.* 2.1) in integrated intensity observed when SP-B$_{1-25}$ is present (see Figs. 2a and 2b). Since GIXD is only sensitive to the ordered phase, this decrease in scattering intensity suggests that a disordered phase occupies *ca.* 50 percent of the monolayer surface area, which is qualitatively consistent with our fluorescence microscopy findings [10, 17].

At 40 mN/m, the mixed film exhibits a single Bragg peak at $Q_{xy} = 1.52$ Å^{-1} with a coherence length of 450 Å. As was observed without the peptide, the condensed phase of the monolayer assumes a hexagonal packing with $\sim 0°$ molecular tilt. The d-spacing is 4.13 Å, with an intermolecular separation of 4.77 Å and an area per molecule of 19.8 Å^2. These lattice parameters are identical (within experimental errors) to the pure PA monolayers at 30 mN/m. This implies that the condensed phase of the film at this elevated pressure is only made up of PA molecules. A direct comparison at 40 mN/m is not possible because the pure PA film collapses below 40 mN/m. Unlike the low surface pressure case, the integrated intensity found in this case is only slightly lower than that in pure PA at 30 mN/m, the ratio being *ca.* 1.3. It should be noted that the two cases are at different pressures (30 mN/m without protein; 40 mN/m with protein), that may also account for the smaller decrease in the scattering intensity. Nonetheless, the smaller decrease is consistent with fluorescence images that show a smaller fluid phase at higher surface pressures [10, 23].

CONCLUSIONS

GIXD measurements were used to investigate the influence of the truncated LS peptide, SP-B$_{1-25}$, on the packing of PA monolayers at the air-water interface. From the GIXD we infer that on pure water subphase, the peptide is incorporated into the disordered phase of the PA monolayer. The insertion of the peptide leads to a reduced amount of ordered phase in the film, as reflected by a decrease in scattering intensity in our GIXD data. This reduction in scattering intensity is particularly apparent at 16 °C. Below the triple point of the PA film, a disordered phase does not exist at non-zero surface pressures in pure PA films; the disordered phase that occurs when peptide is present points to the fluidizing capabilities of SP-B$_{1-25}$. This finding corroborates our fluorescence microscopy data, which show that a disordered phase is created when SP-B$_{1-25}$ is present in the PA monolayer.[10, 17]

Although SP-B$_{1-25}$ induces a disordered phase in the monolayer, our GIXD data show that the peptide does not affect the molecular packing of the ordered phase. Within experimental errors, films with and without SP-B$_{1-25}$ have identical packing parameters. This suggests that the peptide is completely excluded from the ordered phase of the film, and hence must reside in the disordered portion of the monolayer. It should be pointed out that the lattice parameters obtained here for PA and the mixed PA/SP-B$_{1-25}$ monolayers are similar to those found in longer chain fatty acids.

ACKNOWLEDGMENTS

We gratefully acknowledge beam time at HASYLAB at DESY, Hamburg, Germany, and funding by the programs DanSync (Denmark) and TMR of the European Community (Contract ERBFMGECT950059). K.Y.C.L. is grateful for the support from the Camille and Henry Dreyfus New Faculty Award (NF-98-048), the March of Dimes Basil O'Connor Starter Scholar Research Award (5-FY98-0728), the Searle Scholars Program/The Chicago Community Trust (99-C-105), and the American Lung Association (RG-085-N). J.A.Z. and M.M.L. were supported by NIH Grant HL51177. A.J.W. was supported by NIH Grant HL55534. The Manuel Lujan Jr., Neutron Scattering Center is a national user facility funded by the United States Department of Energy, Office of Basic Energy Sciences - Materials Science, under contract number W-7405-ENG-36 with the University of California. This work was also supported in part by the MRSEC Program of the National Science Foundation under Award Numbers DMR-9808595 (The University of Chicago) and DMR96-32716 (University of California, Santa Barbara).

REFERENCES

1. D. L. Shapiro and R. H. Notter, *Surfactant Replacement Therapy*, Alan R. Liss, New York, 1989.
2. A. Waring, W. Taeusch, R. Bruni, J. Amirkhanian, B. Fan, R. Stevens and J. Young, Peptide Research **2**, p. 308-313 (1989).
3. A. Takahashi, A. Waring, J. Amirkhanian, B. Fan and W. Taeusch, Biochim. Biophys. Acta **1044**, p. 43-49 (1990).
4. L. M. Gordon, S. Horvath, M. Longo, J. A. Zasadzinski, H. W. Taeusch, K. Faull, C. Leung and A. J. Waring, Protein Sci. **5**, p. 1662-1675 (1996).
5. M. M. Lipp, K. Y. C. Lee, A. J. Waring and J. A. Zasadzinski, Science **273**, p. 1196-1199 (1996).
6. N. Mathialagan and F. Possmayer, Biochim. Biophys. Acta **1045**, p. 121-127 (1990).
7. A. Cockshutt, D. Absolom and F. Possmayer, Biochim. Biophys. Acta **1085**, p. 248-256 (1991).
8. M. Longo, A. Waring and J. Zasadzinski, Biophys. J. **63**, p. 760-773 (1992).
9. M. Longo, A. Bisagno and J. A. Zasadzinski, Science **261**, p. 453-456 (1993).
10. M. M. Lipp, K. Y. C. Lee, J. A. Zasadzinski and A. J. Waring, Biophys. J. **72**, p. 2783-2804 (1997).
11. J. Als-Nielsen and K. Kjaer in *Phase Transitions in Soft Condensed Matter*, edited by T. Riste and D. Sherrington (Plenum, New York, 1989), p. 113-138.
12. J. Majewski, R. Popovitz-Biro, W. Bouwman, K. Kjaer, J. Als-Nielsen, M. Lahav and L. Leiserowitz, Chem. Eur. J. **1**, p. 302-309 (1995).
13. R. Frahm, J. Weigelt, G. Meyer and G. Materlik, Rev. Sci. Instrum. **66**, p. 1677-1680 (1995).
14. P. Eisenberger and W. C. Marra, Phys. Rev. Lett. **46**, p. 1081 (1981).
15. J. Als-Nielsen, D. Jacquemain, K. Kjaer, F. Leveiller, M. Lahav and L. Leiserowitz, Phys. Rep. **246**, p. 251-313 (1994).
16. K. Kjaer, Physica B **198**, p. 100-109 (1994).
17. M. M. Lipp, K. Y. C. Lee, J. A. Zasadzinski and A. J. Waring, Science **273**, p. 1196-1199 (1996).
18. D. Jacquemain, F. Leveiller, S. Weinbach, M. Lahav, L. Leiserowitz, K. Kjaer and J. Als-Nielsen, J. Am. Chem. Soc. **113**, p. 7684 (1991).
19. K. Kjaer, J. Als-Nielsen, C. A. Helm, P. Tippman-Krayer and H. Möhwald, J. Phys. Chem. **93**, p. 3200-3206 (1989).
20. B. Lin, M. C. Shih, T. M. Bohanon, G. E. Ice and P. Dutta, Phys. Rev. Lett **65**, p. 191-194 (1990).
21. R. M. Kenn, C. Böhm, A. M. Bibo, I. R. Peterson, H.Möhwald, J. Als-Nielsen and K. Kjaer, J. Phys. Chem. **95**, p. 2092-2097 (1991).
22. V. M. Kaganer, H. Möhwald and P. Dutta, Rev. Modern Phys. **71**, p. 779-819 (1999).
23. K. Y. C. Lee, M. M. Lipp, J. A. Zasadzinski and A. J. Waring, Colloids and Surfaces A **28**, p. 225-242 (1997).

RESONANT X-RAY SCATTERING FROM THE SURFACE OF A DILUTE LIQUID Hg–Au ALLOY

E. DiMasi,[1*] H. Tostmann,[2†] B. M. Ocko,[1] P. Huber,[2] O. G. Shpyrko,[2] P. S. Pershan,[2] M. Deutsch[3], and L. E. Berman[4]

[1]Department of Physics, Brookhaven National Laboratory, Upton NY 11973
[2]Div. of Eng. and Appl. Sci. and Dept. of Physics, Harvard University, Cambridge MA 02138
[3]Department of Physics, Bar-Ilan University, Ramat-Gan 52100, Israel
[4]National Synchrotron Light Source, Brookhaven National Laboratory, Upton NY 11973
*Corresponding author: dimasi@bnl.gov
†Present address: Department of Chemistry, University of Florida, Gainesville FL 32611

ABSTRACT

We present the first resonant x-ray reflectivity measurements from a liquid surface. The surface structure of the liquid Hg-Au alloy system just beyond the solubility limit of 0.14at% Au in Hg had previously been shown to exhibit a unique surface phase characterized by a low-density surface region with a complicated temperature dependence. In this paper we present reflectivity measurements near the Au L_{III} edge, for 0.2at% Au in Hg at room temperature. The data are consistent with a concentration of Au in the surface region that can be no larger than about 30at%. These results rule out previous suggestions that pure Au layers segregate at the alloy surface.

INTRODUCTION

Surfaces play an important role in nucleating bulk phases in solid alloys, and because of the reduced atomic coordination, surfaces often exhibit structure and phase behavior distinct from that of the bulk. Liquid metal alloys can be expected to exhibit an even richer variety of surface phases: since the particles are not constrained to lattice sites, the composition at the surface can vary dramatically from that of the bulk. With the application of surface sensitive x-ray scattering techniques to the liquid metal–vapor interface, a number of liquid metal alloys have in fact been shown to exhibit interesting structural phenomena, such as surface segregation and microscopic precursors to wetting, that previously could be only theoretically predicted or indirectly inferred from experiment.[1,2]

Especially interesting behavior was found for liquid Hg-Au alloys near the room temperature solubility limit of 0.14at% Au in Hg.[3] The complicated phase behavior of Hg-Au alloys has long been known, from bulk structural studies,[4] and from studies at the solid amalgam surface.[5] Further motivation for studying liquid Hg alloys comes from experiments in which metal impurities were observed to affect the activation energies of reactions catalyzed by the liquid Hg surface.[6] It is not known whether such effects are due to changes in the electronic properties of liquid Hg, modification of the surface structure, or the formation of intermetallic phases at the surface.

Our study of the x-ray reflectivity of the Hg-Au system at temperatures between −39°C and +25°C revealed two distinct regions of surface phase behavior, depending on whether the bulk solubility limit[7] of Au in Hg is exceeded. At high T and low Au concentrations, surface layering similar to that of pure Hg is observed (Fig. 1, o).[8] By contrast, at low T and comparatively higher Au concentrations, reflectivity measurements revealed a more

183

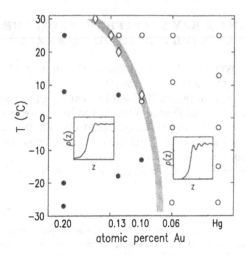

Figure 1: Hg-Au surface phase diagram established in Ref. 3. ◇: Measured solubility limit from Ref. 7 (shaded band is a guide for the eye). ○: Compositions and temperatures for which reflectivity indicates simple oscillatory real-space density profiles, as the inset on the right. •: Compositions and temperatures exhibiting less well defined surface layering and a low-density surface region (inset, left).

complicated surface phase, consistent with a low-density layer appearing at the interface (Fig. 1, •).

Since the new phase was found only for Au-rich samples, it is natural to ask how the Au composition in the surface region differs from that of the bulk. Unfortunately, this elemental specificity is difficult to obtain experimentally. X-ray reflectivity, in which scattered x-ray intensity is measured as a function of momentum transfer q_z normal to the surface, is a sensitive probe of the surface-normal electron density distribution $\rho(z)$, and in the kinematic limit may be taken to be proportional to the Fresnel reflectivity R_F of a homogeneous surface:[9]

$$R(q_z) = R_F \left| (1/\rho_\infty) \int_{-\infty}^{\infty} (\partial\rho/\partial z) \exp(iq_z z) \, dz \right|^2 , \tag{1}$$

where ρ_∞ is the density of the bulk. The electron density variations that appear as modulations in the reflectivity may result either from a compositional change or a mass density change. Thus, direct determination of composition from reflectivity is ambiguous at best. Complementary information is often obtained from electron spectroscopy techniques requiring ultra high vacuum conditions, which cannot be used with high vapor pressure liquid Hg alloys.

These disadvantages can be overcome with the application of resonant x-ray scattering. The apparent electron density of a scattering atom depends on the scattering form factor, which can be taken as $f(q) + f'(E) \approx Z + f'(E)$, where absorption and the weak q dependence have been neglected. When the x-ray energy is tuned to an absorption edge of a scattering atom, the magnitude of f' becomes appreciable. The resulting change in the contrast between unlike atoms provides composition information for the model structure.[10]

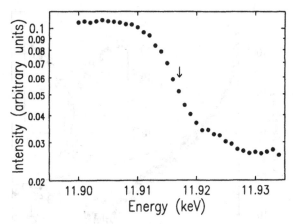

Figure 2: Transmission of x-rays through an Au foil as a function of energy near the Au L_{III} edge (•). The arrow indicates the energy selected for reflectivity measurements.

The starting point for the analysis we will use for this report is a simple real-space density model of the region within 5 Å of the surface:

$$\frac{\rho(z)}{\rho_\infty} = \frac{1}{2}\mathrm{erf}\left(\frac{z}{\sigma_0\sqrt{2}}\right) + \frac{\rho_1/\rho_\infty}{2}\left[\mathrm{erf}\left(\frac{z-z_1}{\sigma_1\sqrt{2}}\right) - \mathrm{erf}\left(\frac{z}{\sigma_0\sqrt{2}}\right)\right] + \frac{h_g/\sigma_g}{\sqrt{2\pi}}\exp\left[\frac{-(z-z_g)^2}{2\sigma_g^2}\right].\ (2)$$

The three terms define, respectively, the liquid–vapor interface at $z = 0$, a region of density ρ_1 at position z_1, and a broad density tail further towards the vapor region. Because of the very small bulk concentration of Au, the reflectivity is calculated as though the system is composed entirely of Hg. A more complete discussion of the modeling is given in Ref. 3.

EXPERIMENT

Au powder was dissolved in pure liquid Hg for a nominal composition of 0.2at% Au. The samples were contained in an ultra high vacuum chamber, with the pressure determined by the Hg vapor pressure at room temperature, with low partial pressures of oxygen and water. The experiment was performed at beamline X25 at the National Synchrotron Light Source using a specialized BNL-designed liquid surface spectrometer. Details of the sample preparation and measurement technique have been given previously,[3] except that for this experiment, a vertically deflecting Si(111) double crystal monochromator upstream of the spectrometer was used to improve the energy resolution to ~ 10 eV.

For the present work, we compare reflectivity measured at 11 keV to measurements made at the Au L_{III} edge at 11.919 keV. The energy choice was based on transmission of the direct beam through an Au foil, shown in Fig. 2.[11] At the inflection point indicated by the arrow, f'_{Au} has its minimum value of -19 electrons, compared to about -6 electrons in the 10–11 keV region.[10] Here the scattering from Au (with $Z=79$) is reduced to about 85% of its 11 keV value. Because of the very small bulk concentration of Au, no Au fluorescence was detected at any energy employed.

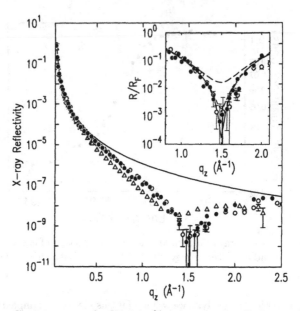

Figure 3: X-ray reflectivity vs. q_z for Hg 0.2at% Au reported in Ref. 3 at 10 keV ($\triangle$), and for the present sample at 11 keV ($\circ$) and 11.919 keV ($\bullet$). Solid line: calculated Fresnel reflectivity R_F of a flat Hg surface. Inset: reflectivity data normalized to R_F at 11 keV ($\circ$) and 11.92 keV ($\bullet$). Lines are calculated from models as described in the text.

RESULTS

X-ray reflectivity obtained from the Hg 0.2at% Au sample at 11 keV (Fig. 3, $\circ$) is similar to that obtained on a sample of approximately the same composition measured previously at 10 keV (Fig. 3, $\triangle$). Compared to the Fresnel reflectivity R_F calculated for a flat Hg surface (Fig. 3, solid line), the experimental data have deep minima for q_z in the range 1.3–1.5 Å^{-1}. This destructive interference comes from a region near the surface where the density is approximately half that of the bulk liquid. Since the previous study showed that the details of the surface structure are very sensitive to Au concentration in this region, the differences between the two samples can be ascribed to a slight difference in the Au concentration.

The normalized reflectivity data have been described by the model profile shown in Fig. 4 (solid line), which produces the calculated reflectivity shown as a solid line in the inset of Fig. 3. The local minimum in the density between the surface layer and the bulk, at $z = -0.5$ Å, is due to the rather small roughnesses used in this simple model, which are required in order to produce the relatively large reflectivity we observe for $q_z > 1.8$ Å^{-1}.

Essentially no difference is seen in the data taken at the Au L_{III} edge. This rules out models having a very high Au composition at the surface. To illustrate the sensitivity of the measurement to the Au composition in the density profiles, we have calculated reflectivity curves for two additional cases. If the entire surface region ($z < 0$) were composed of Au,

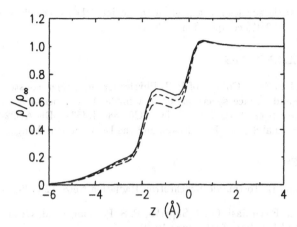

Figure 4: Surface density profiles calculated from the model in Eq. 2. (—): "Pure Hg" profile: $\sigma_0 = \sigma_1 = 0.25$, $z_1 = -1.96$, $\rho_1 = 0.51$, $h_g = 0.8$, $\sigma_g = 1.5$, and $z_g = -2$. (- - -): "40% Au" surface layer, where the apparent density of the step region is reduced by 6% ($\rho_1 = 0.47$, $h_g = 0.75$). (- - -): "100% Au" surface layer, where the apparent density of the step region is reduced by 15% ($\rho_1 = 0.43$, $h_g = 0.68$).

the electron density contributing to the scattering would be reduced by 15% at 11.919 keV. This density profile (Fig. 4, long dashed line) produces a reflectivity curve with a much less pronounced minimum at 1.5 Å^{-1} Fig. 3 (inset, long dashed line). A similar calculation assuming 40% Au in the surface layers, indicated by short dashed lines, lies just outside the experimental error.

A surface composition of 30% Au is compatible with our data (model not shown), and forms an upper bound on the Au concentration, assuming that the energy was tuned precisely to the minimum in f'. To reduce the change in f' by a factor of two, yielding 60% as the upper bound for Au concentration, an error in the energy of > 30 eV would be required.[10] This error is large compared to the width of the energy scan shown in Fig. 2. Further confidence in our experimental sensitivity comes from a subsequent experiment we performed on a liquid Bi-In alloy at the Bi L_{III} edge, where large resonant effects were observed.[12] Our results rule out models in which essentially pure Au layers segregate at the surface of the alloy and produce a coherent contribution to the reflectivity.

CONCLUSIONS

We present the first resonant x-ray scattering measurements from a liquid surface. The surface structure of the liquid Hg-Au alloy system just beyond the solubility limit of 0.14at% Au in Hg had previously been shown to exhibit a unique Au-rich surface phase characterized by a low-density surface region with a complicated temperature dependence. In this paper we present reflectivity measurements near the Au L_{III} edge. The data are consistent with a concentration of Au in the surface region that can be no larger than about 30at%. Our results rule out models in which the surface region consists of pure Au. Since a concentration of $\leq 30at\%$ Au at the surface is consistent with the data, the surface phase may still be quite different from the $0.2at\%$ composition of the bulk.

ACKNOWLEDGMENTS

We are indebted to Scott Coburn of BNL Physics for the design, construction, and preparation of the liquid surface spectrometer that made these experiments possible. We acknowledge support from the U.S. DOE (DE-FG02-88-ER45379, DE-AC02-98CH10886), the U.S.–Israel Binational Science Foundation, and the Deutsche Forschungsgemeinschaft.

REFERENCES

1. E. DiMasi and H. Tostmann, Synchrotron Radiation News **12** (1999) 41.

2. H. Tostmann, E. DiMasi, O. G. Shpyrko, P. S. Pershan, B. M. Ocko, and M. Deutsch, submitted to Phys. Rev. Lett. (June 1999).

3. E. DiMasi, H. Tostmann, B. M. Ocko, P. S. Pershan, and M. Deutsch, J. Phys. Chem. B **103** (1999) 9952.

4. C. Rolfe and W. Hume-Rothery, J. Less-Common Met. **13** (1967) 1.

5. X. M. Yang, K. Tonami, L. A. Tagahara, K. Hashimoto, Y. Wei, and A. Fujishima, Surface Science **319** (1994) L17; C. Battistoni, E. Bemporad, A. Galdikas, S. Kačiulis, G. Mattogno, S. Mickevičius, and V. Olevano, Appl. Surf. Science **103** (1996) 107; R. Nowakowski, T. Kobiela, Z. Wolfram, and R. Duś, Applied Surface Science **115** (1997) 217; J. Li and H. D. Abruña, J. Phys. Chem. B **101** (1997) 2907.

6. G.-M. Schwab, Ber. Bunsenges. **80** (1976) 746.

7. A. S. Kertes, Ed. *Solubility Data Series*, Vol. 25 "Metals in Mercury", Pergamon Press, New York, 1986, p. 378; C. Guminski, J. Less-Common Metals, **168** (1991) 329.

8. E. DiMasi, H. Tostmann, B. M. Ocko, P. S. Pershan, and M. Deutsch, Phys. Rev. B **58** (1998) R13419, and references therein.

9. A. Braslau, P. S. Pershan, G. Swislow, B. M. Ocko, and J. Als-Nielsen, Phys. Rev. A **38** (1988) 2457.

10. G. Materlik, C. J. Sparks, and K. Fischer, eds. *Resonant Anomalous X-ray Scattering: Theory and Applications*, North-Holland, Amsterdam, 1994, p. 47.

11. Error in the monochromator calibration causes the position of the inflection point to appear as 11.917 keV in the energy scan.

12. E. DiMasi *et al*, to be published.

Experimental Investigations of the Interaction of SO₂ with MgO

ANDREA FREITAG, J. A. RODRIGUEZ AND J. Z. LARESE
Chemistry Department, Brookhaven National Laboratory, Upton, NY 11973-5000

Abstract

High resolution adsorption isotherms, temperature programmed desorption (TPD), x-ray diffraction (XRD) and x-ray absorption near edge spectroscopy (XANES) methods were used to investigate the interaction of SO_2 with high quality MgO powders. The results of these investigations indicate that when SO_2 is deposited on MgO in monolayer quantities at temperatures near 100K both SO_3 and SO_4 species form that are not removed by simply pumping on the pre-dosed samples at room temperature. TPD and XANES studies indicate that heating of pre-dosed MgO samples to temperatures above 350 °C is required for full removal of the SO_3/SO_4 species. XANES measurements made as a function of film thickness indicate for coverages near monolayer completion that the SO_4 species form first.

Introduction

Sulfur dioxide is one of the main pollutants released into the atmosphere as a result of volcanic activity, and the burning of sulfur bearing fossil fuels in automobile engines, industrial complexes, power plants and households. The subsequent interaction of the sulfur dioxide with air and atmospheric moisture results in the formation of "acid rain" leading to the corrosion of metals and degradation of stone buildings and statuary [1]. To diminish the environmental effects of sulfur dioxide emissions one must either dissociate or remove the SO_2 from the effluent. MgO and CaO are two materials that are widely used as commercial scrubbers in industry[2]. We describe our recent investigations using adsorption isotherms, TPD, XANES and XRD investigations of the interaction of SO_2 with MgO.

Experimental Results

A novel process recently developed in our laboratory [3] was used to produce the MgO powders used in the present experiments. Transmission electron microscopy (TEM) indicates that the MgO consists of uniform cubic particles, approximately 2000Å on a side with predominantly the (100) surface exposed(see Fig.1). The quality of the substrates was judged using a methane adsorption isotherm performed volumetrically at 77 K, a typical example of which is shown in Fig.2. The isotherms are performed using an automated isotherm apparatus which has been described elsewhere[4]. Typical adsorption areas of about 10 m²/gm are obtained.

In order to quantify the adsorption characteristics of sulfur dioxide with the

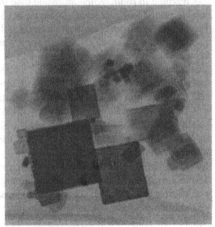

Fig. 1 TEM image of MgO Powder. Largest cube is about 2000Å on an edge.

MgO(100) surfaces we performed adsorption isotherms. Before use, the MgO powder was heat-treated in a quartz oven *in vacuo* at about 950°C for about 36 hours and subsequently transferred in a inert-gas filled glove box into a sample cell that was mounted on a closed-cycle refrigerator. Fig.3 shows the typical adsorption behavior of SO_2 on MgO at 200 K when two successive isotherms are performed. The second SO_2 isotherm was performed after the sample was warmed to room temperature while simultaneously evacuating the adsorption cell with a turbo pumped based pumping station (base pressure ~10-7 torr). Notice that during the second adsorption isotherm there is nearly a fifty percent reduction in the apparent amount of SO_2 adsorbed.

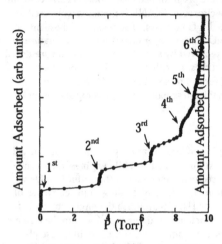

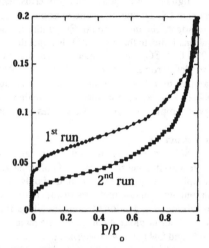

Fig. 2 CH_4 on MgO isotherm at 77K.   Fig. 3 SO_2 on MgO isotherm at 200K.

In order to understand the isotherm results shown in fig. 3 temperature programmed desorption of the SO_2 was performed. A MgO sample predosed with about 2 layers of SO_2 near 77K was warmed to room temperature and then transferred in an argon filled glove bag to a quartz sample cell. The quartz cell containing the SO_2 loaded MgO was then placed inside of a furnace. The evolution of mass products from the cell was monitored using a quadrupole mass spectrometer while a linear temperature ramp was applied. Fig. 4 illustartes the SO_2 signal (64 amu) as a function of time (temperature noted). It is quite clear that an increase in the SO_2 evolution starts at temperatures above of 100°C that decreases above about 350°C. This clearly establishes that significant quantities of SO_2 are still adsorbed on the MgO (100) surface until well above 100°C reconciling the observed decrease in the adsorption capacity shown in fig. 3 above.

To determine the chemical nature of the adsorbed SO_2 species XANES at the sulfur K-edge was performed on MgO samples loaded at 77K with about two monolayers of SO_2. The samples were handled in the following way. First, the SO_2 loaded MgO was warmed to room temperature and evacuated (similar to the way the isotherm samples were handled). A small alloquot of the dosed sample was removed from the quartz cell in an argon filled glove bag and set aside for XANES analysis. The quartz cell containing the remaining sample was then heated to 100°C while being evacuated. Once the sample reached 100°C it was quenched to room temperature and another alloquot was removed in a glove bag and set aside for analysis. This process was repeated at 100°C increments up to 400°C. The five alloquots were examined using XANES on the X19-A beamline at the NSLS in the "fluorescence yield mode" with a Stern-Heald-Lytle detection scheme. Spectral locations of the S, SO_3 and SO_4 species were calibrated using standards ZnS, $NaSO_3$ and $MgSO_4$. The results are plotted in fig. 5. It is clear that both SO_3 and SO_4 species are present on the MgO (100) surface with the SO_4 species being the most stable (i.e. present

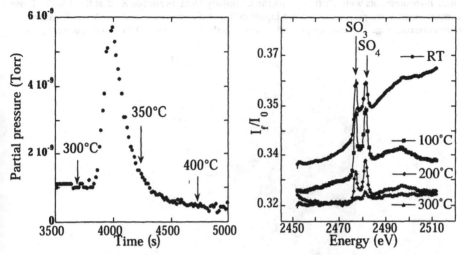

Fig. 4 TPD of monolayer SO_2 on MgO. Approximate temperature indicate arrows.

Fig. 5 Temperature dependence of Sulfur K-edge XANES for ~2 monolayer SO_2 on MgO.

until the highest temperatures). Furthermore, the temperature dependence of the XANES signals is consistent with the TPD results shown in fig. 4. The experiment described above monitors the relative concentration of SO_3/SO_4 species on the MgO as a function of temperature, but it is also important to determine the coverage dependence. Hence, another set of SO_2 dosed MgO samples were prepared by depositing one, two and three layers of SO_2 on the MgO at 77K. The SO_2 loaded samples were warmed to room temperature and evacuated as before. Fig. 6 shows the

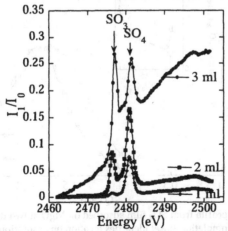

Fig. 6 Coverage dependence of Sulfur K-edge XANES at Room Temperature for SO_2 on MgO

resulting XANES spectra. Once again, these results establish the presence of both SO_3 and SO_4 on MgO(100) and furthermore, that the SO_4 species are the first to form.

Powder x-ray diffraction (XRD) has also been used to study the adsorption of SO_2 on MgO. These measurements were performed on the Chemistry Dept. beamline X7B at the NSLS. These measurements were performed using a highly collimated x-ray beam and a MAR 345 image plate detector. Fig. 7 shows the general experimental arrangement. Difference patterns (gener

EXPERIMENTAL SETUP

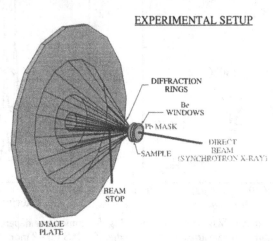

Fig. 7 Schematic diagram of X-ray Diffraction Setup for SO_2/MgO studies on X7b at NSLS.

ated by subtracting a diffraction pattern of a clean MgO powder from one of the same sample after SO_2 is adsorbed) are presented in fig. 8 for two different SO_2 coverages indicated by the arrows labelled "A" and "B" inset from fig. 3 above. Several comments can be made concerning

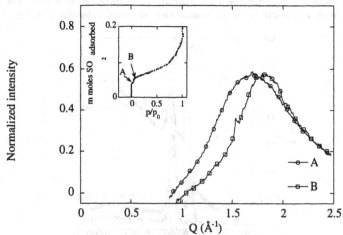

Fig. 8 X-ray diffraction profile from SO_2 film adsorbed on MgO at two different coverages near monolayer completion (inset identifies location on adsorption isotherm).

the diffraction measurements. First the broad diffraction feature recorded between 1.25-2.35 Å^{-1} in Q indicates that no long range order appears in the SO_2 films. Second, the movement of the peak of the diffraction pattern to higher Q values with increasing SO_2 coverage indicates that the mean density of the adsorbed species increases. Third, no evidence for the formation of magnesium sulphate or other sulfur based magnesium compounds is recorded. These diffraction data are still in the preliminary state and future studies will be aimed at obtaining a more comprehensive set of data as a function of temperature and SO_2 coverage.

Discussion

Numerous experimental studies of the interaction of SO_2 with MgO have been performed. The EPR work of Lin and Lunsford [5] found that SO_2 is photochemically oxidized to SO_3^- on the surface of MgO in the presence of water vapor while infrared work by Bensitel *et al* [6] found that bulk-like sulfate (SO_4^-) surface species were formed on a high surface area MgO. Waqif et al.[7] have also examined the interaction of SO_2 with MgO and find that sulfite, SO_3, species are formed and Schoonheydt and Lunsford [8] reported that sulfate (SO_4) species form upon heating. The interaction of SO_2 with acid and basic sites of clean, fully dehydroxylated MgO have been investigated by Pacchioni, Clotet and Ricart [9]using *ab initio* cluster model calculations. They find that SO_2 adsorbs molecularly at five-coordinated Mg^{2+} sites and evidence for SO_3 formation at basic O^{2-} ions. However, they rule out the possibility of SO_4 formation at an unreconstructed MgO surface.

Fig. 9 schematically illustrates how the formation of both SO_3 and SO_4 can take place on the MgO (100) surface. Note that the SO_4 species formation requires that the sulfur sits in a bridging location between adjacent surface oxygen atoms(i.e. with the C_{2v} axis parallel to the surface normal). However, the SO_4 formation requires either the oxygen-oxygen distance at the surface of the MgO be less than that found in the bulk lattice or that a distorted sulfate species forms such that the sulfur-oxygen bond length is greater than that found in the isolated molecule.

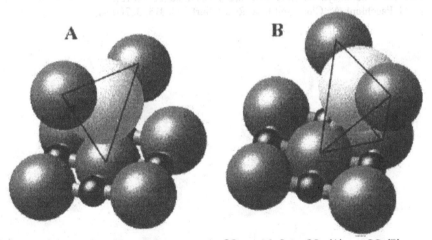

Fig. 9 Schematic view of possible configurations for SO_2 on MgO for SO_3 (A) and SO_4(B) species. Sulfur, Oxygen and Magnesium atoms in order of size from largest to smallest.

Conclusion

As we have indicated above, understanding the interaction of SO_2 with MgO is an important step in identifying better ways to remediate its environmental impact. We have demonstrated that exposure of monolayer quantities of SO_2 to MgO results in the formation of surface adsorbed SO_3 and SO_4 species that are stable to temperatures in excess of 300°C. No evidence of atomic sulfur deposition on the MgO surface is recorded during the sulfite/sulfate decomposition process. Future work will be aimed at understanding what effects coadsorption of other simple molecules like water and carbon monoxide and the addition of controlled amounts of dopants to the MgO powders have on the adsorption and dissociation properties of SO_2.

Acknowledgement

This work was supported by LDRD funds from Brookhaven National Laboratory and the U.S. Department of Energy, Materials Science Division under Contract No. DE-AC02-98CH10886.

References

[1] A. C. Stern, R.W. Boubel, D.B. Turner and D. L. Fox, Fundamentals of Air Pollution, 2nd edition (Academic Press, 1997).

[2] Slack, A. V; Hollidan, GA *Sulfur Dioxide Removal from Waste Gases*, 2nd Ed.;Noyes Data Corp.:Park Ridge NJ 1975.

[3] J. Z. Larese and W. Kunnmann, Patent pending.

[4] Z. Mursic, M. Y. M. Lee, D. E. Johnson and J. Z. Larese, Rev. Sci. Instr. **67**, 1886 (1996).

[5] M. J. Lin and J. H. Lunsford, J. Phys. Chem. **79**, 892 (1975).

[6] M. Bensitel, M. Waqif, O.Saur, and J. C.Lavalley, J. Phys. Chem. **93**, 6581(1989).

[7] M. Waqif, A.M. Saad, M. Bensitel, J./ Bachellier, O. Saur and J. C. Lavalley, J. Chem. Soc. Faraday Trans. **88**, 2931(1992).

[8] R. A. Schoonheydt and J. H. Lunsford, J. Catal. **26**, 261(1972).

[9] G. Pacchioni, A. Clotet and J. M. Ricart, Surf. Sci. **315**, 337(1994).

SYNCHROTRON X-RAY SCATTERING STUDY ON OXIDATION OF AlN/SAPPHIRE

H. C. KANG, S. H. SEO, and D. Y. NOH
Department of Materials Science and Engineering,
Kwangju Institute of Science and Technology, Kwangju, KOREA 500-762, dynoh@kjist.ac.kr

ABSTRACT

We present an x-ray scattering study of the oxidation of AlN/sapphire films into γ-Al$_2$O$_3$ upon annealing. Epitaxial AlN/Sapphire(0001) transforms into nano-crystalline epitaxial γ-Al$_2$O$_3$ during annealing at temperatures above 800 °C in air. The crystalline orientational relation between the γ-Al$_2$O$_3$ and AlN are $<111> \, // <0001>$ in the film normal direction, and $<1\bar{1}0> \, // <11\bar{2}0>$ in the film plane direction. The domain size of the spinel γ-Al$_2$O$_3$ crystalline is smaller than 50 Å in both out-of-plane and in-plane directions. The XPS depth profiles of the oxide film showed that the film is composed of aluminum and oxygen, and the atomic concentration ratio is about 2:3.

INTRODUCTION

Due to their thermal stability and mechanical strength, aluminum oxide thin films are widely investigated for various applications such as catalysis, coating, microelectronics, and composites[1,2]. High quality insulators are also in demand for gate insulators in ultra small field effect transistors as well as for silicon-on-insulator(SOI) devices[2-6]. It has been suggested that crystalline aluminum oxides thin films be fabricated by oxidizing AlN films[5,6] grown on silicon. The crystal quality of the oxide films, however, would be limited by the quality of the host AlN films grown on silicon.

Recently, it has been reported that high quality epitaxial AlN films could be grown on sapphire (0001) substrates[7,8]. To understand the nature of the formation of the crystalline aluminum oxide, using a high quality single crystalline AlN film is crucial, although growth on sapphire is less desirable, from a technological perspective, than on silicon. The goal of this study is to reveal the oxidation process of AlN/sapphire, and to provide structural and chemical information for the resulting aluminum oxide film.

In this paper, we present a synchrotron x-ray scattering investigation of the structural transformation of epitaxial AlN/sapphire films into epitaxial γ-Al$_2$O$_3$ film during thermal oxidation. The nano-size γ-Al$_2$O$_3$ was initially formed on the surface of the AlN film at about 800°C. As the oxidation temperature increases to 1000°C, all the AlN film transformed to a γ-Al$_2$O$_3$ film which became epitaxial to sapphire substrate.

EXPERIMENTAL PROCEDURE

Epitaxial AlN(0001) films were prepared on single crystal sapphire(0001) substrates by radio frequency(RF) magnetron sputtering deposition. As the sputter target, pure Al(99.999%) was used, and 5×10^{-3} torr of pure N$_2$ gas was used as the carrier gas. The substrate temperature was held at 300°C and the RF power was set at 50W. The AlN films thus obtained were epitaxial with high crystallinity. The $<0001>$ direction of AlN is parallel to the substrate normal direction, the <0001> direction of sapphire, while the in-plane $<10\bar{1}0>$ of AlN is rotated 30° with respect to that of sapphire. The detailed process of the AlN growth and the structural properties of AlN are reported elsewhere[7].

The AlN films are oxidized by thermal annealing to temperatures higher than 800°C in air and

195

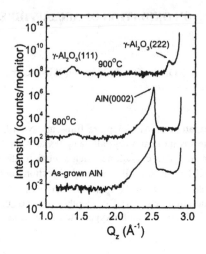

Figure 1. The x-ray diffraction profiles of AlN/sapphire during annealing. As the annealing proceeds, the AlN transform into γ-Al_2O_3 as indicated by the growth of the small γ-Al_2O_3 (111) peak.

in an oxygen atmosphere. Annealing in vacuum or in nitrogen does not oxidize the AlN films. For in-situ studies of the oxidation, a specially designed heating stage that can be mounted on a four-circle goniometer has been used. The oxidation temperature was monitored by a K-type thermocouple that is calibrated by an optical pyrometer.

The synchrotron x-ray scattering experiment was carried out at beamline 5C2 at Pohang Light Source (PLS) in Korea. The incident x-rays were vertically focussed by a focusing mirror, and monochromatized to 9 keV by a double bounce Si(111) monochromator. For detector resolution, two-pairs of slits are used. The diffraction profile along the substrate normal direction was normally investigated during annealing. An off-specular Bragg peak is also investigated to study the epitaxial relationship of the oxide film.

RESULTS

Figure 1 shows the diffraction profile along the substrate normal direction before and after annealing. The diffraction profile before annealing is peaked at the AlN(0002) Bragg peak position, $2.52\,\text{Å}^{-1}$. The asymmetric tail in the low q-side indicates that the strain distribution of the AlN film is not uniform. The asymmetric profile is often observed in high quality AlN films grown by sputter deposition. The detailed structural properties of the AlN films used in this experiment have been reported[7]. The AlN film is epitaxially grown on sapphire as discussed previously.

As the AlN film is annealed to 800°C, the AlN peak starts to disappear and a small, broad peak appears near $q=1.40\,\text{Å}^{-1}$. The small peak grows as the annealing time or the annealing temperature increases, indicating the transformation is kinetically limited. The AlN film is completely transformed to the new phase upon annealing longer than one hour at 800°C or higher than 1000°C. We attribute this transformation to the oxidation of the AlN film. The nitrogen in the AlN film is replaced by oxygen resulting in a form of aluminum oxide. We note that the oxidation starts from the surface of the AlN film rather than from the interface between the film and the substrate sapphire. The oxygen source is the ambient oxygen in air rather than the oxygen in sapphire. The oxidation does not occur when the film is annealed in vacuum or in nitrogen environment.

The peak position of the new peak is close to the expected position of the (111) Bragg reflection of the cubic γ-Al₂O₃ phase, although the peak position decreases slightly during oxidation. It is, however, difficult to determine the phase of the oxide since the aluminum oxynitride (AlON) yields a Bragg peak very close to the observed peak position. To determine the chemical composition of the oxide, we performed a depth profiling x-ray photoemission spectroscopy (XPS) analysis of a film annealed at 800°C for one hour. Figure 2 shows the result. The XPS spectra have peaks at aluminum 2p position and at oxygen 1s position. Near the nitrogen 1s position, however, there is no detectable signal indicating that the nitrogen content is minimal. The ratio of atomic concentration of aluminum and oxygen turns out to be about 2 to 3. We, therefore, conclude that the film transformed into crystalline Al₂O₃.

There are two crystalline phases of Al₂O₃. One of them is well-known, sapphire, which is also denoted as hexagonal corundum α-Al₂O₃. The positions of the Bragg peaks from sapphire are, however, very far from the observed peak position of the annealed film. Incidentally the substrate used in this experiment is sapphire. The other crystalline form of Al₂O₃ is the cubic γ-Al₂O₃ phase. Since the chemical composition and the Bragg peak position of the annealed film match to those of γ-Al₂O₃, we conclude that the AlN transforms into γ-Al₂O₃ during annealing. As a result of the oxidation, a crystalline γ-Al₂O₃ film formed on the α-Al₂O₃, sapphire substrate.

Since the γ-Al₂O₃ phase is formed by replacing nitrogen in the epitaxial AlN by oxygen, it is

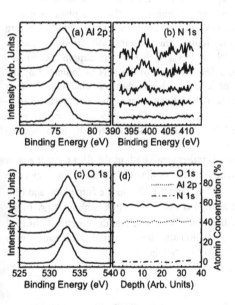

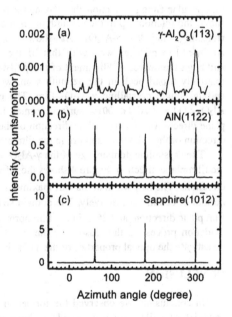

Figure 2. The XPS spectra near (a) Al 2p, (b) N 1s, (c) O 1s, during depth profiling. (d) Relative amount of aluminum, oxygen, and nitrogen.

Figure 3. Azimuth scan of (a) γ-Al₂O₃<1$\bar{1}$3>, (b) AlN<11$\bar{2}$1>, (c) sapphire<10$\bar{1}$2>.

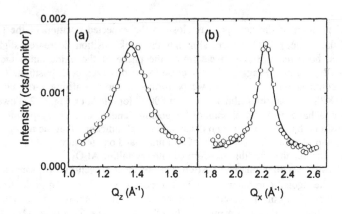

Figure 4. The diffraction profile of γ-Al$_2$O$_3$ < 1$\bar{1}$3 > (a) along the surface normal direction, (b) in the film plane direction. The lines are results of a fit to Lorentzian.

likely that γ-Al$_2$O$_3$ is also epitaxial. To study the epitaxial relation, we have measured the (1$\bar{1}$3) off-specular Bragg reflection that has non-zero momentum transfer in the film plane. Figure 3 shows the relationship in the azimuth orientation between the sapphire substrate, the AlN before annealing, and the γ-Al$_2$O$_3$. As shown in the figure, the azimuth direction of the γ-Al$_2$O$_3$ < 1$\bar{1}$3 > matches well with that of the AlN < 11$\bar{2}$1 > and that of the sapphire < 10$\bar{1}$2 >. Although the degree of alignment of the γ-Al$_2$O$_3$ is less than that of the AlN as indicated by the relatively broad peak in the azimuth scan, we believe that this is the first observation of an epitxial single crystalline γ-Al$_2$O$_3$ film. The epitaxial relationship may be summarized as γ-Al$_2$O$_3$ < 111 > // AlN < 0001 > (surface normal direction), and γ-Al$_2$O$_3$ < 1$\bar{1}$0 > // AlN < 11$\bar{2}$0 > (in-plane direction). The crystalline direction of the AlN is rotated by 30° relative to the respective direction of the substrate sapphire in the film plane.

The crystalline domain size of the γ-Al$_2$O$_3$ film is rather small both in the film normal and in the film plane direction. Figure 4 shows the diffraction profile of the (1$\bar{1}$3) peak in the respective directions. The half width at the half maximum is 0.07 Å^{-1} and 0.12 Å^{-1} in the film plane and film normal directions, respectively. The domain size estimated from the widths is about 40 Å in the film plane direction, and 25 Å in the film normal direction. We believe that the kinetically limited oxidation process is the reason for the small crystal domain size. It would be interesting to investigate the optical properties of the nano-sized γ-Al$_2$O$_3$ crystals.

CONCLUSION

In conclusion, we observed the formation of a cubic γ-Al$_2$O$_3$ film by thermally oxidizing epitaxial AlN film grown on sapphire above 800 °C in air using synchrotron x-ray scattering technique. The film is epitaxial with γ-Al$_2$O$_3$ < 111 > // AlN < 0001 > and γ-Al$_2$O$_3$ < 1$\bar{1}$0 > // AlN < 11$\bar{2}$0 >. The crystal domain was smaller than 50 Å in both out-of-plane and in-plane direction indicating that the γ-Al$_2$O$_3$ is in the form of nano-crystals. The chemical composition was confirmed by an XPS depth profiling analysis. We believe that this is one of the first reports of the observation of epitaxial crystalline aluminum oxide film. Further studies on the detailed kinetics of the oxidation process are necessary to elucidate the nature of the formation of γ-Al_2O_3 film. Optical and electronic properties of the epitaxial nano-crystalline γ-Al_2O_3 film would be interesting.

ACKNOWLEDGEMENT

This work has been supported by KOSEF through ASSRC(1999), and by Korean Research Foundation made in the program year of 1998. PLS is supported by Ministry of Science and Technology in KOREA

REFERENCES

1. D. R. Clarke, Phys. Stat. Sol. (a) 166, 183 (1998)

2. T. Kimura and M. Ishida, Jpn. J. Appl. Phys. 38, 853 (1999)

3. J. T. Zborowski, T. D. Golding, R. L. Forrest, D. Marton, and Z. Zhang, J. Vac. Sci. Technol. B 16(3), 1451 (1998)

4. N. Yu, T. W. Simpson, P. C. McIntyre, M. Nastasi, and I. V. Mitchell, Appl. Phys. Lett. 67(7), 924 (1995)

5. J. Kolodzey, E. A. Chowdhury, G. Qui, Olowolafe, C. P. Swann, K. M. Unruh, J. Suehle, R. G. Wilson, and J. M. Zavada, Appl. Phys. Lett. 71(26), 3802 (1997)

6. E. A. Chowdhury, J. Kolodzey, J. O. Olowolafe, G. Qiu, G. Katulka, D. Hits, M. Dashiell, D. Weide, C. P. Swann, and K. M. Unruh, Appl. Phys. Lett. 70(20), 2732 (1997)

7. H. C. Kang, S. H. Seo, D.Y. Noh, Jpn. J. Appl. Phys 38, Suppl. 38-1, 187 (1999)

8. K. H. Shim, J. Myoung, O. Gluschenkov, K. Kim, C. Kim, and I. K. Robinson, Jpn. J. Appl. Phys. 37, L313 (1998)

STUDY OF THE BURIED INTERFACE BEHAVIOR OF LIQUID CRYSTAL THIN FILMS USING SYNCHROTRON RADIATION AND GRAZING INCIDENCE X-RAY SCATTERING MODE

Y. Hu, L.J. Martínez-Miranda
Department of Materials and Nuclear Engineering, University of Maryland, College Park, MD 20742, yufeihu@eng.umd.edu

ABSTRACT

We have used the intensity and tunability of a synchrotron X-ray source in order to access the buried interface between a glass substrate and a liquid crystal thin film. We find that for energy of 9.4kev, the X-rays can penetrate a 0.22mm substrate. Grazing Incidence X-ray Scattering has been used to study the alignment of the films as a function of depth and temperature. Our results indicate the presence of both a chevron structure and a structure similar to the helical twist-grain-boundary (TGB) phase. Some films have a disordered interfacial layer. This technique can be applied in the study of semiconductor devices as well as surfactant film interfaces.

INTRODUCTION

With highly collimated beam and good cross sections, bright synchrotron sources are particularly suited for the technique of surface diffraction such as grazing incidence X-ray scattering (GIXS). [1]It is possible to acquire otherwise low intensity diffraction data. GIXS is a powerful tool for the analysis of thin films and surfaces. In the situation where X-rays strike the surface at a grazing angle near or within the range for which total external reflection occurs, only the near surface layer is illuminated and the information thus obtained reveals structure of the surface or interface region in preference to that of the interior. At higher incidence angle, regions in the bulk and buried interfaces of films can be studied. It has been widely used for the study of surface reconstruction [2] and adsorption, [3] as well as structural depth profiling of thin films. [4] We find GIXS is very useful in study of liquid crystal interfacial behavior.

The promising future of surface-stabilized ferroelectric liquid crystal displays (SSFLCD) has created much interest in the interfacial behavior of smectic liquid crystals under different conditions. Rieker et al. [5] observed a chevron structure in liquid crystal sandwiched between two glass plates instead of the assumed bookshelf geometry. Patel [6] found that non-chiral smectic liquid crystals may also exhibit a TGB structure under twisted boundary conditions, which was previously reported only among chiral smectic liquid crystals. There are also many theoretical studies modeling and explaining experimental results. [7]

In this paper, we present the results of a depth profile study of smectic-A octylcyanobiphenyl (8CB) films deposited on grated glass substrates, which indicate the presence of a TGB structure in coexistence with a chevron structure inside the films.

EXPERIMENTAL

Octylcyanobiphenyl (8CB) was chosen for these experiments because of its chemical stability and room temperature smectic A phase. It exists in the SmA phase below 306.6K, in the nematic phase above that temperature and in the isotropic liquid phase above 313K. The smectic phase has a layer spacing of 31.6 Å.

Substrate Preparation

The grating on glass was made at the University of Maryland engineering clean room. The grating period is 10 μm and its depth is approximately 0.1μm over an area 1.2cm×1.2cm. We used a modified photolithography method to prepare these gratings. First, the glass substrate was cleaned with trichloroethylene (TCE), manufactured by Alfa Aesar, acetone and methanol to remove all contaminants. Then, a positive photoresist, Hoechst's AZ1512, was spun on with thickness ~1μm. After softbaking at 90°C for 30 min., the substrate was exposed to UV light for about 10s. We use AZ400 developer to develop the pattern and then postbake it at 100°C for 30 min. Finally the substrate was etched in 10:1 hydrofluoric acid for 2 mins. After using acetone to strip off photoresist the grating substrate is done. The grating was etched on a 0.2mm glass slide.

GIXS

Our X-ray studies were conducted at the National Synchrotron Light Source at Brookhaven National Laboratory, beamlines X-18A and X-22B. The X-ray energy used was 9.4keV, which allowed the beam to penetrate through the glass slide in order to probe the substrate-LC interface. The experimental resolution is $0.003q_0$ with a slit size of $2mm^2$. The experimental setup is shown in Fig.1. The grating's long axis was placed perpendicular to the beam at the nominal $\phi = 0°$ position where ϕ is the azimuthal angle in the plane of the film. In-plane scattering at $\phi = 0°$ is due to molecules aligned along the long axis of the gratings. Scattering at $\phi = 90°$ originates from those molecular layers lying perpendicular to the grating direction.

The GIXS technique exploits the fact that the index of refraction for X-rays is slightly less than 1; this means that there exists a critical angle, α_c, below which the X-rays are totally externally reflected. For $\alpha < \alpha_c$, the X-ray wave in the film is evanescent and is limited to the top layer of materials. The penetration depth can be controlled by varying incident angle α and so depth profile information can be obtained. The penetration depth is given by [4]

$$D(\alpha) = \lambda/4\pi q \tag{1}$$

Where

$$q = [\sqrt{(\alpha^2 - \alpha_c^2)^2 + 4\delta_i^2} + \alpha_c^2 - \alpha^2]^{1/2}/\sqrt{2} \tag{2}$$

$$\lambda = 1.308\text{Å}, \quad \delta_i = \lambda\mu/4\pi.$$

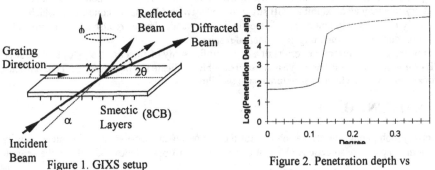

Figure 1. GIXS setup

Figure 2. Penetration depth vs incidence angle

Fig. 2 shows calculated penetration depth vs different incident angle.
The calculated penetration depth of X-ray in glass is shown in Fig. 4. Based on this Figure, we chose X-rays 1.308Å in wavelength. At this energy, the penetration depth for glass is of the order of 280μm, which allows the beam to penetrate through a glass slide 0.22 mm thick. [8] The sample is aligned such that part of the beam passed through the glass by its side, as shown in Fig. 3. The incidence angle is very small and the X-rays are almost parallel to the interface. Performing the GIXS experiment as described above allows us to directly access the glass-interface and to study any variations in the structure of the liquid crystal.

RESULTS AND DISCUSSION

There is a geometrical relationship between the angle χ and α given by

$$\sin \alpha = \sin \theta \sin \chi \qquad (3)$$

Therefore, we can change the α angle by controlling the angle χ. Each α corresponds to a different depth inside the sample.

Fig. 5 shows a typical result in smectic liquid crystals with thickness 70μm and at temperature T = 305.3K. We can see clearly that a TGB structure existed at this sample as shown by the variation in the ϕ angle . At different χ angles, i.e. at different penetration depths, the ϕ angle changed continuously from 56° to -14°. It means that grains rotated with to each other at different layers through the interface. Renn and Lubensky first postulated the TGB structure theoretically, [9] which consists of regularly spaced grain boundaries of screw dislocations that are parallel to each other within the boundary, but are rotated by a fixed angle with respect to the screw dislocation in adjacent grains. However, most previous experiments demonstrate it only in chiral materials. Our results are due to two competing boundaries: near the glass boundary, the grated substrate aligns the LC along the grating direction. At the LC-air interface, the molecules may tilt away from the normal to the surface in any direction as observed in focal conic textures. This observation is consistent with our previous results. [10]

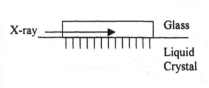

X-ray

Glass

Liquid
Crystal

Figure 3. Cross section of substrate

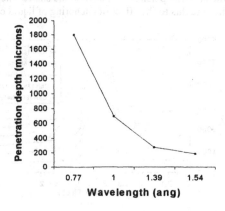

Figure 4. Penetration depth vs X-ray wavelength

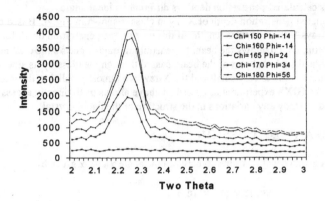

Figure 5. 2θ Scan at T=305.3K   70μm LC

A second set of results is shown in Fig. 6 for a 106μm thick liquid crystal at T=305.7K. We observe that as the χ angle increase, the 2θ peak position shifted toward high angle then shifted back to low angle. The variance of the peak position implies a change in the thickness of smectic layer or d spacing. This observation suggests the existence of chevron structure as shown in Fig.7. At first, the smectic liquid crystal layer becomes thicker at the glass-liquid crystal interface due to the tilt of the layer. After higher penetration depth, it reached the tip of chevron structure which is thinner. The cause of formation of chevron structure may be the mismatch between the smectic layer and the grating as Rieker et al [5] have suggested.

Fig.8 is a plot of intensity vs temperature at different penetration depths with film thickness 104μm. Generally, the intensity becomes lower as the temperature increases which shows that molecules were more active and random and that smectic ordering weakened. We find that smectic peak persists well beyond the smectic-nematic transition temperature T=306.3K. We attribute this to the effect of anchoring of liquid crystal molecules.

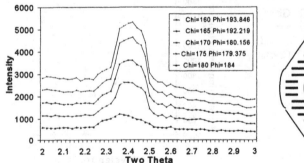

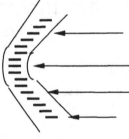

Figure 6. 2θ (deg) scan at T=305.7K

Figure 7. Chevron structure

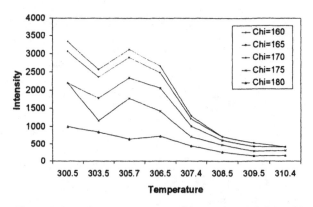

Figure 8. Intensity vs temperature with 104μm thick LC phi~180°

CONCLUSIONS

We have performed GIXS studies of the buried glass interface in hybrid 8CB liquid crystal films. Our results suggest the presence of both a chevron structure and TGB structure near the bottom of the grating glass substrate under these competing boundary conditions. These structures persist above the smectic - nematic transition temperature. More results and detailed analysis will be presented in a later publication.

ACKNOWLEDGEMENT

This work was supported by NSF Grant No. ECS-95-30933. Work at the NSLS is partially supported by the U.S. Department of Energy.

REFERENCES

1. W. C. Marra, P. Eisenberger, and A. Y. Cho, *J. Appl. Phys.* **50**, 6927(1979).
2. J. Bohr, R. Feidenhans'l, M. Nielson, M. Toney, R.Johnson, and I. Robinson, *Phys. Rev. Lett.* **54**, 1275(1985).
3. M. Marra, P. Fuoss, and P. Eisenberger, *Phys. Rev. Lett.* **49**, 1169(1982).
4. M. Toney and T. Huang, *J. Mater. Res.* **3**, 351(1988).
5. T. P. Rieker, N. A. Clark, G. C. Smith, D. S. Parmar, E. B. Sirota & C. Safinya, *Phys. Rev. Lett.* **59**, 2658(1987).
6. J.S. Patel, *Phys. Rev. E* **49**, R3594(1994).
7. S. Kralj and S. Zumer, *Phys. Rev. E* **54**, 1610(1996).
8. L. J. Martinez-Miranda, Y. Hu and T. Misra, *Mol. Cryst. Liq. Cryst.***329**, 121(1999).
9. T.C. Lubensky and S.R.Renn, *Phys. Rev. A*, **41**, 4392 (1990).
10. Y. Hu and L. J. Martinez-Miranda, MRS 1998 symposium.

INVESTIGATION OF INHOMOGENEOUS IN-PLANE STRAIN RELAXATION IN SI/SIGE QUANTUM WIRES BY HIGH RESOLUTION X-RAY DIFFRACTION

Y. Zhuang*, C. Schelling*, T. Roch*, A. Daniel*, F. Schäffler* G. Bauer*, J. Grenzer**, U. Pietsch**, S. Senz***
*Institut für Halbleiterphysik, Universtät Linz, A-4040 Linz, Austria, yzhuang@hlphys.uni-linz.ac.at
**Institut für Physik, Universtät Potsdam, D-14115 Potsdam, Germany
***Max Plank Institut für Mikrostrukturphysik, D-06120 Halle, Germany

ABSTRACT

The structural properties of Si/SiGe quantum wires, which were grown by local solid source molecular beam epitaxy through a Si_3N_4/SiO_2 wire-like shadow mask, were investigated by means of high resolution x-ray coplanar and x-ray grazing incidence diffraction ,as well as by transmission electron microscopy. High resolution x-ray coplanar diffraction was used to obtain the average in-plane strain in Si/SiGe wires before and after removing the Si_3N_4/SiO_2 shadow mask. X-ray grazing incidence diffraction measurements were performed to obtain information on the shape of the wires and on the depth-dependent strain relaxation. A finite element method was used to calculate the strain distribution in the Si/SiGe wires and in the Si substrate which clearly show the influence of the Si_3N_4/SiO_2 shadow masks on the strain status of the Si/SiGe wires in agreement with the experimental data.

INTRODUCTION

Si/SiGe multi quantum wells with finite lateral size have attracted a lot of interest for their potential application in high frequency electronic and optoelectronic devices. Selective epitaxial growth on patterned substrates is a promising technique offering possibilities for (i) in situ fabrication of integrated circuits [1] and (ii) improvement of the structural and optical properties of wires [2, 3]. Due to the 4% lattice mismatch between Si and Ge, the reduction of the lateral size causes a non-uniform strain relaxation. X-ray coplanar diffraction [4-7] and grazing incidence diffraction (GID) [8-10] have been widely used to investigate this non-uniform strain relaxation on dry or wet-etched wires and dots as well as on self-organized wire and dot structures. To the best of our knowledge no work has been done so far on the strain relaxation of Si/SiGe wires grown by selective epitaxial growth through shadow masks. In this work, we present investigations on the inhomogeneous strain relaxation in Si/SiGe wires grown by solid source molecular beam epitaxy through Si_3N_4/SiO_2 shadow masks. The average in-plane strain in Si/SiGe wires is determined using x-ray coplanar diffraction. The depth-dependent in-plane strain relaxation and information on the shape of the Si/SiGe wires are obtained from x ray grazing incidence diffraction.

EXPERIMENT

Si wafers with a 105nm thick thermal SiO_2 layer capped with a 120 Å thick Si_3N_4 layer were patterned laterally by holographic lithography and reactive ion etching. For the holographic lithography an Ar-ion laser operating at a wavelength of 458nm was used and the etching was performed in a parallel plate reactor with a mixture of CF_4 and H_2 for the

SiO$_2$ and Si$_3$N$_4$ layers. The lateral period of wires is of about 800nm and their height was around 900 Å. The orientation of the wires was along the [$\bar{1}$10] direction. After the reactive ion etching, the SiO$_2$ layer was selectively etched by a diluted HF solution to create an undercut and thus lead to a Si$_3$N$_4$ shadow mask. Through this shadow mask, a six-period Si/Si$_x$Ge$_{1-x}$ multi quantum well structure (MQW) was deposited on the Si (001) substrate, using solid source molecular beam epitaxy. The structural parameters were the following: Si layer thickness 100 Å, SiGe layer thickness 30 Å, Ge concentration 30%. Prior to growth, a RCA cleaning procedure with subsequent HF-removal of the oxide was employed. Before MBE growth the wafers were subjected to a thermal cleaning step at 650°C. The common oxide-desorption step raising the substrate temperature above 900° was not performed, in order to avoid the change of the Si surface profile and of shadow mask by Si mass transport [11].

Coplanar x-ray diffraction measurements were done on a Philips MRD diffractometer with a four-crystal Ge(220) monochromator. Cu Kα_1 (λ=1.5405 Å) radiation was used with $\Delta\lambda/\lambda$ ~10^{-4}. X-ray GID measurements had been performed at BW2 in HASYLAB (Hamburg, Germany) using a wavelength of 1.36 Å with a wavelength spread of $\Delta\lambda/\lambda$ ~10^{-5}. A series of slits and a Si(111) analyser were mounted in front of a scintillation detector to achieve a high resolution of about 0.005nm^{-1} within the plane (in x-y plane) and 0.001 nm^{-1} for out of plane (x-z plane) in reciprocal space.

For reference a six-period Si/Si$_x$Ge$_{1-x}$ MQW structure was grown on an unpatterned Si(001) wafer and its structural parameters were determined employing rocking curves along the q$_z$- axis, recorded around the (004) reciprocal lattice point (RLP) in coplanar diffraction geometry. The thicknesses of the Si and SiGe layers were determined to be 94.1 Å and 24.9 Å, respectively, and the Ge concentration x to be 25.3%.

RESULTS: STRAIN ANALYSIS

The strain status of Si/SiGe wires depends on the Si and SiGe layer thicknesses, the Ge content of the MQW structure, the lateral dimension of the wires and on the presence or absence of the Si$_3$N$_4$/SiO$_2$ shadow mask. The in-plane strain relaxation of Si/Si$_x$Ge$_{1-x}$ wires was detected by measuring asymmetrical reciprocal space maps (RSM) as shown in Fig.1 where the scattered intensity was shown as a function of reciprocal space coordinates along the growth ([001], q$_z$-) and an in-plane ([110], q$_x$-) direction. For comparison Fig.1(a) shows a (224) asymmetric RSM measured on the reference sample. The same q$_x$ values for the Si substrate reflection and for the 0th order Si/SiGe superlattice peak SL$_0$ indicate the pseudomorphic growth of Si/Si$_x$Ge$_{1-x}$ MQW on the Si substrate. Figs.1(b) and (c) represent the (224) asymmetrical RSMs before (b) and after (c) removing the Si$_3$N$_4$/SiO$_2$ shadow mask. The lateral periodicity of the intensity maxima reflects the period of the Si/Si$_x$Ge$_{1-x}$ wires Λ, and Λ is determined by $\Lambda = 1/\Delta q_{x,wires}$ of about 800 nm (where $\Delta q_{x,wires}$ denotes the spacing of adjacent lateral peaks in reciprocal space).

In order to obtain quantitative values for the average in-plane strain, several q$_x$ line scans were generated by projecting the scattered intensity in a selected area in the vicinity of the Si substrate peak and of the SL$_0$ peak in the (224) asymmetrical RSM onto the q$_x$ axis. From the shift of the envelope maxima of the wire samples with respect to the substrate peak before and after removing the Si$_3$N$_4$/SiO$_2$ shadow mask, the mean in-plane strain $<\varepsilon_{xx}>$ is obtained to be $<\varepsilon_{xx,a}> = \Delta q_{x,a}/q_0$=4.8×10^{-4}, $<\varepsilon_{xx,b}>= \Delta q_{x,a}/q_0$=7.5×10^{-4}, respectively; q$_0 = 4\pi\sin\theta_B/\lambda$ and θ_B is the Bragg angle of the substrate. These results show that an increase of the in-plane lattice relaxation occurs in the Si/SiGe wires after the removal of

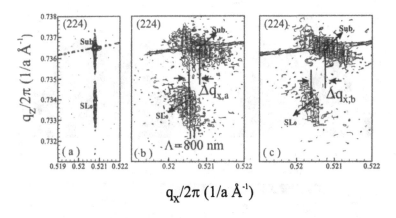

$$q_x/2\pi \ (1/a \ \text{Å}^{-1})$$

Figure 1: (224) asymmetrical reciprocal space maps measured on (a) reference sample, i.e. Si/Si$_x$Ge$_{1-x}$ MQWs grown on an unpatterned Si substrate, (b)Si/Si$_x$Ge$_{1-x}$ wires before removing the Si$_3$N$_4$/SiO$_2$ shadow mask, (c) Si/Si$_x$Ge$_{1-x}$ MQWs after removing the Si$_3$N$_4$/SiO$_2$ shadow mask. Λ denotes the lateral period of the Si/Si$_x$Ge$_{1-x}$ wires.

the Si$_3$N$_4$/SiO$_2$ shadow mask.

Quantitative calculations of the strain distribution presented in the Si/SiGe wires structure were performed by using a finite element method (FEM), as shown in Fig.2. Both the Si and SiGe layers were chosen as three-dimensional anisotropic linear elastic solids. The detailed boundary conditions were described in Ref. [7]. The Si/SiGe wires structure was assumed to be of rectangular shape with a width of about 5600 Å and a total height 780 Å. Since the thickness of the Si$_3$N$_4$ layer was much smaller than that of the SiO$_2$ layer, we neglected the presence of the Si$_3$N$_4$ layer in our FEM calculations. Thus the height of the SiO$_2$ wires is equal to the total thickness of the SiO$_2$ layer, i.e. 1050 Å, their width is 1000 Å. Consequently, there remains a gap of about 200 Åbetween Si/SiGe wires and the SiO$_2$ wires (see Fig. 2). In order to consider the variation of the strain distribution caused by the presence of the amorphous SiO$_2$ wires, in the FEM calculations we dealt with the specific volume mismatch between Si and SiO$_2$ rather than a lattice constant mismatch [7]. The degree of volume mismatch was set to $f = 100\%$. As shown in Fig. 2, the presence of the SiO$_2$ shadow mask changed the strain distribution both in the Si substrate and in the Si/SiGe wires. There, in the top part, the in-plane strain was larger than at the bottom, reflecting the larger lattice relaxation at the top. For comparison with the experimental x-ray data, an averaging of the in-plane strain within the whole Si/SiGe wires was performed and the mean in-plane strain $<\varepsilon_{xx}>$ was obtained to be $<\varepsilon_{xx}> = 1.2 \times 10^{-3}$ if the SiO$_2$ mask is present and $<\varepsilon_{xx}>=1.4 \times 10^{-3}$ without the SiO$_2$ mask. Apparently the SiO$_2$ mask exerts a tensile force on the Si substrate underneath the SiO$_2$ mask, and thus leads to a reduction of the in-plane lattice relaxation in the Si/SiGe wires. Removing the SiO$_2$ mask, the tensile force disappearred and consequently made a further lattice relaxation in the Si/SiGe wires possible.

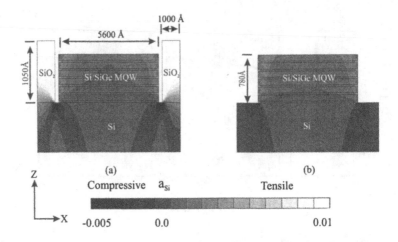

Figure 2: Contour plot of ε_{xx} calculated by finite element method, (a) with SiO$_2$ mask, (b) without SiO$_2$ mask.

X-ray grazing incidence diffraction experiments were performed in order to obtain information on the depth-dependent in-plane strain relaxation. Figure 3(a) shows a sketch of GID measurement. By varying the incidence angle α_i and the exit angle α_f, the depth-dependent information was obtained from the different penetration depth probed by the x-rays. In all the GID measurements, we set $\alpha_i = \alpha_f$. Since the wires were orientated along the [$\bar{1}$10] direction, so the scan along the x ([110]) direction, i.e. q_x scan, reflected information on the wires. The two possibilities to realize the q_x scan were shown in the inset of Fig. 3(a); h denotes the diffraction vector and u is the displacement field within the wires. Transversal line scans reflected the shape of the wires and their lateral periodicity. Longitudinal scans are shown in Fig. 3(b) for several angles α_i corresponding to information depths of 50 Å, 100 Å, 2000 Å, 3000 Å, 4000 Å for the Si/SiGe wires and to 30 Å, 50 Å, 600 Å, 2000 Å and 3000 Å for Si substrate, respectively. For the smallest penetration depth (curve (a)), the diffracted intensity stems mainly from the top part of the Si/SiGe wires and its modulation reflects the lateral period of the wires. The maximum of the envelope curve of the lateral peaks yields an average in-plane strain of 1.0×10^{-3} (Fig. 3(b), curve a) and the corresponding value obtained from FEM calculations within the top 50 Å is 1.6×10^{-3}. Increasing the penetration depth, i.e. curves (b) and (c), a broadening of the peak appears. The x-rays probe the complete Si/SiGe wires structure of about 780 Å, and the top part of the Si substrate of about 600 Å from which the compressively strained parts (see Fig. 2(b)) cause the apparent modulation at q_x values larger than 0.5208 Å^{-1}. For curves (d) and (e), the x-rays probe much deeper into the Si substrate with a penetration depth above 2000 Å. The broadening of the peaks becomes weaker (curve d)) and disappears in curve (e). The lateral wire peaks are recovered and the maximum of their envelope curve shows a small shift towards the Si substrate peak as compared to curve (a), indicating a smaller in-plane lattice constant. The mean in-plane strain is then determined to be of 8.9×10^{-4} while the FEM value for the whole Si/SiGe wires structure is 1.4×10^{-3}.

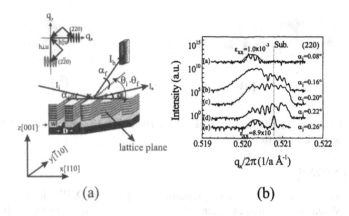

(a)   (b)

Figure 3: (a) Sketch of a grazing incidence diffraction (GID) set-up. I_i, I_s and I_h denote the incidence beam, the scattered beam and the diffracted beam, θ_i and θ_f are the in-plane angles between I_i, I_f and the diffraction plane, respectively. The inset shows a sketch of the longitudinal scans in reciprocal space. (b) GID longitudinal line scans for different incidence angle $\alpha_i = \alpha_f$.

In order to understand the origin of the broad intensity maximum, a cross-section view by using transmission electron microscopy (TEM) was made, as shown in Fig. 4. The single crystalline Si/SiGe MQW appears between the Si_3N_4/SiO_2 mask, whereas it is polycrystalline on top of the mask. At the interface between the Si/SiGe layers and the Si substrate defects appear, possibly due to remaining natural oxide in the 650°C cleaning step. These defects are clearly present in Fig. 4(b), which is a micrograph with a larger magnification factor. In order to avoid significant Si mass transport, we did not employ the standard 900°C thermal desorption step prior to the MBE growth, and apparently a certain amount of chemical contamination remained on the sample surface, which served as the source for these defects.

The presence of these defects gives rise to the broadening of the GID peaks as shown in Fig. 3(b) curves (b) and (c). For the wavelength of λ=1.36 Å, the critical angle of the total external reflection for the Si substrate α_c^{Si} is 0.197°, while that for the Si/SiGe wires α_c^{wire} is about 0.14° since it is dependent on the average electron density [10]. Usually the diffracted intensity of GID measurements shows a maximum at the incidence angle close to the critical angle α_c and above the critical angle, the intensity drops dramatically. The total scattered intensity is the sum of the intensity scattered from the Si/SiGe wires and the distorted top part of the Si substrate in a ratio weighted by the respective linear absorption coefficients. For α_i =0.16° and α_i =0.20°, the scattering from the distorted top part of the Si substrate is quite important and yields this broadening in Fig. 3(b) curves (b) and (c). By a further increase of the penetration depth, the x -rays probe very deeply into the Si substrate, which leads to a smaller relative contribution to the total scattered intensity from the distorted region. Thus both the Si/SiGe wire peaks as well as the substrate peak become clearly visible for curve (e). These data show that by using GID, information on the presence of interface defects can be obtained by a non-destructive method.

(a)   (b)

Figure 4: TEM cross-section micrograph (a) overview and (b) image with a larger magnification factor of the growth interface.

CONCLUSIONS

The strain relaxation of Si/SiGe quantum wires grown by MBE through a Si_3N_4/SiO_2 shadow mask was investigated by means of high resolution x-ray diffraction. The presence of the Si_3N_4/SiO_2 mask exerts a tensile force on the Si substrate, which restricts in-plane lattice relaxation within the Si/SiGe wires. A larger lattice relaxation was found for the top part as compared to the bottom of the Si/SiGe wires. For numerical simulations, strain calculations using the finite element method were performed, which are in good agreement with the x-ray measurements. Furthermore, the presence of defects at the interface between the Si/SiGe wires and the Si substrate could be detected by x-ray grazing incidence diffraction, and their presence was proven by cross-section transmission electron microscopy.

This work was supported by the FWF, and the GMe, Vienna, Austria. We thank Austria Microsystem International for providing the thermal oxides and nitrides. Y. Z. thanks OeAD for the financial support.

References

[1] G. W. Neudeck et al, J. Vac. Sci. Technol. B 17, 994 (1999)

[2] J. Brunner, et al., J. Crystal Growth, 150, 1060 (1995)

[3] M. Kim et al., J. Crystal Growth, 167, 508 (1995)

[4] L. Tapfer et al., Appl. Phys. A 50, 3 (1990)

[5] V. Holy et al., Phys. Rev. B 52, 8348 (1995)

[6] Q. Shen et al., Phys. Rev. B 53, R4237 (1996);Rev. B 54, 16381 (1996)

[7] Y. Zhuang et al., J. Phys. D: Appl. Phys. 32, A224 (1999)

[8] U. Pietsch et al., J. Appl. Phys. 74, 2381 (1993)

[9] N. Darowski et al., Physica B, 248, p. 104 (1998);Appl. Phys. Lett. 73, 806 (1998)

[10] Y. Zhuang et al., Physica B in print

[11] M. E. Keeffe et al., J. Phys. Chem. Solids, 55, 10, p. 965 (1994)

STRUCTURE OF THE NEAR-SURFACE WAVEGUIDE LAYERS PRODUCED BY DIFFUSION OF TITANIUM IN LITHIUM NIOBATE

Y. AVRAHAMI* , E. ZOLOTOYABKO*, W. SAUER**, T. H. METZGER**, J. PEISL**
*Department of Materials Engineering, Technion-Israel Institute of Technology, Haifa 32000, Israel, zloto@tx.technion.ac.il
**Sektion Physik der Universität München, Geschwister-Scholl-Platz 1, D-80539 München, Germany

ABSTRACT

Titanium-induced structural modifications in thin waveguide layers of lithium niobate have been investigated by grazing incidence diffraction and complementary thin film techniques. The study was focused on the high-temperature phase transformation in this system and its influence on the lattice parameter changes, depending on the annealing time.

INTRODUCTION

Ti-diffusion at $1000\,^{0}C$ is used in fabricating lithium niobate ($LiNbO_3$)-based optoelectronic devices in order to increase the refractive index of the near-surface waveguide layer. An optical barrier to the propagation of light into the crystal bulk arises from the replacement of Li or Nb cations by Ti ions and the related changes in structural parameters and strain levels. In this paper, the phase formation processes accompanying high-temperature Ti-diffusion in thin waveguide layers of $LiNbO_3$, and their influence on the structural characteristics of the $LiNbO_3$ matrix were studied by means of grazing-incidence x-ray diffraction (GID) on a synchrotron beam line, conventional high-resolution (HRXRD) and powder x-ray diffraction. Complementary techniques, such as transmission (TEM) and scanning (SEM) electron microscopy as well as secondary ion mass spectrometry (SIMS) were also used. The focus was on the development and temporal decay of a rutile-like (RL) phase, which is of great importance to device technology.

EXPERIMENT

Samples were prepared by Ti sputtering at room temperature and 4 Torr Ar pressure on the polished surface of Y-cut $LiNbO_3$ wafers, 3 inch diameter and 500 μm thick. A 30-35 nm thick Ti-layers were observed in high-resolution scanning SEM cross-sections (see Fig. 1). The Ti-deposited wafers were cut into $10x15\ mm^2$ pieces which were then heat-treated in air at 995 °C for various durations between 0.5 and 6 hours. Ti-penetration into the $LiNbO_3$-bulk was monitored by SIMS. The heat treatment resulted in an approximately 1 μm thick Ti-diffused waveguide layer after 1 hour of annealing.

At high temperature and in oxygen environment, the metal Ti is rapidly reduced to a rutile form (TiO_2). The replacement of Li or Nb cations by Ti ions during Ti diffusion also facilitates the formation of lithium triniobate ($LiNb_3O_8$) [1, 2]. As was first suggested in Ref. [3], TiO_2 and $LiNb_3O_8$ create a solid solution,

213

$(Li_{0.25} Nb_{0.75} O_2)_{1-x}(TiO_2)_x$ which plays an important role in the fast Ti diffusion through the LiNbO$_3$-lattice.

Fig. 1. High-resolution SEM cross-section of an as-deposited lithium niobate (LN) sample.

Fig. 2. TEM microdiffraction pattern showing the presence of a tetragonal RL phase within lithium niobate (LN). Diffraction spots, indicated by (101), (301) and (-101), originate from the RL-phase. For better resolution, diffraction spots belonging to the LN-matrix are marked here by four digits.

RESULTS

In this study, by means of TEM, we directly observed the presence of the RL-phase in LiNbO$_3$ at early stages of annealing. A microdiffraction pattern (zone axis [0001]LiNbO$_3$) taken from the sample annealed for 0.5 h is shown in Fig. 2. Selected diffraction spots originating from the rhombohedral LiNbO$_3$ -matrix are indicated by four digits, while several diffraction spots originating from the tetragonal RL phase, by three digits. Below, throughout the paper we will use only three digit notations. The analysis of TEM microdiffraction patterns allowed us to establish that the RL-phase is formed epitaxially with respect to the matrix, and orientation relations between the tetragonal RL-phase and the rhombohedral LiNbO$_3$ phase are: (301)||(010) , (101)||(110), and (010)||(001), respectively. Thus, for our Y-cut LiNbO$_3$ samples, the (301) atomic planes of the RL phase will be parallel to the sample surface, while the (010)RL-planes will be perpendicular to the sample surface. This conclusion has been confirmed by conventional x-ray diffraction measurements performed in para-focusing Bragg-Brentano geometry, in which a diffraction signal from the (301)RL atomic planes parallel to the sample surface was detected (see Fig. 3). Integrated (301)RL diffraction intensities, as a function of annealing time, are presented below, in Fig.7.

In order to investigate the layers located closer to the surface we performed GID measurements using the Hasylab beam line D4 at the DORIS storage ring of the DESY synchrotron (Hamburg, Germany). An x-ray beam (λ = 0.15453 nm) from a

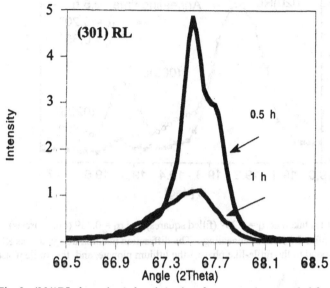

Fig. 3. (301)RL intensity (arb.units) taken from samples annealed for 0.5 h and 1 h.

Ge(111)crystal-monochromator was used to take diffraction profiles in the vicinity of the (006)LiNbO$_3$ surface reflection.

Measurements were performed at several incident angles, α_i, above and below the critical angle of the total external reflection, α_c = 0.25°, providing a wide range of x-ray penetration depths into the LiNbO$_3$. For each α_i-value, the diffraction profiles were recorded in the exit-angle-resolved mode (α_f - spectra), and within the $\omega/2\Theta$–interval containing (020)-rutile, (020)-RL and (006)-LiNbO$_3$ reflections . Diffraction intensities integrated over α_f-range were then plotted as functions of ω at different α_i.

Two such profiles taken at α_i = 0.20° < α_c (i.e. an x-ray penetration of only 5 nm into the LiNbO$_3$) and at α_i = 0.35° > α_c (70 nm x-ray penetration into the LiNbO$_3$) from a sample heat-treated for 0.5 h, are shown in Fig. 4. The profiles contain three peaks, corresponding to the RL-phase, the LiNbO$_3$ matrix, and a small amount of the non-reacted rutile phase. It can be seen that after 0.5 h of annealing, the RL-phase is dominant in the thin near-surface layer. The fact that the diffraction peak from non-reacted rutile is only barely visible in the spectrum taken at α_i = 0.20° < α_c, implies that the rutile phase remains not exactly at the crystal surface, but is located in a deeper layer. Islands of rutile phase at the RL / LiNbO$_3$ interface were indeed observed in the TEM cross-sections (see Fig. 5). After more than 0.5 h of annealing, the peak of the rutile phase disappears, and the GID profiles consist of the contributions from the RL and LiNbO3 phases only (see Fig.6).

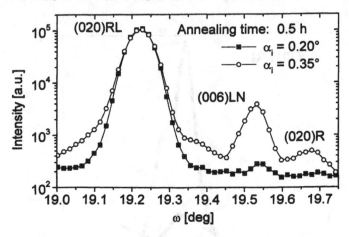

Fig. 4. GID profiles taken at $\alpha_i = 0.20^o$ (filled squares) and $\alpha_i = 0.35^o$ (open circles) from the sample after 0.5 hour of annealing. The diffraction maxima designated as RL, LN and R correspond to the rutile-like phase, the lithium niobate, and the rutile phase, respectively.

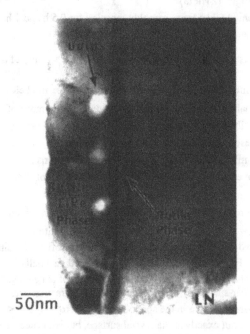

Fig. 5. TEM cross-section taken from a sample annealed for 0.5 h. The micrograph shows a coexistence of the RL-phase, the non-reactive rutile and the lithium niobate (LN) matrix. Several voids are visible at the reaction front due to the fast TiO_2 diffusion.

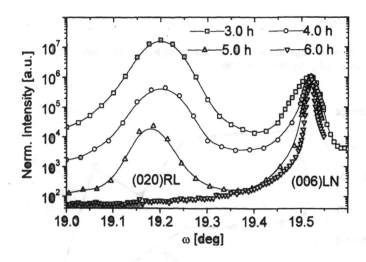

Fig. 6. GID profiles taken at $\alpha_i = 0.20^o$ from the samples annealed for different periods. In order to focus on the relative contribution of the RL-phase, the lithium niobate (LN) diffraction peak is artificially kept constant in all spectra.

From a comparison between the relative peak intensities in Fig. 6 one may infer the reduction of the RL-phase with increasing annealing time. After 4 h, the RL diffraction signal becomes smaller than the $LiNbO_3$ contribution, and after 6 h of annealing only the $LiNbO_3$ diffraction peak remains in the GID profiles, independently of the α_i – value. The temporal decay of the RL-phase during annealing is clearly visible in Fig. 7, in which the diffraction intensities (in a logarithmic scale), taken from the RL-phase by GID as well as by conventional x-ray diffraction, are plotted.

The influence of the RL phase on the structural parameters in the waveguide layer was studied by HRXRD. In fact, by using HRXRD we were able to extract the depth-resolved profiles of the $LiNbO_3$ lattice parameter, modified by Ti-diffusion [4]. By comparing those to the Ti-concentration profiles measured by SIMS, a numerical factor, K, was obtained which allowed to characterize quantitatively the extent of lattice contraction due to Ti-incorporation in the $LiNbO_3$ crystal. The K-factor was found to increase with annealing time, indicating a variable strain contribution to the structural parameters of the waveguide layer due to the gradual disappearance of the rutile-like phase, as observed by GID and other methods.

In fact, a detailed analysis [4] showed that lattice mismatches and strain components acting on the interface between the RL phase and the $LiNbO_3$ bulk, are significantly smaller than those between the Ti-diffused layer and the $LiNbO_3$ matrix. This is the reason for the relatively small K-values during the first 3-4 h of annealing. The complete disappearance of the RL-phase after 6 h annealing leads to the 1.7-times increase in the K-value due to the additional lattice contraction in the Y-direction caused by the interfacial strain components. This means the much stronger effect of Ti incorporation on the structural parameters and the refraction index of $LiNbO_3$.

CONCLUSION

These findings explain why 6 h annealing is used in industrial processing with a 30-40 nm initially deposited Ti layer.

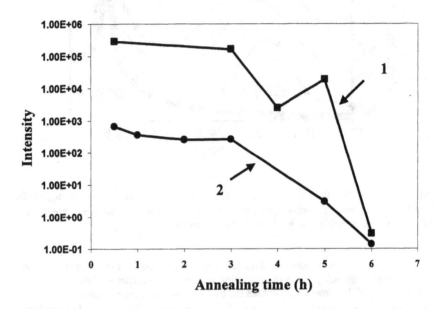

Fig. 7. Dependence of diffraction intensity (in a logarithmic scale) recorded from the RL-phase, as a function of annealing time: 1- (020)RL –reflection, GID data; 2 – (301)RL-reflection, conventional x-ray diffraction data.

ACKNOWLEDGEMENT

This research was supported by the German-Israeli Foundation (GIF) for Scientific Research and Development under contract # I-0406-006.07/95 . Dr. W. D. Kaplan (Technion) is gratefully acknowledged for his help in TEM.

REFERENCES

[1] M. A. McCoy, S. A. Dregia, and W. E. Lee, J. Mater. Res., 9, 2029 (1994).

[2] E. Zolotoyabko Y. Avrahami, W. Sauer, H. Metzger, and J. Peisl, Appl. Phys. Lett. , 73, 1352 (1998).

[3] C. E. Rice and R. J. Holmes, J. Appl. Phys., 60 , 3836 (1986) .

[4] Y. Avrahami and E. Zolotoyabko, J. Appl. Phys., 85, 6447 (1999).

STRUCTURAL CHARACTERIZATION USING SYNCHROTRON RADIATION OF OXIDE FILMS AND MULTILAYERS GROWN BY MOCVD

C. DUBOURDIEU*, J. LINDNER*, M. AUDIER*, M. ROSINA*, F. WEISS*, J.P. SÉNATEUR*, J.L. HODEAU**, E. DOORYHEE***, J.F. BÉRAR***
* Laboratoire des Matériaux et du Génie Physique, CNRS UMR 5628, INPG, ENSPG BP 46, 38402 St Martin d'Hères, France
** Laboratoire de Cristallographie, CNRS, BP 166, 38042 Grenoble cedex 9, France
*** European Synchrotron Radiation Facility, BP 220, 38043 Grenoble cedex, France

ABSTRACT

Synchrotron radiation at the European Synchrotron Radiation Facility has been used to characterize oxide thin films and multilayers grown by metal organic chemical vapor deposition (MOCVD). Reflectometry measurements were performed in the low angle region to get information on the quality of film/substrate and film/film interfaces of the heterostructures. The experimental data were compared to simulated spectra. High angle diffraction experiments were performed on superlattices of $(BaTiO_3/SrTiO_3)_{15}$ and $(La_{2/3}Sr_{1/3}MnO_3/SrTiO_3)_{15}$. The multilayers are oriented with the [001] direction perpendicular to the substrate plane. The 00l diffraction peaks were recorded up to the 008 one. The satellite peaks observed for both types of multilayers indicate the good coherence over the whole stacking. For the $(BaTiO_3/SrTiO_3)_{15}$ systems, the diffraction peaks were particularly well resolved even for the highest order (008 peak), showing excellent epitaxial quality with abrupt interfaces for periods up to 16 nm.

INTRODUCTION

The progress gained in oxide growth over the last decade allows now to grow heterostructures (multilayers, superlattices) combining differents oxides with a nanometer scale control. The properties of an individual oxide (superconductivity, piezoelectricity, ferroelectricity, magneto-resistivity...) can be modified when assembled with another functional oxide due to strain effects, interface effects (reduced dimensions), or coupling effects. In this paper, we report on two kinds of superlattices : magnetic superlattices and ferroelectric superlattices.

Manganite compounds ($La_{1-x}A_yMnO_3$, where A is a vacancy or a divalent ion such as Sr) exhibit a colossal magnetoresistance that can be of interest for magnetic sensors. However, the sensitivity is quite low (magnetic fields of the order of few tesla are required). In the search for better sensitivity, we investigate the properties of artificial structures such as $(La_{0.67}Sr_{0.33}MnO_3/SrTiO_3)_n$ superlattices, where manganite films are separated by insulating non magnetic layers.

Barium titanate $BaTiO_3$, together with its solid solution $Ba_{1-x}Sr_xTiO_3$ is a promising material for future dynamic random acces memories (DRAM) and tunable microwave devices. In this perspective, it is interesting to study the dielectric response (as a function of temperature and frequency) of $(BaTiO_3/SrTiO_3)_n$ superlattices.

In both cases, the crystalline quality of the layers as well as the quality of the film/substrate and film/film interfaces (roughness, interdifussion...) is very important. In this paper, we investigate the structure of the superlattices with synchrotron radiation. Reflectometry measurements in the low angle region and diffraction measurements in the high angle region were performed at the European Synchrotron Radiation Facility (ESRF). The dielectric properties and magneto-transport properties of the superlattices will be presented elsewhere.

EXPERIMENT

Thin films and multilayers were grown by metal organic chemical vapor deposition (MOCVD) using a liquid-injection system for the precursors delivery. The magnetic heterostructures (based on manganite) and the ferroelectric heterostructures (based on BaTiO$_3$) were prepared in two similar reactors described elsewhere [1]. The deposition temperature was 800°C for the ferroelectric heterostructures and 700°C for the magnetic heterostructures. The total pressure was 6.66 kPa and the oxygen partial pressure was 3.33 kPa. After deposition, the heterostructures where annealed under 1 atm of oxygen. Details about the preparation can be found in references [2,3].

BM2 [4] and BM16 [5] are two ESRF stations both equipped with two long-curved mirrors for vertical focusing and $\lambda/3$ harmonic rejection, and a Si[111] two-crystal monochromator with a resolution $\Delta E/E$ of about 2.10^{-4}. The reflectometry experiments were performed at BM16 at a wavelength of 0.4 Å on the 2-circles diffractometer. High angle diffraction at 0.775 Å was performed at BM2 (CRG-D2AM) on the 7-circles goniometer. The X-ray beam was focused in the horizontal plane by the sagittally bent second crystal of the monochromator.

The multilayers are oriented with the [001] direction perpendicular to the substrate plane. Therefore, only the 00l diffraction peaks were recorded up to the 008.

RESULTS

Magnetic heterostructures

The reflectivity curve of a La$_{1-x}$MnO$_3$ film (x~ 0.2) deposited on (001) SrTiO$_3$ is shown in figure 1. Such films have a magnetoresistance typically of the order of 20% at room temperature (290 K) under 1 T. Simulations of the system give a good agreement with the experimental curve for a film thickness of 102 nm and a roughness of about 0.8 nm (less than three unit cells). For comparison, the simulated curve corresponding to a 1.2 nm roughness is indicated. The calculated thickness is in good agreement with the thickness calculated from the reflectivity curve (110 nm) recorded *in situ* during deposition (using a 670 nm emitting diode laser and an index of 2.2 for the film).

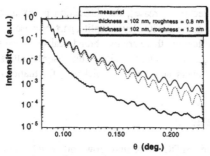

Figure 1 : Reflectivity curve ($\lambda=0.4$ Å) of a manganite $La_{0.8}MnO_3$ film deposited on $SrTiO_3$.

Multilayers alternating manganite and $SrTiO_3$ films were grown on $LaAlO_3$ substrates. The diffraction spectrum recorded for the 002 reflection on a standard diffractometer (Cu Kα radiation) for a $(La_{2/3}Sr_{1/3}MnO_3/SrTiO_3)_{15}$ multilayer of period $\Lambda = 32$ nm is shown in figure 2(a). The presence of satellite peaks reveal a good coherence over the stacking. It is generally not possible to record valuable data over higher orders (004, 005...) on a standard diffractometer owing to the lack of intensity and resolution. Orders up to 008 were recorded at BM2 line. Figures 2(b) and 3 show the diffraction patterns for the 002, 004 and 008 reflections.

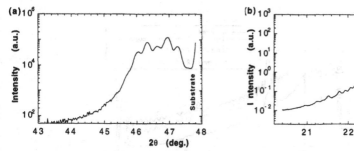

Figure 2 : θ/2θ scan on the 002 reflection of a $(La_{2/3}Sr_{1/3}MnO_3/SrTiO_3)_{15}$ multilayer grown on $LaAlO_3$. a) standard diffractometer ($\lambda=1.54$ Å) - b) using synchrotron radiation ($\lambda=0.775$Å).

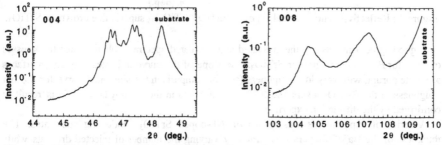

Figure 3 : θ/2θ scans on the 004 and 008 reflections using synchrotron radiation ($\lambda=0.775$Å) of a $(La_{2/3}Sr_{1/3}MnO_3/SrTiO_3)_{15}$ multilayer grown on $LaAlO_3$.

As the order of the reflection increases, the diffraction peaks arising from the manganite layers and those arising from the SrTiO$_3$ layers split apart because of the difference in lattice parameters. For the 008 reflection, there are two broad peaks with no structure clearly visible, which is indicative of roughness or interdiffusion. This disorder alters the film/film interfaces quality in both types of layers and leads to thickness fluctuations throughout the whole stacking.

Ferroelectric heterostructures

Thin films of BaTiO$_3$ and SrTiO$_3$ were first grown on SrTiO$_3$ and LaAlO$_3$ substrates. From the θ/2θ scans performed on a standard diffractometer (Cu Kα wavelength) BaTiO$_3$ films are found to cristallize in a cubic structure on both substrates, with a lattice parameter of 3.995 Å. Both BaTiO$_3$ and SrTiO$_3$ films are strongly oriented and only 001 lines show up on the diffractograms. On LaAlO$_3$ substrates, the full width at half maximum of the 002 rocking curves are 0.45° and 0.16° for BaTiO$_3$ and SrTiO$_3$ respectively. The roughness of the films was studied by atomic force microscopy. The root mean square value of the roughness is 1 nm and 2 nm for BaTiO$_3$ and SrTiO$_3$ respectively.

(BaTiO$_3$/SrTiO$_3$) multilayers were grown on SrTiO$_3$, SrTiO$_3$:Nb (for dielectric measurements) and LaAlO$_3$ substrates. The reflectivity curve of a (BaTiO$_3$/SrTiO$_3$)$_{15}$ multilayer grown on SrTiO$_3$ is shown in figure 4. The desired thicknesses were 7.5 nm for each layer.

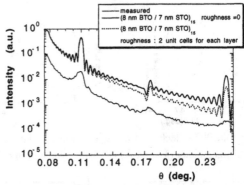

Figure 4 : Reflectivity curve (λ=0.4 Å) of a (BaTiO$_3$/SrTiO$_3$)$_{15}$ superlattice grown on SrTiO$_3$.

A good agreement between the measured and simulated spectra is obtained for thicknesses of 8 nm for BaTiO$_3$ and 7 nm for SrTiO$_3$. The slope of the curve indicates that the interfaces are quite abrupt, with very low roughness (see for comparison the simulated curve for 0.8 nm roughness in the BaTiO$_3$ layers and 0.78 nm roughness in the SrTiO$_3$ layers). This result is confirmed by the diffraction experiments.

A series of superlattices was grown on Nb-doped SrTiO$_3$ conducting substrates. The thickness of the BaTiO$_3$ layers was varied by varying the number of injected droplets while keeping the thickness of the SrTiO$_3$ layers constant (~ 8.0 nm). The θ/2θ scans on the 002 line obtained on a standard diffractometer are shown in figure 5. The respective calculated periods are reported in figure 5. A linear relationship is found between the period of the superlattices and the number of injected droplets for the BaTiO$_3$ layers. It shows that it is possible with our CVD process to grow superlattices with few nanometers interlayers thicknesses by controlling the amount of injected liquid. In figure 6, 004 and 008 diffraction spectra obtained using synchrotron radiation are shown for three of these multilayers. In the case of the thickest period (31.8 nm) the diffraction peaks do not exhibit a well resolved fine structure for the 008 order. This is indicative of defects in the interlayers structures, or roughness at the interfaces which leads to thickness fluctuations and shortens the coherence length. A broadening of the satellites peaks was also observed in the case of the (La$_{2/3}$Sr$_{1/3}$MnO$_3$/SrTiO$_3$)$_{15}$ superlattice of period 32 nm (see discussion above).

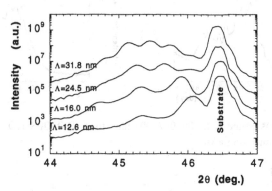

Figure 5 : θ/2θ scans obtained on a standard diffractometer(λ=1.54 Å) on the 002 peak for a serie of (BaTiO$_3$/SrTiO$_3$)$_{15}$ multilayers of different periods deposited on Nb-doped SrTiO$_3$.

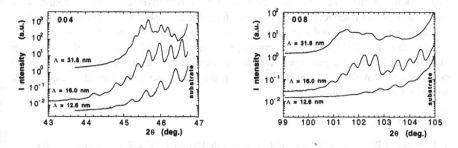

Figure 6 : θ/2θ scans using synchrotron radiation on the 004 and 008 peaks for a serie of (BaTiO$_3$/SrTiO$_3$)$_{15}$ multilayers of different periods deposited on Nb-doped SrTiO$_3$.

The quality of the superlattices obtained on LaAlO$_3$ is as good as on SrTiO$_3$ substrates. As an example, the θ/2θ scans obtained on the 004, 006, 007 and 008 lines of a superlattice consisting of ~ 10 nm BTO layers alternated with ~ 10 nm STO layers are shown in figure 7. The satellites peaks are very well resolved, even for the highest orders (007, 008), which reveals sharp interfaces and low interdiffusion.

223

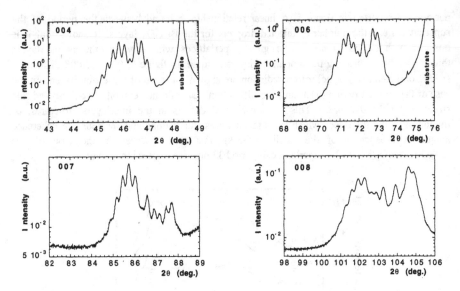

Figure 7 : θ/2θ scans on the 004, 006, 007, 008 lines for a $(BaTiO_3/SrTiO_3)_{15}$ superlattice of period 20 nm deposited on $LaAlO_3$

CONCLUSIONS

Synchrotron radiation was used to characterize the quality of $(BaTiO_3/SrTiO_3)_{15}$ and $(La_{2/3}Sr_{1/3}MnO_3/SrTiO_3)_{15}$ multilayers grown by liquid-injection MOCVD. The reflectivity measurements together with the diffraction data obtained up to the 008 reflection show a very good quality of these superlattices: the coherence is kept over the whole stacking, with particularly abrupt interfaces in the case of the ferroelectric superlattices. As the period of the superlattice increases, disorder appears (roughness) that leads to thickness fluctuations.

In the case of $(BaTiO_3/SrTiO_3)_n$ multilayers, anomalous diffraction at the strontium absorption edge will be performed. Diffraction anomalous fine structure analysis (DAFS) could indeed provide additional information on the local order given the site- and element specificity of this technique.

REFERENCES

1. F. Felten, J.P. Sénateur, F. Weiss, R. Madar, J.de Physique IV, **C5**, p. C5-1079, (1995)

2. J. Lindner, F. Weiss, W. Haessler, G. Köbernik, J.P. Sénateur, S. Oswald, J. Santiso, A. Figueras, Mat. Res. Soc. Symp. Proc. **541**, p. 501, (1999)

3. C. Dubourdieu, M. Audier, J.P. Sénateur, J. Pierre, to appear in Journal of Applied Physics

4. J.L. Ferrer, J.P. Simon, J.F. Bérar, B. Caillot, E. Fanchon, O. Kaïkati, S. Arnaud, M. Guidotti, M. Pirocchi, and M. Roth, J. Synchrotron Rad. **5**, p. 1346 (1998)

5. A.N. Fitch, Mater. Sc. Forum **228**, p. 219 (1996)

Microbeam Diffraction, Microtomography, Topography

PLASTIC DEFORMATION AND RECRYSTALLIZATION STUDIED BY THE 3D X-RAY MICROSCOPE

D. JUUL JENSEN[1], Å. KVICK[2], E.M. LAURIDSEN[1], U. LIENERT[2], L. MARGULIES[1+2], S.F. NIELSEN[1] and H.F. POULSEN[1]

[1]Materials Research Department, RISØ National Laboratory, Frederiksborgvej 399, DK-4000 Roskilde, Denmark

[2]ESRF, BP 220, F-38043 Grenoble Cedex, France

ABSTRACT

A newly developed synchrotron instrument – the so-called 3D X-ray microscope – is presented. The instrument is placed at the Materials Science beamline at ESRF and dedicated to local µm scale structural characterization within bulk materials. In this paper, emphasis is on *in-situ* studies of thermomechanical processing. The potential of the instrument for characterization of single nuclei and grains is described and discussed based on both first results and planned experiments.

INTRODUCTION

Recently significant progress has been achieved in the understanding of basic metallurgical processes such as plastic deformation and recrystallization, see e.g. [1 - 4]. Important for this progress has been the use of several advanced characterization techniques. However, such techniques are typically restricted to either detailed studies near surfaces (e.g. microscopy) or more average studies deeper below the surface or in the bulk (e.g.standard X-ray or neutron diffraction). With today's available techniques it is not possible to characterize non-destructively local processes occurring on a µ-scale in the bulk of a metallic sample. This means that despite of the cited progress in understanding, essential questions are still unanswered or only addressed theoretically. Within the fields of plastic deformation and recrystallization examples of such questions are:

- How do the single grains in a polycrystalline sample subdivide and rotate in orientation space during deformation?

- Where in a deformed microstructure and with which crystallographic orientations do nuclei form during recrystallization?

- How do individual nuclei grow during recrystallization?

With the 3D X-ray microscope [5,6], developed in a collaborative effort between Risø and ESRF, it may become possible to address experimentally these and, of course, also many other questions. Following a short description of the experimental technique, some examples are given of experimental investigations which cannot be carried out with other techniques. These examples are limited to non-destructive characterization of grain structures in 3D, grain subdivision during plastic deformation and growth of individual nuclei during recrystallization. Further applications within the same scientific framework can be foreseen covering characteri-

227

zation of dislocation densities and distributions, recovery, development of nuclei, phase transformations and grain boundary migration. For a status on local strain and stress analysis using the 3D X-ray microscope the reader is referred to [7].

THE 3D X-RAY MICROSCOPE

The 3DXRD microscope was commissioned during the Summer 1999 in a new experimental hutch at the Materials Science beamline at the ESRF synchrotron in France. The underlying principle of the instrument is the focusing of high energy X-rays (50 – 100 keV) by means of a bent Laue crystal and/or multilayer mirrors to one or two-dimensional spots of 1-5 or 5x10 micron size. In this way a unique combination of high flux (10^{11} cts/sec), high spatial resolution and high penetration power (in the mm-range) is obtained. With a focal distance of 2 meters the divergence of the monochromatic beam is tailored to the applications (0.1 – 1 mrad), while leaving ample space for large sample environments. For technical details, see Lienert *et al.* [6,8].

The instrument itself – sketched in Fig 1 - is basically a heavy-duty horizontal two-axis set-up, designed for holding 200 kg with an absolute accuracy of a few microns [6]. Two CCDs are available, a wide-range, low spatial resolution one (pixel size of 67 µm), and a narrow-range, high spatial resolution one (pixel size of 5 µm). Both can be translated in all directions, and may even be used simultaneously as the high resolution one is semi-transparent. Thanks to the low scattering angles at high energies, experiments can in general be performed with an x-y-z- ω setup, that is without the need for Eulerian cradles and the associated sphere-of-confusion problems. A photo of the sample-detector arrangement as used for X-ray tracing experiments (cf. below) is reproduced as Fig 2. Sample auxiliaries include 2 furnaces and a 25 kN Instron tensile machine. All movements are motorized, and all motors, detectors and auxiliaries are controlled via SPEC, the ESRF control software.

Four different methods have been developed for depth-resolved studies: x-ray tracing with line [9] or point focus, conical slits [10], and focusing analyzer optics [11]. In all cases the diffracted beam is transmitted through the sample and the illuminated area is determined by the focusing. A schematic presentation of the four principles is given in Fig 1, and a comparison of the essential features in Table 1. Automatic data analysis for the 3 former set-ups – including extensions to stress analysis - are presently under development.

APPLICATIONS

3D Grain Mapping

Recently much focus has been paid to grain boundary characterization [e.g. 12]. One of the fields emerging from this is termed grain boundary engineering [13–16]. The aim of grain boundary engineering is to optimize properties of a material by tailoring the grain boundary characteristics. It has been shown that a high fraction of special boundaries, typically low-angle and/or low Σ coincidence boundaries can significantly improve specific bulk properties. A well-known example of grain boundary engineering is the reduction of intergranular corrosion and stress corrosion cracking in steels by increasing the fraction of low Σ coincidence boundaries.

Grain boundary characterization and grain boundary engineering require statistical characterization of very large numbers of grain boundaries in typical polycrystalline materials. Out of the nine parameters describing a grain boundary, five are important for the properties, namely the crystallographic misorientation across the boundary (three parameters) and the grain boundary normal relative to the macroscopic sample axes (two parameters) [17].

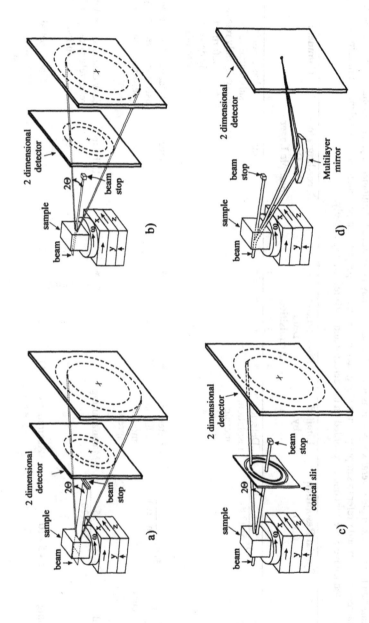

Fig 1. Sketches of the experimental set-up with 4 different principles for 3D beam definition inside the sample. a) X-ray tracing (line focus), b) X-ray tracing (point focus), c) conical slit, d) focussing optics.

TABLE 1

Characteristics for identification of individual nuclei/grains of four techniques for 3D beam definitions. Depending on the type of beam definition, data from for example several x,y positions inside the sample may be achieved simultaneously. This is marked by * in the table. The parameters x, y, η and r refer to sample position along incident beam, sample position perpendicular to incident beam, azimuthal angle along diffracted Debye-Scherrer cone as recorded on the detector and (hkl) reflex, respectively.

Type	Spatial Resolution (μm)	Limited by	Simultaneous Data Acquisition				Recording Time For One Sample Layer	Comment
			x	y	η	r		
X-Ray tracing (line focus)	$x \cdot 5 \cdot \dfrac{5}{Sin(2\theta)}$	Detector resolution, mosaic spread	*	*	*	*	1 min.	Fast, requires substantial software analysis, only relevant for undeformed or weakly deformed materials
X-Ray tracing (point focus)	$5 \cdot 5 \cdot \dfrac{5}{Sin(2\theta)}$	Detector resolution, mosaic spread	*		*	*	1 hour	As above, but can be used for more heavily deformed materials
Conical slit	$5 \cdot 5 \cdot \dfrac{25}{Tan(2\theta)}$	Manufacturing			*	*	12 hours	Easy data analysis, but relatively poor spatial resolution
Focussing optics	$5 \cdot 5 \cdot 10$	Aberrations in optics	(*)				-	Good spatial resolution, but slow

Fig. 2. Photo showing parts of the set-up for X-ray tracing. A motorized slit system is seen to the right, the sample in a small vacuum furnace in the middle and the CCD is seen to the left in the picture.

At present, several microscopy-based techniques are used to determine these parameters. The misorientation across a large number of grain boundaries is easily determined by the electron back scattering pattern (EBSP) technique in the scanning electron microscope (SEM). This technique allows large sample areas to be inspected, simple sample preparation and use of automatic data analysis procedures for determination of the crystallographic orientation [18, 19]. The determination of the grain boundary normal is more troublesome. Using the EBSP technique this requires the tedious and destructive sample sectioning [20, 21] in small steps. With transmission electron microscope (TEM) techniques the grain boundary normal may be determined by tilting the sample in the microscope [e.g. 22], but the TEM technique is not well suited for statistical characterizations because of the very limited sample area.

Synchrotron Measurements. The 3D X-ray microscope is capable of precise identification of grains in a polycrystalline sample. This identification includes i) x, y, z positions of the grains inside the sample and ii) the crystallographic orientations of the grains. The precision with which these parameters need to be determined depends on the actual application. Sometimes it is enough to know the center of mass for the individual grains, whereas in other cases the exact locations (with μm-precision) of for example grain boundaries and triple junctions are needed.

The first series of synchrotron measurements performed with the aim of a 3D characterization of the grain structure made use of the conical slit system. For the results to be reported below focussing was not yet available so the incoming monochromatic beam was defined by a Tungsten slit to $50 \times 50 \ \mu m^2$. The conical slit system was placed 50 mm behind the sample and the openings in the slit system were 20 μm wide. The energy of the incident beam was 87.2 keV, tuned to match the conical slit, such that the (200) and the (331) reflections from the actual copper sample could be recorded on the CCD.

After alignment of the conical slit system [10], the crystallographic orientations of the selected grains in the polycrystalline sample were determined by translating the sample (x, y, z) until the gauge volume was within the grain of interest and acquiring images while rotating the sample from $\omega = -90°$ to $90°$. Then the grain boundaries in a given z plane were determined by translating the sample in x,y relative to the gauge volume and recording the x,y positions where the intensity of the diffraction spot had decreased to half of the maximum value.

An example of the result is shown in Fig. 3b. The same z-plane was subsequently characterized using the EBSP technique, see Fig. 3a. A comparison between Figs. 3a and b shows a reasonable match, but the precision with which the grain boundary positions are determined varies from boundary to boundary. The cause of this was mainly the choise of manufacturing process, machining, for the prototype conical slit in use. By means of spark cutting a superior device – with precise gaps corresponding to 6 reflections of an FCC material – has now been produced and tested [10]. Also an improved data analysis method is being developed which makes use of several reflections for a given grain.

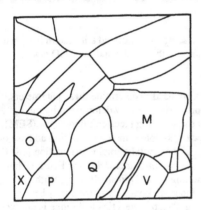 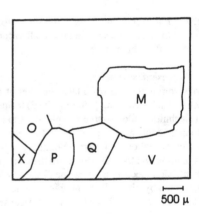

500 μ

Fig. 3.   Grain structure in a given section of an OFHC copper sample a) characterized by EBSP, b) characterized by the 3D X-ray microscope (conical slit).

In the second series of synchrotron measurements for 3D grain structure characterization, the X-ray tracing method was used to define the third dimension. With this method the diffraction pattern is recorded for three distances between the sample and the CCD and for 22 ω settings (-26° $\le \omega$ <16°). While recording, ω is scanned $\pm 1°$ in order to cover all ω angles. By an advanced analysis procedure including information about the diffraction peak position as well as the diffraction peak shape, it has been possible to back calculate the full 3D grain structure [9]. An example for a single grain is shown in Fig. 4. The grain structure seen in colour is determined by EBSP and the line profile is the synchrotron result. Improved analysis procedures are presently being developed which is believed to lead to even more precise boundary determination.

Depending on the actual application, either of the two approaches, the conical slit or the X-ray tracing can be selected.

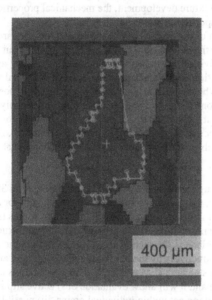

Fig. 4.   A selected grain in a 99.996% pure aluminium sample characterized by EBSP and the 3D X-ray microscope (X-ray tracing-line focus, see Fig. 1a).

Plastic Deformation

A new trend in quantitative characterization of deformation microstructures is to combine the purely morphological studies with measurements of crystallographic orientations in selected local (μm sized) volumes of the deformed microstructure. The latter is generally done by Kikuchi line technique in the TEM [23, 24] or by the EBSP technique in the SEM (18, 19, 25). The investigations have revealed that the original grains subdivide in typical metals and alloys during plastic deformation. Subdivision by cell formation is a well-known phenomena in metals with medium to high stacking fault energy, but the subdivision may also take place on larger length scales by: i) Formation of single-walled dense dislocation walls (DDWs) and/or double-walled microbands (MBs). The DDW/MBs are seen as elongated, nearly straight, vertically parallel dislocation boundaries which delineate several cells, thus bonding cellblocks. Typical examples are shown in Fig. 5. ii) Subdivision on a grain scale, for example due to grain-grain interactions or due to the formation of transition bands separating matrix bands. This subdivision is typical for both single crystals and polycrystals. For a general description and recent overview, see [2, 4].

The investigations have also revealed that the grain subdivision depends on the orientation of grains. The morphology may appear different in grains of different orientations [26] and the DDW/MBs can be parallel or inclined to a slip plane [27]. The dislocation boundaries developed during the deformation are associated with changes in crystallographic orientation. It is generally found that the misorientation across the dislocation boundaries increases with increasing strain. The increase is being more rapid for the DDW/MBs than for the cell boundaries. At medium and high strains, misorientations greater than 15° are observed across many of the DDW/MBs, and the DDW/MBs become indistinguishable from the original grain boundaries. This has important implications for the overall texture development, the mechanical properties, and for recrystallization occurring upon subsequent annealing.

The observed subdivision of the original grains means that the standard hypothesis [e.g. 28] that an original grain during the deformation continues as a deformed but "unbroken" grain, cannot be correct. The subdivision has to be incorporated. For example, applying crystal plasticity, such an incorporation requires a detailed analysis of the actual subdivision in relation to the active slip systems. Whereas such an analysis is experimentally possible for single crystals [e.g. 29 - 31], it cannot be done for polycrystalline materials. For the polycrystals only a snapshot of the subdivision at a given strain can be recorded and the evolution cannot be followed (except at the surface which is considered non-representative). A full analysis of active slip systems is thus not possible, as the previous crystallographic orientations of a given grain are not known.

Synchrotron Plans. Although the synchrotron technique allows a full 3D characterization of crystallographic orientations, the spatial resolution is not fine enough for a mapping of the cell structure of typical metals or alloys deformed to medium or high strains (see Fig. 5). Instead, it is planned to use the 3D X-ray microscope to characterize the 3D grain structure in the bulk of an undeformed sample (as described in the section above). Then deform the sample and characterize it, both non-destructively by the 3D X-ray microscope and after sectioning to an appropriate sample plane by standard SEM and TEM. This procedure is considered experimentally fairly simple and it will allow evaluation of i) active slip systems in relation to morphology, ii) texture development within individual grains, iii) possible effects of grain rotation on morphology, and iv) grain-grain interactions.

Recrystallization

The modernization and automation of the EBSP technique have lead to a break-through in characterization of recrystallization and recent international recrystallization conferences have been full of EBSP data [e.g. 1]. Reasons why EBSP has become such an important technique in recrystallization studies are that it is easy to use and that the spatial ($\geq 0.5 \ \mu m$) and angular ($\geq 1^{\circ}$) resolution of the technique match that of the topic – nuclei are typically a few micrometer in diameter and the crystallographic orientations generally only needs to be determined with a precision of a few degrees.

Many of the EBSP recrystallization investigations have been dealing with nucleation trying to identify where nuclei with certain orientations form in the deformed microstructure. A notable result has been the observation that in hot deformed aluminium alloys, cube oriented nuclei develop in so-called cube bands which are bands of deformed materials with cube or near cube

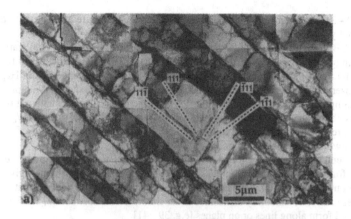

Fig. 5.   TEM micrographs from the longitudinal (RD-ND) plane showing grain subdivision .
a) Pure Al cold rolled 10%. DDWs and MBs are seen at an angle of approximately
40° to the rolling direction RD and parallel to a trace of a (111) plane [31]. b) Commercially pure Al cold rolled 90%. Some lamellar boundaries are observed to be almost
parallel to the rolling plane.

orientation [32 - 34]. Another important observation is that the nuclei frequently are partly surrounded not by high angle but rather by low angle boundaries [35].

Also many EBSP investigations have been dealing with growth during recrystallization. Here the EBSP technique has allowed an extension of the standard Cahn-Hagel analysis of true average growth rates [36] to include also orientations [37]. With this extended Cahn-Hagel method, the observed nuclei/grains are classified according to their orientations into pre-selected classes, for example if a nucleus/grain is within 15° from the ideal cube orientation, it is classified as a cube nucleus/grain. Then average growth rates for each class of nuclei/grains are calculated. The results have shown that in many cold-rolled samples, nuclei/grains of certain orientations grow faster than others [e.g. 35, 38].

The experimental data necessary for an extended Cahn-Hagel analysis of growth rates, also form the basis for microstructural path modelling [39] which, besides growth rates, give information about the nucleation. For several fcc metals such modelling indicates that the nucleation sites cannot be randomly distributed in the deformed microstructure, but the nuclei must preferential form along lines or on planes [e.g. 39 - 41].

There are, however, many investigations for which the EBSP technique is not sufficient because of its limitation to 2D. It can for example not be experimentally determined how the nuclei are arranged in 3D space, what the nucleation rate is or how individual nuclei/grains grow in the bulk. All these three topics have been addressed by the use of the 3D X-ray microscope and some of the first results on growth of individual nuclei/grains are summarized below. For more information [42].

Synchrotron Measurements. The point focus variant of the X-ray tracing method was used with the sample being placed 20 cm off-focus. A slit with a 70 x 70 μm opening was mounted in front of the sample. The sample material was aluminium AA1050 cold rolled to 90% reduction. The sample thickness parallel to the beam was 1 mm. The gauge volume in which recrystallization was followed was thus 70 x 70 x 1000 μm. The sample was annealed in-situ at 270°C in a small furnace (see Fig. 2) on the 3D X-ray microscope.

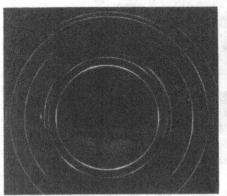 

Fig. 6.   Images recorded on the 2D detector (see Figs. 1 and 2) of commercially pure aluminium. a) Cold rolled to 90%. b) As a) and annealed at 270°C for 9720s.

The signal recorded on the 2D detector for the deformed sample before annealing is seen in Fig. 6a). Each of the rings corresponds to different reflections (111), (200) etc., and the intensities along the rings are intensities along an almost straight diameter in the corresponding pole figures. Upon annealing spots appear in the ring pattern, see Fig. 6b. Each spot is a recrystallization nucleus. The intensity in the spots increases with increasing annealing time as the nuclei grow. By integrating the intensity within a spot and plotting it as a function of annealing time, growth curves for the individual nuclei are obtained. It is assured that only nuclei originating and growing within the selected 70 μm x 70 μm wide channel are recorded. This is done by increasing the width of the channel to 100 μm x 100 μm and eliminating all spots for which the intensity increases. If the intensity of a given spot increases this would mean that the nucleus either was formed or have grown outside the selected channel and the recorded growth curve would, therefore, not be the true growth curve for the entire nucleus.

So far, only 6 nuclei have been analyzed [41]. Three of the growth curves are shown in Fig. 7. The results reveal that the growth curves for the six inspected nuclei are all very different; none of them has the same nucleation rate nor incubation period and some show a very smooth development, whereas others show growth with different rates at various times during the anneal.

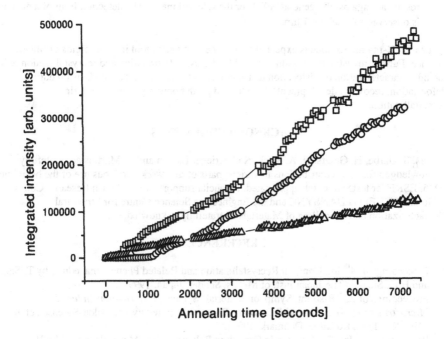

Fig. 7. Growth curves for 3 individual nuclei. The graphs reveal that each nucleus has its own recrystallization kinetics.

237

SUMMARY AND OUTLOOK

- A new high energy X-ray synchrotron instrument – the 3D X-ray microscope has been developed. It allows non-destructive characterizations on a local (μ-sized) scale in the bulk of typically millimeter sized metal samples.

- Some of the first experiments using the 3D X-ray microscope have been dealing with characterization of individual nuclei or grains.

  - It has been shown that typical polycrystalline grain structures can be characterized in 3D non-destructively. The necessary acquisition time is of the order of hours for a full 3D picture.

  - The nucleation and growth of individual nuclei has been followed in-situ during recrystallization. It is found that the nucleation rate and growth rate vary significantly from nucleus to nucleus. It is furthermore observed that some nuclei have a distinct incubation period before nucleation and that some nuclei vanish again after some growth during recrystallization. Such observations have never been possible before and the latter two results are against the general belief for the selected metal. The detection limit of a nucleus is observed to be about 1 μm.

The 3D X-ray microscope is expected to become a powerful tool in many fields of Material Science. Further investigations within the fields of plastic deformation and recrystallization will include characterization of dislocation distributions and densities, grain subdivision during deformation, recovery, development of nuclei and grain boundary migration during recrystallization.

ACKNOWLEDGEMENTS

Drs. S. Garbe, H. Graafsma, A. Koch, N.C. Krieger Lassen and R. Martins are gratefully acknowledged for their contributions to various parts of this work. Gracious use of the beamline ID15, ESRF for tests are much appreciated. Financial support by the Danish Research Counsils (SNF and STVF) via DANSYNC and the Engineering Science Centre for Structural Characterization and Modelling of Materials is gratefully acknowledged.

REFERENCES

1. Proceedings of 4th Int. Conf. on Recrystallization and Related Phenomena, edited by T. Sakai and H.G. Suzuki (The Japan Inst. of Metals, Sendai, Japan, 1999).
2. Proceedings of 20th Risø Int. Symp. on Materials Science, *Deformation-Induced Microstructures: Analysis and Relation to Properties,* edited by J.B. Bilde-Sørensen et al. (Risø Nat. Lab., Roskilde, Denmark, 1999).
3. Proceedings 3rd Int. Conf. on Grain Growth in Polycrystalline Materials, edited by H. Weiland, B.L. Adams and A.D. Rollett (The Minerals, Metals and Materials Society, 1998).
4. N. Hansen and D. Juul Jensen, Phil. Trans. R. Soc. London **357**, p. 1447, 1467 (1999).
5. H.F. Poulsen, S. Garbe, T. Lorentzen, D. Juul Jensen, F.W. Poulsen, N.H. Andersen, T. Frello, R. Feidenhans'l, H. Graafsma, J. Synchrotron Rad. **4**, p. 147-154 (1997).

6. U. Lienert, H.F. Poulsen and Å. Kvick in Proceedings of the 40th Conference of the American Institute of Aeronautics and Astronautics on Structures, Structural Dynamics and Materials (St. Louis, USA, 1999) **A99-24795**, p. 2067-2075.

7. U. Lienert, R. Martins, S. Grigull, M. Pinkerton, H.F. Poulsen, Å. Kvick, these proceedings.

8. U. Lienert, C. Schulze, V. Honkimäki, T. Tschentscher, S. Garbe. O. Hignette, A. Horsewell, M. Lingham, H.F. Poulsen, N.B. Thomsen, E. Ziegler, J. Synchrotron Rad. **5**, p. 226, 231 (1998).

9. H.F. Poulsen, S.F. Nielsen, E.M. Lauridsen, U. Lienert, R.M. Suter and D. Juul Jensen. Submitted for publication.

10. S.F. Nielsen, A. Wolf, H.F. Poulsen, M. Ohler, U. Lienert and R.A. Owen, in print J. Synchrotron Rad.

11. U. Lienert, H.F. Poulsen, V. Honkimäki, C. Schulze, O. Hignette, J. Synchrotron Rad. **9**, p. 979, 984, (1999).

12. Proceedings 6th Conf. on Frontiers of Electron Microscopy in Materials Science, Ultramicroscopy **67** (1997).

13. T. Watanabe in Proc. Textures of Materials ICOTOM 11, edited by Z. Liang et al. (Int. Academic Publishers, Beijing 1996) p. 1309 – 1318.

14. G. Palumbo, P.J. King, K.T. Aust, U. Erb and I.c. Lichtenberger, Scripta metall. mater. **25**, p. 1775, 1780 (1991).

15. P. Lin, G. Palumbo, U. Erb and K.T. Aust, Scripta metall. mater. **33**, p. 1387, 1392 (1995).

16. G. Palumbo and K.T. Aust in Proc. *Grain Growth in Polycrystalline Materials III*, edited by H. Weiland et al. (The Minerals, Metals & Materials Society, 1998) pp. 311-320.

17. B.L. Adams and C.T. Wu in Proc. Textures of Materials ICOTOM 11, edited by Z. Liang et al., Int. Academic Publishers, Beijing, 1996, pp. 23-30.

18. N.C. Krieger Lassen, D. Juul Jensen and K. Conradsen, Scan. Microscopy **6**, p. 115, 121 (1992).

19. S.I. Wright and B.L. Adams, Metall. Trans. **23A**, p. 759, 767 (1992).

20. M.V. Kral and G. Spanos, Acta mater. **47**, p. 711, 724 (1999).

21. A.D. Rollett et al. in *Microscopy and Microanalysis*, Springer, 1999, pp. 230 – 231.

22. X. Huang and Q. Liu, Ultramicroscopy **74**, p. 123, 130 (1998)

23. Q. Liu, Ultra microscopy **60**, p. 81, 89 (1995).

24. N.C. Krieger Lassen in *Proc. 16th Risø Int. Symp. on Mat. Sci.*, edited by N. Hansen et al. (Risø, Roskilde, Denmark, 1995), p. 405 – 411.

25. D.J. Dingley, Scan. Electron Microscopy **11**, p. 569, 575 (1984).

26. X. Huang and N. Hansen, Scripta mater. **37**, p. 1, 7 (1997).

27. G. Winther, D. Juul Jensen and N. Hansen, Acta mater. **45**, p. 5059, 5068 (1997).

28. G.I. Taylor, J. Inst. Met. **62**, p. 307, 324 (1938).

29. J.H. Driver in Proc. 16th Risø Int. Symp. on Mat. Sci. edited by N. Hansen et al. (Risø, Roskilde, Denmark, 1995) p. 25 – 36.

30. Å. Godfrey, D. Juul Jensen and N. Hansen, Acta mater. **46**, p. 823, 833 and p. 835, 848 (1998).

31. Q. Liu and N. Hansen, Proc. R. Soc. Lond. **A454**, p. 1, 19 (1998).

32. H. Weiland and J.R. Hirsch, Textures and Microstructures **14-18**, p. 647, 652 (1991).

33. I. Samajdar and R.D. Doherty, Scripta metall. mater. **32**, p. 845, 850 (1995).

34. H.E. Vatne, O. Daaland and E. Nes, Mat. Sci. Forum **157 – 162**, p. 1087, 1094 (1994).

35. D. Juul Jensen, Acta mater. **43**, p. 4117, 4129 (1995).

36. J.W. Cahn and W.C. Hagel in Decomposition of Austenite by Diffusional Processes, edited by V.F. Zackay and H.I. Aaronson (Interscience 1962), p. 131 – 196.

37. D. Juul Jensen, Scripta metall. mater. **27**, p. 533-538 (1992).
38. O. Engler, Acta mater. **46**, p. 1555, 1568 (1998).
39. R.A. Vandermeer in Proc. 16th Risø Int. symp. on Mat. Sci., edited by N. Hansen et al. (Risø, Roskilde, Denmark, 1995), p. 193 – 213.
40. R.A. Vandermeer and D. Juul Jensen, Acta metall. mater. **42**, p. 2427, 2436 (1994).
41. R.A. Vandermeer and D. Juul Jensen, Met. Trans. A **26**, p. 2227, 2235 (1995).
42. E.M. Lauridsen, D. Juul Jensen, U. Lienert and H.F. Poulsen, submitted for publication.

HIGH SPATIAL RESOLUTION STRAIN MEASUREMENTS WITHIN BULK MATERIALS BY SLIT-IMAGING

U. LIENERT *, R. MARTINS *, S. GRIGULL *, M. PINKERTON **, H.F. POULSEN ***, Å. KVICK *

* European Synchrotron Radiation Facility, 6 rue Jules Horowitz, B.P. 220, 38043 Grenoble Cedex, France, lienert@esrf.fr
** Manchester Materials Science Centre, Grosvenor St., M17HS, Manchester, England
*** Materials Research Department, Risø National Laboratory, 4000 Roskilde, Denmark

ABSTRACT

High energy synchrotron radiation is employed for residual strain measurements from local gauge volumes within the bulk of polycrystalline materials. The longitudinal spatial resolution is defined by placing a narrow imaging slit behind the sample and recording the intensity distribution on a position sensitive detector. It is shown that the sample to slit distance can be increased without sacrificing longitudinal resolution by applying a reconstruction technique. Hence, space is provided for large samples and sample environments. The reconstruction technique is described and validated by measuring the residual strain profile of a shot-peened Al sample. A longitudinal gauge length of 95 μm is achieved at 52 keV with a sample to slit distance of 10 cm.

INTRODUCTION

Third generation high energy synchrotron facilities like the European Synchrotron Radiation Facility (ESRF) in Grenoble, France, provide high energy X-ray beams (E > 50 keV) of unprecedented brilliance. Penetration depths of several millimeters are obtained for many metals and ceramics of technological interest, enabling non-destructive structural characterization within the bulk. A novel experimental station, referred to as 3-Dimensional X-Ray Diffraction microscope (3DXRD), has been constructed at the Materials Science Beamline ID11 of the ESRF [1]. Here, micrometer sized gauge volumes are achieved by means of focussing optics. The utilization of Position Sensitive X-ray Detectors (PSD) provides fast data acquisition and allows *in-situ* investigations during thermo-mechanical processing [2].

The gauge volume is confined perpendicular to the incoming beam by the dimensions of the focal spot. The longitudinal resolution on the other hand is conventionally defined by a crossed beam technique. A collimator, usually made out of two slits, is placed behind the sample and selects diffracted beams from a confined intersection length with the incoming beam. The longitudinal resolution degrades with increasing distance between sample and collimator due to the finite opening divergence of the collimator. The collimator therefore needs to be placed close to the sample restricting the available space for samples and sample environment.

Here we describe a novel technique for strain measurements that provides a good longitudinal resolution but does not require the placement of an optical element closely behind the sample.

EXPERIMENT

Set up and coordinate system

The experiment was performed on the 3DXRD station. The set up is sketched in Fig. 1 showing the vertical scattering plane. An energy band of $\Delta E/E = 0.34$ % was selected at 52 keV by a bent Laue crystal and focussed to a 1 mm wide and $\Delta foc = 6$ µm high line. The samples were positioned at the line focus and an imaging slit was installed behind the sample in the diffracted beam. The imaging slit was made out of 2 mm thick rounded and polished WC blades spaced by 10 µm thick foils. An effective slit opening Δs of 15 µm was inferred from scanning the slit through the line focus. The positions of the diffracted beams perpendicular to the incoming beam were measured by a PSD with a spatial resolution below 100 µm.

Fig. 1: Experimental set-up. A horizontal line focus is obtained by a bent Laue monochromator placed in the direct beam. Due to the crystal bending an energy gradient is caused along the crystal as indicated by the short and long dashed lines, symbolizing beams of short and long wavelength, respectively. Diffracted beams are selected by a narrow imaging slit and detected on a position sensitive detector.

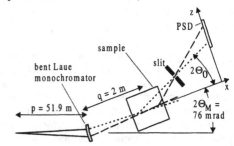

The coordinate systems are shown in Fig. 2. A laboratory coordinate system is defined by taking the incident beam as x-axis. The PSD is placed parallel to the z axis. The zero positions x_0, z_0, and the reference scattering angle $2\Theta_0$, are defined by a calibration sample as described below. A sample system, designated by a dash ', is defined by a shifted x' axis such that the zero position coincides with the center of the sample. The sample position relative to the laboratory system is described by the parameter Δx.

Fig. 2: Ray diagram and coordinate system describing the slit imaging technique. The diffracted beam indicated by the solid line represents the reference beam as diffracted by a strain free calibration sample. d_{SD} was about 2050 mm.

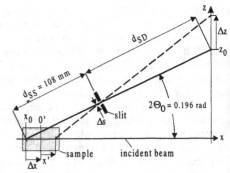

Strain measurements by slit imaging

The aim is the measurement of the difference between the actual atomic lattice spacing d at position x' within the sample and the strain free lattice parameter d_0. The differential form of Braggs law gives

$$\frac{d(x')-d_0}{d_0} = -\frac{2\Theta_c(x')-2\Theta_0}{2\tan\Theta_0} = -\frac{\Delta 2\Theta(x')}{2\tan\Theta_0} \quad . \quad (1)$$

242

which for the present case amounts to 10 μm for δε = 10⁻⁴. The sample should therefore be positioned with an accuracy of 10 μm. The factor b in (2) should be determined to an accuracy of δx'/ Δξ = 2 % and the term in the square brackets can be neglected. It is in general sufficient to measure strain to an accuracy of a few percent so that d_{SD} and Θ_0 should be measured within 1 %.

Calibration and longitudinal resolution

The determination of the geometric parameters is best achieved by translating a thin strain free sample along the incident beam. Such a sample was not available when the experiment was performed so a 28 μm thick cross rolled Al foil was used instead. Fig. 3 shows the center position and integrated intensities of the peaks observed on the PSD when the foil was translated along x. The factor b is given by the slope of the center positions plotted as function of foil position Δx. The width Δξ over which a signal is obtained is in good agreement with the instrumental terms in (4).

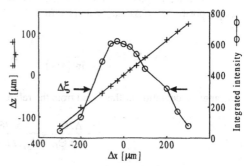

Fig. 3: Calibration scan by translating a 28 μm thick Al foil along the incoming beam. The peak centers together with a linear fit and the integrated intensities of the detector signal are plotted.

The longitudinal resolution, Δx_{long}, is measured as the width of the peak observed in one independent channel of the PSD when a thin foil is translated along x. Three geometric factors contribute to the achievable resolution: the opening of the slit Δs, the height of the incident beam Δfoc and the width of an independent PSD channel Δpsd. Assuming all contributions to be Gaussian one obtains

$$\Delta x_{long} = \frac{1}{\tan(2\Theta_0)} \sqrt{\left(\frac{d_{SS}+d_{SD}}{d_{SD}}\Delta s\right)^2 + \left(\Delta foc\right)^2 + \left(\frac{d_{SS}}{d_{SD}}\Delta psd\right)^2} \qquad (6)$$

A longitudinal resolution Δx_{long} = 90 μm is calculated from the experimental parameters. The contribution from the slit opening Δs dominates but the large value of d_{SS} does not substantially increase Δx_{long} as long as d_{SS} is small compared to d_{SD}. Fig. 4 shows the experimental result obtained by scanning the Al calibration foil along the beam. The overall peak shape is well approximated by a Gaussian. The calculated and observed values are in good agreement considering the finite thickness of the Al foil and the simplifying assumption of Gaussian contributions.

Fig. 4: Longitudinal resolution function as obtained by translation of a 28 μm thick Al foil along the beam and recording the signal in a single independent detector channel . The experimental data points are fitted to a Gaussian.

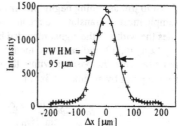

$2\Theta_c(x')$ is the angular center position of the diffraction peak from the gage volume centered at x'. It is clear from Fig. 2 that beams impinging on the PSD for one sample setting Δx come from different positions x' within the sample. The center position $\Theta_c(x')$ can therefore not be obtained directly as center position of the intensity distribution on the PSD. Instead, the diffraction peaks for each sample position x' are reconstructed from the measured Δx and Δz positions before finding the center.

The source point x' of a diffracted beam detected at a position Δz on the PSD for a sample positioned at Δx can be derived from Fig. 2 as

$$x' = b\Delta z\left[1 + \frac{\Delta z}{d_{SD}\sin(2\Theta_0)}\right]^{-1} - \Delta x, \quad b = \frac{d_{SS}}{d_{SD}\tan(2\Theta_0)} \quad .(2)$$

The angular deviation of this beam from the reference scattering angle $2\Theta_0$ is obtained as

$$\frac{\Delta 2\Theta}{2\tan\Theta_0} = \frac{\cos 2\Theta_0}{d_{SD} 2\tan\Theta_0}\Delta z \tag{3}$$

Equations (2) and (3) establish the transformation from the experimental parameters, Δx and Δz, to the desired strain as function of position within the sample.

Peak width and tolerances

The correct mean scattering angle $2\Theta_c$ is only obtained if the complete diffraction profiles are measured. Therefore intensity profiles $I(\Delta z)$ must be measured over a sufficiently large range of sample translations Δx (*cf.* Fig. 3). A signal is only observed over a distance $\Delta\xi$ around the position x_0 which is taken as the center of the integrated intensity measured on the PSD as function of Δx. z_0 is defined as the center of the peak on the PSD if the sample is at x_0. The width $\Delta\xi$ of this distribution depends on the energy bandwidth $\Delta E/E$ and divergence of the incident beam and on the intrinsic width of the sample reflection $\Delta\Theta_{hkl}$, which is caused by size effects and microstrains. For the actual set-up $\Delta\xi$ is calculated as

$$\Delta\xi = \frac{d_{SS}}{\sin(2\Theta_0)}\sqrt{\left(\frac{\Delta E}{E}\right)^2\left[2\tan\Theta_0 + 2\tan\Theta_M\left(1 + \frac{q\sin\Theta_{IN}}{p\sin\Theta_{OUT}}\right)\right]^2 + (\Delta\Theta_{hkl})^2} \tag{4}$$

where Θ_M is the monochromator Bragg angle and Θ_{IN} and Θ_{OUT} the angles between the incoming/exiting beams and the surface normal of the Laue crystal. The instrumental term leads to $\Delta\xi = 490 \ \mu m$ assuming negligible size effects. Equation (4) provides the distance over which the sample must be translated to reconstruct a complete diffraction peak. For homogeneous samples the width of the signal recorded by the PSD, $\Delta\xi$, is given by $\Delta\xi / b$.

Equation (2) indicates that the finite distance between sample and slit d_{SS} introduces a coupling between the position within the sample and the local strain. The strain causes a shift of the source point by an amount $\delta x'$ of

$$\delta x' = \frac{d_{SS} 2\tan\Theta_0\delta\varepsilon}{\sin(2\Theta_0)} \approx d_{SS}\delta\varepsilon \tag{5}$$

Sample description and measurements

The shot peened Al sample consisted of two pieces of an area of 10×10 mm² and 5 mm thickness. Each plate was shot peened on one side and the plates were then mounted to form a 10 mm cube such that the shot peened surfaces faced each other resulting in an internal interface to cancel out both surface and absorption effects in strain measurements [3]. The in-plane strain was measured, *i.e.* the scattering vector of the investigated [3 1 1] reflection was parallel to the interface.

For a reference measurement the interface was aligned parallel to the incoming beam and the depth dependent strain profile was recorded by translating the sample perpendicular to the interface. In this case the depth resolution was given by the lateral beam width of 6 µm and no reconstruction needed to be applied as the sample is considered homogeneous along the beam. The strain measurement was performed by an analyzer crystal.

Then, the sample was rotated by 90° such that the interface was perpendicular to the incident beam. The sample was translated along the beam probing the depth strain gradient now with the longitudinal gauge length by applying the slit imaging technique. The sample was oscillated parallel to the interface by ±1 mm during the latter measurements.

As no strain free reference sample was available, the absolute zero position of the strain profile was chosen such that the average strain over the full sample thickness was zero. The b parameter as obtained from the calibration foil was corrected to account for the different d spacings.

RESULTS

The measured strain profiles are plotted in Fig. 5. The reference profile outside the plastically deformed region indicates a small number of detected diffracting grains due to the fact that the sample was not oscillated and an analyzer crystal was used. The symmetric shape reflects the mounting of two shot peened surfaces against each other, the interface being at x' = 0. As intended, the plastic deformation produces a compressive layer of 200 µm thickness which should inhibit surface crack propagation. The strongest compression is situated slightly below the surface resulting in a narrow bump in the strain profile at x' = 0. This feature provides a sensitive test of the spatial resolution. It is almost completely resolved if the spatial resolution is below 50 µm and not observed at all above 200 µm.

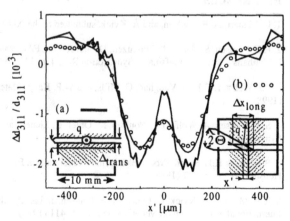

Fig. 5: In-plane [3 1 1] strain profile of the shot-peened Al sample. q indicates the scattering vector. Plotted are: (a) experimental reference profile (beam parallel to shot-peened surfaces)(bold), (b) experimental profile from the reconstruction technique (beam perpendicular to the shot-peened surfaces) (circles) and the experimental reference profile convoluted with a Gaussian resolution function of 100 µm FWHM (thin). The insets show the scattering geometry. The shot-peened regions are hatched.

Good agreement is found between the reconstructed strain profile and the reference profile convoluted with a Gaussian of 100 μm FWHM which is close to the resolution obtained from the translation of the calibration foil. The center bump is not resolved without application of the reconstruction procedure.

The plastically deformed region also leads to an increase of the width of the diffraction peaks from $\Delta d/d = 3.5 \times 10^{-3}$ to 5.5×10^{-3}, the former being due to the energy bandwidth of the focussed beam whereas the latter reflects size effects.

CONCLUSIONS

A novel technique is presented for the local measurement of strain fields within polycrystalline bulk materials. A longitudinal resolution of 95 μm was achieved by mounting a narrow slit 10 cm behind the sample and back tracing of the diffracted beams as recorded by a position sensitive detector. Substantial space is therefore provided for samples and environment. The technique may also be used to define 3 dimensional confined gauge volumes by additional horizontal focussing of the incident beam providing a gauge volume of $5 \times 10 \times 100$ μm^3. Averaging over a sufficient number of grains can be achieved by oscillating the sample perpendicular to local strain gradients.

The slit imaging technique can be extended to observe simultaneously several complete diffraction rings by conical apertures [4]. In combination with a 2 dimensional detector a bi-axial strain state can be measured by a single exposure [5] and the complete strain tensor and the local texture are obtained by rotating the sample around a single axis.

Potential applications where high spatial resolution is required include systems of inhomogeneous plastic deformation [6], deep strain and texture gradients within mechanically treated surfaces and thick coatings and strainfields at buried interfaces in composite materials.

ACKNOWLEDGMENTS

The authors are grateful to Prof. P.J. Webster for providing the shot-peened Al sample. Support for this work was provided by the Danish Reseach Councils, STVF and SNF (via Dansync).

REFERENCES

1. U. Lienert, H. F. Poulsen, and Å. Kvick, submitted to the AIAA Journal.

2. H.F. Poulsen, S. Garbe, T. Lorentzen, D. Juul Jensen, F.W. Poulsen, N.H. Andersen, T. Frello, R. Feidenhans'l, and H. Graafsma, J. Synchrotron Rad. **4**, p. 147 (1997).

3. P.J. Webster, G.B.M. Vaughan, G. Mills, and W.P. Kang, Mater. Sci. Forum vol. 278-81, p. 323 (1998).

4. S. F. Nielsen, A. Wolf, H. F. Poulsen, M. Ohler, U. Lienert, and R. A. Owen, submitted to J. of Synchrotron Rad. .

5. B.B. He & K.L. Smith, Proc. SEM Spring Conference on Experimental and Applied Mechanics, Houston, Texas, p. 217 (1998).

6. R.V. Martins, Å. Kvick, U. Lienert, H.F. Poulsen, and A. Pyzalla, Proceedings of the 20th Risø International Symposium on Materials Science, p. 411 (1999).

3-D MEASUREMENT OF DEFORMATION MICROSTRUCTURE IN Al(0.2%)Mg USING SUBMICRON RESOLUTION WHITE X-RAY MICROBEAMS

B.C. Larson, N. Tamura, J.-S. Chung, G.E. Ice, J. D. Budai, J. Z. Tischler,
W. Yang, H. Weiland,[b] and W.P. Lowe,[c]
Oak Ridge National Laboratory, Oak Ridge, TN 37830
[b]Alcoa Technical Center, Alcoa Center, PA
[c]Howard University, Washington D.C.

ABSTRACT

We have used submicron-resolution white x-ray microbeams on the MHATT-CAT beamline 7–ID at the Advanced Photon Source to develop techniques for three-dimensional investigation of the deformation microstructure in a 20% plane strain compressed Al(0.2%)Mg tri-crystal. Kirkpatrick-Baez mirrors were used to focus white radiation from an undulator to a 0.7 x 0.7 μm^2 beam that was scanned over bi- and tri-crystal regions near the triple-junction of the tri-crystal. Depth resolution along the x-ray microbeam of less than 5 microns was achieved by triangulation to the diffraction source point using images taken at a series of CCD distances from the microbeam. Computer indexing of the deformation cell structure in the bi-crystal region provided orientations of individual subgrains to ~0.01°, making possible detailed measurements of the rotation axes between individual cells.

INTRODUCTION

High-temperature mechanical deformation is an important industrial process for the generation of desired polycrystalline grain structure and forming of structural materials. The disoriented crystalline grains that form the structural units of polycrystalline materials are defined by grain boundaries and grain boundary intersections. These grain boundaries and triple-junctions severely restrict deformation induced slip processes and, hence, significantly increase the strength of polycrystalline materials relative to that of single crystals.[1,2] Although highly refined methods have been developed for the control of polycrystalline microstructure, little fundamental understanding exists regarding the details of mechanisms that lead to the observed microstructures.

X-ray microbeam diffraction represents a powerful new technique for the investigation of orientation and tri-axial strain for individual grains at sub-micron levels.[3-5] As this technique is developed, it will provide a means for non-destructively studying the local microstructure and stress/strain in bulk materials. In this paper we describe methods for producing submicron x-ray beams of white x-rays using Kirkpatrick-Baez (K-B) mirrors, and we illustrate their application to 3-D microstructure investigation using a CCD detector. White microbeams are used to obtain 3-D microstructural information through diffraction measurements on a bi-crystal region of an Al(0.2%)Mg tri-crystal deformed 20% in plane strain.

EXPERIMENT

A microbeam diffraction system with interchangeable white and monochromatic radiation has been constructed on the MHATT-CAT beam line at the Advanced Photon Source (APS).[3,4] ORNL and Howard University have fabricated a (small-offset) monochromator that can be translated in or out of the APS undulator beam to produce co-linear monochromatic or white radiation, respectively.[6] These (interchangeable) white or monochromatic beams are subsequently focused by crossed K-B mirrors with elliptical figures. As depicted in Fig. 1, white/monochromatic undulator radiation is incident on a vertical deflecting mirror with a 130 mm focal length and a horizontal deflecting mirror with a 60 mm focal length. Assuming 50x50 μm^2 beam defining slits before the mirrors, the converging microbeam has angular spreads of less than a milliradian (~0.05/130 and 0.05/60) in the vertical and horizontal directions, respectively. Diffraction patterns are collected on a 1242 x 1152 pixel CCD coupled to a phosphor screen by a 1:1 (untapered) fiber optics bundle. The ~22.5 micron pitch pixels of the CCD corresponds to a nominally 28 x 26 mm^2 active area that is typically positioned a distance of ~ 30-60 mm from the sample.

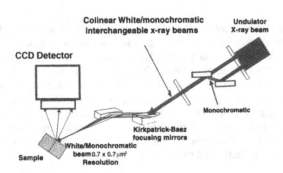

Fig. 1. Schematic overall view of microbeam components. A removable double-crystal monochromator provides white or monochromatic x-rays to crossed K-B mirrors, which focus a submicron beam onto the sample that diffracts into the CCD detector.

Data collection, reduction, and analyses are performed using interactive software that is the subject of ongoing development. Automatic and manual peak searches with two-dimensional Gaussian, Lorentzian, or Pearson VII fitting functions are used to identify and tabulate diffraction features; typically, sub-pixel precision is obtained for diffraction peak positions. Angular and positional calibration of the diffraction system components is accomplished through white beam diffraction measurements on a perfect Ge crystal. Typically, more than 40 reflections are collected from a Ge crystal for ~3-5 detector positions; these measurements are used to iteratively refine the positional and angular orientation parameters of the detector relative to the microbeam. By fitting these data to the undistorted diamond cubic structure of Ge, the refinement yields detector orientations and positions such that the diffraction patterns correspond to perfect Ge to strain values of < 5 x 10^{-5}; the height of the detector relative to the microbeam is typically determined to a precision of ~3 μm in this process, which does not change with sample.

For thin film samples, the experimental techniques discussed above provide micron resolution measurements simply by the use of a high-resolution translation stage to scan the two-dimensional sample under the submicron beam. Three-dimensional samples, on the other hand, present an additional challenge because they require spatial resolution along, as well as perpendicular to, the direction of the beam.[8] Figure 2 illustrates schematically the use of geometrical triangulation to achieve spatial resolution along the x-ray beam. By collecting measurements at detector positions 1 and 2 (without moving the sample), it is possible to determine the origin along the beam direction of diffraction peaks or features. With the use of a high precision detector translation stage, it is possible to achieve sub-pixel precision depth-triangulation along the microbeam. It is important to notice that only the position of the sample surface or sub-surface diffracting grains along the microbeam changes in switching from a Ge

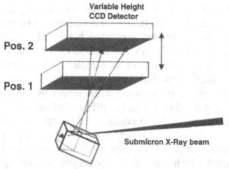

Fig. 2. Illustration of triangulation for obtaining position resolution along the beam..

standard to a sample crystal. When desired, the actual surface position has been determined by triangulation measurements on small Si flakes/grains sprinkled onto the sample surface.

The Al(0.2%)Mg tri-crystal sample used in the measurements presented in this paper was grown at the Alcoa Technical Center and deformed in plane strain.[7] The tri-crystal had near columnar grains with [001] orientations near the surface normal. The (compression) deformation of the ~10x15x20 mm^3 sample was carried out at 200° C in a channel die with the columnar and triple-junction directions along the zero strain direction of the channel..

MICROBEAM DIFFRACTION MEASUREMENTS

Figures 1and 3 illustrate schematically the overall microbeam configurations used to perform diffraction measurements on the deformed sample. Figure 3 contains an optical photograph of

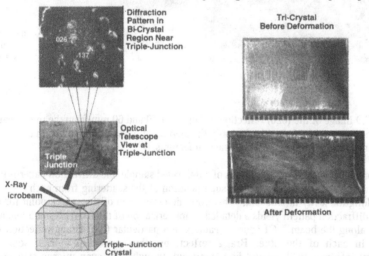

Fig. 3. Schematic view of microbeam diffraction from a deformed tri-crystal. The optical pictures at the right are before and after 20% plane strain; the pictures at the left depict the beam hitting the grain boundary producing a CCD diffraction pattern from both grains.

the sample before and after deformation.[7] The schematic drawing of the tri-crystal in the lower left corner of Fig. 3 indicates the orientation of the columnar-like grain structure and the triple-crystal junction in the measurements presented here. The optical telescope image is a photograph from a long focal-length telescope focused on the area immediately surrounding the position where the x-ray microbeam strikes the sample. This telescope image contains an ~800 μm wide view of the sample area during the x-ray measurement, which is used as a rough indicator of the position of the micro-beam with respect to optically visible features on the sample; precise positioning is inferred from the diffraction features.

The CCD image at the top of Fig. 3 contains a diffraction pattern collected with the x-ray beam traversing the bi-crystal grain boundary to the left of the triple junction point, as indicated schematically in the lower left part of the figure. Diffraction spots from both grains in the bi-crystal region are present in the image. At this beam position, the upper grain produces the arc-shaped patterns and the lower grain gives rise to the shorter, vertically elongated spot pattern. For the measurements reported here, the sample normal was oriented at an angle of 45° toward the beam; therefore, each diffraction pattern provides information on the deformation microstructure and sub-grain orientations in the deformed sample along a line with 0.7 x 0.7 μm² cross-section penetrating ~0.25-0.5 mm into the sample at a 45° angle.

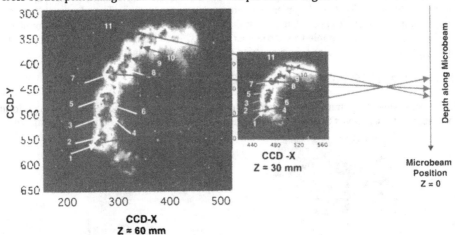

Fig. 4. CCD image of the (026) reflection in Fig. 3 for 30 and 60 mm from the microbeam height. The two images show schematically the method of triangulation to obtain the source point of the scattering for the diffraction features.

For the kinematical scattering conditions of a deformed sample, the diffraction patterns in the CCD image can be assumed to represent a superposition of the scattering from each crystallite illuminated along the x-ray microbeam. Therefore, determination of the source point for each feature in the diffraction pattern yields a detailed characterization of the local crystal orientations and positions along the beam. Of course, features in a particular (hkl) Bragg reflection have counterparts in each of the other Bragg reflections, thus providing a rich source of microstructural information. It should be recognized, though, that even micron size sample movements with respect to the beam lead to diffraction changes for micron size substructure.

The 3-D triangulation measurements are performed by positioning the detector at varying distances above the micro-beam and collecting diffraction pattern images without sample motion. Figure 4 shows an expanded view of the (026) diffraction spot in Fig. 3 for detector distances of 60 and 30 mm from the beam. The deformation microstructure produces a diffraction pattern composed of ~15 major cluster-like structures distributed in the form of an arc. Simultaneous analysis of the diffraction features for each of the (hkl) reflections will provide detailed, spatially resolved microstructural information along the x-ray microbeam. Techniques for such a comprehensive analysis are under development, but in this paper we restrict consideration to determination of the position (along the beam) of the origin of the diffraction features to demonstrate the triangulation technique.

The diffraction peak positions indicated by the arrows in Fig. 4 were determined by fitting two-dimensional Lorentzians to the peaks in diffraction images for 10 detector heights. These peak positions (in pixels) for grain C are plotted in Fig. 5 for each of the arrow positions in Fig. 4, and similarly for four peaks corresponding to the lower grain A of the bi-crystal. Figure 5 shows the locus of the peak positions as a function of detector distance projected onto the plane containing the x-ray microbeam (i.e. $Z_{Det} = 0$). The expanded view in the right hand panel indicates the position along the beam that is identified as the source position for each of the diffraction peaks. Noting that the beam direction is implied to be from top-to-bottom in the figure, it can be inferred that the beam traverses ~50 μm through the lower grain in Fig. 3 and ~250 μm into the upper grain (11th arrow) in Fig. 4. It is important to note that the x-position is defined by the microbeam position, so it is significant that the inferred x-positions from the triangulation form an essentially straight line. Since the x-position at $Z_{Det} = 0$ is determined in the same manner as the y-position, the deviations of the x-position from a straight beam line provide an indicator of the uncertainties in the determination of the y-position (along the beam).

Using the deviations of the x-position and the fact that the two grains neither overlap nor leave a gap at the boundary, we estimate the uncertainty along the beam to be less than 5 microns, and in the 2-3 micron range in many cases. Triangulation on weak peaks or peaks that become multiple peaks for larger detector distances do not achieve such precision, of course, and

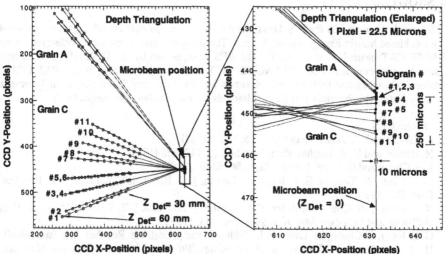

Fig. 5. Projection of diffraction peak positions for detector distances varying from 30 – 60 mm onto the microbeam. Extrapolation of the lines to $Z= 0$ determines the positions of the diffracting grains along the beam with an enlargement at right. The x-direction is perpendicular to the beam, the y-direction is along the beam, and the z-direction above the beam.

these cases will require fitting multiple peaks or additional techniques. The data presented indicate the precision for representative data that do not subdivide at large detector distances. The series of discrete clusters of diffraction intensity in Figs. 3,4 is consistent with the anticipated presence of layers of semi-aligned Al crystallites separated by dislocation walls with small angular rotations between layers.[1,2] By identifying the corresponding peaks in a number of (hkl) reflections, the orientation matrices[3,4] of each of the individual diffraction features in Fig. 3 can be determined. As a result, it is possible to determine the angular separation and the axis of rotation between individual Al grains in their respective positions along the x-ray beam. For example, we have determined that the peaks at arrow positions 1 and 10 (i.e. bottom of arc to top of arc) have a relative rotation of 4.38° around the ~[-2,3,3] direction. We comment that both the direction and magnitude of the angular rotations for sample positions far removed from the grain boundary (say ~ 1mm) were observed to be significantly different from those in Fig. 3, reflecting the impact of the grain boundary on active slip systems during deformation.

CONCLUSIONS

A fundamental understanding of the mechanisms involved in the production of the well-known two-dimensional dislocation wall microstructure is of high interest both technologically and fundamentally. Continued development of the x-ray microbeam techniques discussed here, including local stress/strain measurements, will provide new information regarding underlying issues involved in microstructural evolution. Microbeam x-ray techniques provide mesoscale information complementary to that presently available by TEM and electron backscattering OIM techniques. In particular, microbeam x-ray diffraction provides non-destructive bulk measurements and they provide the possibility of measuring local stress distributions under loaded and/or dynamic conditions.

ACKNOWLEDGMENTS

Research sponsored by the U.S. Department of Energy under contract DE-AC05-96OR22464 with Lockheed Martin Energy Research Corp. The x-ray measurements were performed on the MHATT-CAT beam line at the APS. The APS is supported by the DOE Office of Energy Research under contract W-31-109-ENG-38. N. Tamura and J.-S. Chung were ORNL-ORISE postdoctoral associates; their present addresses are the Advanced Light Source at Lawrence Berkeley National Laboratory and UIUC-Materials Research Laboratory, respectively.

REFERENCES
1. P. Hähner and M. Zaiser, Acta Mater. **45**, 1067 (1997).
2. N. Hansen and D. A. Hughes, Phys. Stat. Sol. (b) **149**, 155 (1995).
3. J.-S. Chung and G.E. Ice, J. Appl. Phys., **86**, 5249 (1999).
4. J.-S. Chung, N. Tamura, G. E. Ice, B. C. Larson, J. D. Budai, and W. Lowe, Mat. Res. Soc. Proc. **563**, 169 (1999).
5. N. Tamura, J.-S. Chung, G. E. Ice, B. C. Larson, J. D. Budai, J. Z. Tischler, M. Yoon, E. L. Williams, and W. Lowe, Mat. Res. Soc. Proc. **563**, 175 (1999).
6. G. E. Ice, J.-S. Chung, W. Lowe, E. L. Williams, and J. Edelman, Rev. Sci. Inst. (submitted).
7. H. Weiland and R. Becker, in *Proceeding 20th Riso Intern. Symp. on Mat. Sci.: Deformation-induced Microstructures: Analysis and Relation to Properties*, edited by T. Leffer and O.P. Pederson, Riso Intern. Laboratory, Roskilde, Denmark 1999, p. 213.
8. D.P. Piotrowski, S.R. Stock, A. Guvenilir, J.D. Haase, and Z.U. Rek, Mat. Res. Soc. Symp. **437**, 125 (1996).

NONDESTRUCTIVE DETERMINATION OF THE DEPTH OF DIFFERENT TEXTURE COMPONENTS IN POLYCRYSTALLINE SAMPLES

C.R. PATTERSON II[1,2], K.I. IGNATIEV[1], A. GUVENILIR[1,3], J.D. HAASE[1], R. MORANO[1,4], Z.U. REK[5], S.R. STOCK[1]
[1]School of Materials Sci. & Eng. & Mechanical Properties Research Lab., Georgia Institute of Technology, Atlanta, GA 30332-0245 USA   stuart.stock@mse.gatech.edu
[2]presently at Pratt & Whitney, West Palm Beach, FL, USA
[3]presently at Motorola, Ed Bluestein Blvd, Austin, TX, USA
[4]presently with US Navy
[5]Stanford Synchrotron Radiation Laboratory, SLAC, Stanford Univ., Stanford, CA, USA

ABSTRACT

The surface and the center average textures (macrotexture) of plates of many alloys differ substantially, and nondestructive methods for determining the depth of different texture components would be very useful in various applications. This report evaluates one method based on recording microbeam transmission Laue patterns as a function of sample-detector separation and on tracing the diffracted rays back to their physical origin. Polychromatic synchrotron x-radiation and absorption edge filters are used. Results from sections of Al-Li 2090 T8E41 plates are reported, and limitations of the ray tracing technique are discussed

INTRODUCTION

Ray tracing with x-ray microbeam diffraction is one approach to mapping grain orientations as a function of three-dimensional position. Use of polychromatic synchrotron x-radiation is particularly advantageous because all crystalline material in the beam will diffract; unlike the situation with monochromatic radiation, data for the entire irradiated crystalline volume is obtained simultaneously. In this approach, the pattern of intensity diffracted by a polycrystalline sample is recorded on a two-dimensional detector as a function of separation between the sample and the detector. Using linear regression, the various diffracted beams (i.e., rays) can be traced to their origins within the sample.

The principles of the ray tracing technique are outlined elsewhere for the transmission Laue setting [1-3] used in the work reported below. Earlier work on model 5.0 mm diameter samples comprised of ~ 1 mm chunks of single crystals revealed diffracted beam origins could be fixed with uncertainties (standard errors) in depth (i.e., along the axis of the microbeam) somewhat smaller than ± 0.1 mm [3, 4]. It is important to note the following experimental details as these affect depth resolution: a 0.1 mm diameter pinhole collimator defined the irradiated volume, the two-dimensional detector was an image plate read with 0.1 mm pixels and the pixels with maximum diffracted intensity in each diffraction spot were used for ray tracing. High resolution x-ray computed tomography confirmed the ray tracing results for the "large-grained" samples [3, 5].

Many interesting polycrystalline samples, however, have much smaller grain sizes along at least one direction and have much more complex textures than studied before. It is unclear whether ray tracing with polychromatic beams will work as well for this class of samples as it did with the model samples, and pieces of 12.7 mm thick plate with very different textures are used in the present study to examine how precisely grains can be located in these more complex samples.

SAMPLE AND EXPERIMENTS

Two 22 mm x 38 mm x ~1.1 mm pieces were cut from a 12.7 mm thick plate of Al-Li 2090 T8E41; piece 1 was from the surface and piece 5 from the surface of the plate. Pole figures reveal very different macrotextures at the plate center and surface [6]. Pieces 1 and 5 were mounted on either side of a 3 mm thick acrylic plate using two pins (Fig. 1), and the pieces' L (rolling) and T (long transverse) directions were horizontal and vertical, respectively. This simulated the separation between pieces 1 and 5 within the as-received plate and the grain orientations encountered in nondestructive texture mapping of compact tension samples.

The pieces and acrylic plate were mounted perpendicular to the x-ray beam axis at bending magnet Beamline 2-2 of SSRL (Stanford Synchrotron Radiation Laboratory, 3.0 GeV storage ring energy and beam currents between 100 and 20 mA). A 0.03 mm diameter pinhole collimator mounted ~550 mm from the sample was used to define the irradiated volume. Transmission Laue patterns were recorded on image storage plates with both pieces 1 and 5 in the beam, for piece 1 only in the beam and for piece 5 only in the beam. The image plates were clamped to a holder mounted perpendicular to the incident beam, were removed for reading after each exposure and were read on a Fuji BAS-2000 system with 0.1 x 0.1 mm^2 pixels. Because the Laue patterns consisted of radial streaks spanning a considerable range of wavelengths, a Pd filter was used to change the streaks' contrast at the K-absorption edge wavelength [7], and the midpoint of the edge transition was used for ray tracing. A Pb beam stop, nearly opaque to the incident beam, was used to prevent overexposure and fogging of the image plate while allowing the highly attenuated incident beam to be recorded.

Three sets of diffraction patterns were recorded at eight sample-detector separations by translating the image plate holder in increments of 40.0 mm; precision of the translator was considerably better than 0.1 mm. All positions are reported in terms of the TS scale, that is, in terms of the numerical values inscribed on the translator: the numerical values of the TS positions of the plate holder range between 260 and 540 mm and of the origins of the diffraction streaks between roughly 669 and 675 mm. Diffraction patterns recorded with piece 1 only in the beam, with piece 5 only in the beam and with both pieces in the beam were compared; while this was not necessary for the ray tracing analysis, it was a helpful check on the results. Linear regression determined where the separation r between incident beam and diffracted beam (i.e., absorption edge position) extrapolated to zero; this was the position of the volume originating the feature used in ray tracing.

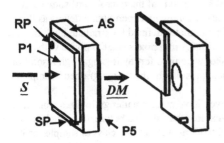

Fig. 1. Mount with pieces 1 and 5 (P1, P5) in the beam (left) and only P5 in the beam (right). The hole in the acrylic separator (AS) allowed the x-ray beams (S and DM) to pass without encountering the plastic. One pin (RP) allowed either piece to be rotated out of the beam. The other pin (SP), on which the samples rested, enabled the pieces to be re-positioned precisely.

RESULTS AND DISCUSSION

Figure 2 shows diffraction patterns with only piece 5 (left) and with both piece 5 and piece 1 (right) in the beam; the former (and data derived from it) is designated 'P5', the latter 'P1+ P5' and data from piece 1 by itself 'P1'. Diffraction from 111, 200, 220, 311 and 222 can be seen, and careful comparison of 'P1', 'P5' and 'P1+P5' patterns reveals the origin of the streaks, at least those which do not overlap. Streaks D1, D2, C, R, Q, G, T and S1 are clear of intereference from piece 1 streaks, appear in both diffraction patterns of Fig. 2 and are therefore from piece 5.

The streaks identified in Fig. 2 are 111 and 200 reflections and are typical. The large number of streaks observed in Fig. 2 is not surprising nor are the systematic differences in streak orientations between pieces 1 and 5. The outer portion of 12.7 mm thick plates of Al-Li 2090 T8E41 consist of partially recrystallized material while the centers are unrecrystallized [6, 8]; along the L,T and S (short transverse) plate directions, the grain dimensions are ~2 mm, 0.5 mm and 0.05 mm, respectively [8]. Thus, within the ~1.1 mm thickness of either piece, one expects 20 - 25 grains to be traversed in piece 1 and the same number in piece 5.

If streaks from P5 and P1 were close to each other, their origin could be determined by inspection (i.e., the edge position of one was farther from the incident beam position). The enlargements of two groups of streaks in Fig. 3 are examples where origin of diffracted intensity could be differentiated directly; the shifts are so small, however, that simple inspection or even a single precise measurement of r, the separation between edge and incident beam positions, will not suffice for depths much smaller than 2 or 3 mm.

Fig.2. Diffraction patterns 'P5', i.e., from piece 5 only, (left) and 'P1+P5', from pieces 1 and 5, (right) recorded with the Pd filter in the beam. The brighter the pixel, the greater the diffracted intensity. 'G' results from 111 diffraction at the Pd K-edge wavelength and 'S1' from 200 diffraction. The dark diamond and small light point in the middle of the image are the beam stop and incident beam, respectively.

A total of eight streaks from 'P1' and eight from 'P5' were traced back to their origins. The origins for the streaks identified from piece 5 were within 669.3 < TS (mm) < 670.8, and streaks from piece 1 lay within 673.2 < TS (mm) < 674.5. The range of TS origins for piece 5 was 1.5 mm and for piece 1 was 1.3 mm; these values were somewhat larger than the measured thickness of each piece (~1.1 mm). The separation between the closest origins of pieces 1 and 5 is 2.4 mm; this is somewhat lower than the 3.0 mm thickness of the separator. In light of the uncertainties associated with the extrapolation (discussed in the following paragraph), the data is in good agreement with the physical extent of the sample.

The minimum uncertainty (as expressed by the standard error of the estimated origin) was about 0.5 mm, the largest was 2.3 mm, the median was 1.1 and the mean was 1.2. The uncertainty was between 0.5 and 1.0 mm for six streaks, between 1.0 and 1.5 mm for six streaks, between 1.5 and 2.0 mm for two streaks and between 2.0 and 2.5 mm for two streaks. No pattern was evident in the uncertainties.

The uncertainties in the location of the depth of the diffracting volumes are about an order of magnitude larger than those obtained earlier with two large-grained model samples [3, 4]. The main difference between the present and earlier experiments was the diffraction features used for ray tracing. One model sample consisted of chunks of Si broken from 0.5 mm thick wafers and produced transmission Laue patterns of sharply-defined spots; the foci of different zones (ellipses) were traced back to their origins. The second model sample contained single crystal blocks of Al, deformation within the blocks produced a pattern of streaks and the pixel with maximum diffracted intensity for each streak was used in ray tracing. Usually there was more than one pixel with the maximum value, and the average pixel position was used for ray tracing. In both model samples multiple pixels were "averaged" in defining the feature used

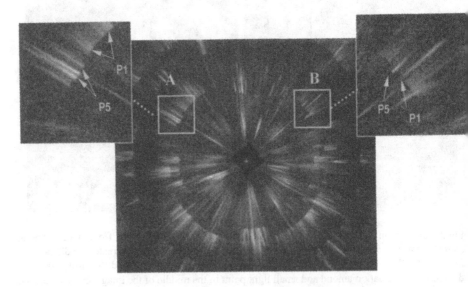

Fig. 3. Diffraction pattern 'P1+P5' and enlargements of areas A and B (left and right, respectively) showing clear offsets of edge positions for 200 diffraction streaks from pieces 1 and 5 (labeled 'P1' and 'P5', respectively). Clearly, piece 1 is farther than 5 from the image plate.

for ray tracing. This procedure is quite robust; noise from scatter has little influence. The feature used in the present work, the midpoint of the change in contrast across the absorption edge position, provides a much less robust marker for extrapolation. Two-dimensional fitting/smoothing routines (considerably more sophisticated than the linear fit used here) might be able to improve the robustness of ray tracing using absorption edges but probably not more than a factor of two or three.

The intensity profiles across absorption edges vary considerably for different streaks. While scatter and other sources of noise contribute to the uncertainty in fixing the edge position, sample-related effects appear to be more important. Inhomogeneous deformation within the volume sampled by the columnar beam can contribute significantly: the diffracted intensity is a convolution of the distribution of wavelengths in the beam with the volume-weighted distribution of orientations and of microstrain. This affects different streaks and makes automatic analysis problematic but will not be a factor for any one streak.

In highly textured samples such as Al-Li 2090 T8E41 the strong texture can cause problems in ray tracing. Fully 40% of the volume of the center 3 mm of 12.7 mm thick plates consists of groups of 5-10 adjacent, pancake-shaped grains [9, 10] with nearly identical orientations; this near single crystal mesotexture leads to streaks which lie close together or overlap. The overlap may be difficult to resolve, except at very large sample-image plate separations, and changes with TS position (e.g., Fig. 4); this subtle effect, if unrecognized, can cause large variability in the separation between incident beam and edge position. Further, the vicinity of the change in contrast across the absorption edge position is where multiple closely-spaced streaks will be most difficult to detect, and this appears to be the largest contribution to the uncertainty in origin of diffracting volumes seen in the present work.

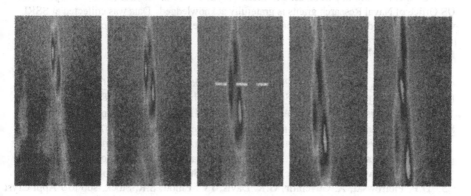

Fig. 4. Images of the same set of diffraction streaks recorded at five sample-to-detector separations. The farther right, the greater the separation. The contrast was adjusted to produce maximum visibility of the streaks' maxima in diffracted intensity (i.e., the white centers of the streaks). Surrounding the maxima are intermediate intensities (shown in black), and the fuzzy gray to black areas outside the black envelopes show lowest intensities. The position along the streak image, the greater the diffracted wavelength. If the absorption edge cut the middle of the left-most streak (white dashed line in the middle image), for example, the experimentally determined edge position on the left side of the composite streak would differ substantially from that on the right side.

Decreasing the diameter of the column of material irradiated can improve the precision of ray tracing results; use of a 0.01 mm diameter pinhole collimator placed much closer to the sample has become routine since the present data were collected. Capability for reading image plates with 0.05 mm pixels also has become available and should help improve the results. Finally, use of maximum intensity pixels of streaks and careful scanning of streaks for unanticipated substructure can improve reliability of ray tracing. It is unlikely that ray tracing in the transmission Laue geometry and using image plates will be able to reliably identify the depth of adjacent 0.05 mm thick grains, but it seems reasonable that depth resolutions between 0.2 and 0.3 mm should eventually be achievable with polychromatic radiation (i.e., one-third to one-quarter of the piece thickness used in the present experiments.

CONCLUSIONS

Microbeam x-ray tracing with polychromatic synchrotron radiation and using absorption edge filters was used to investigate the precision and accuracy with which the depths of different texture components could be determined inside thick plates of commercial Al alloys. A model sample consisting of two ~1 mm thick pieces of Al-Li 2090 T8E41, one cut from the surface and the second from the center of a 12.7 mm thick plate, spaced about 3 mm apart simulated an intact plate. The 16 diffraction streaks investigated could be followed back to the correct piece, but the standard error in estimating the origins of the streaks averaged 1.1 mm.

ACKNOWLEDGMENTS

This paper is based in part on the MS thesis [11] of one of the authors (CRP). Support from US Office of Naval Research grants is gratefully acknowledged. Data was collected at SSRL which is supported by US Department of Energy, Office of Basic Energy Sciences.

REFERENCES

1. S.R. Stock, A. Guvenilir, D.P. Piotrowski, Z.U. Rek, in Applications of Synchrotron Radiation Techniques to Materials Science II, MRS Vol. 375 (1995) 275-280.
2. D.P. Piotrowski, S.R. Stock, A. Guvenilir, J.D. Haase, Z.U. Rek, in Applications of Synchrotron Radiation Techniques to Materials Science III, MRS Vol. 437 (1996) 125-128.
3. D.P. Piotrowski, "Synchrotron Polychromatic X-ray Diffraction Tomography of Large-grained Polycrystalline Materials," MS Thesis, Georgia Institute of Technology, 1996.
4. D.P. Piotrowski, A. Guvenilir, Z.U. Rek, S.R. Stock, sub. to J Appl Cryst 1999.
5. D.P. Piotrowski, A. Guvenilir, G.R. Davis, J.C. Elliott, S.R. Stock, sub to J Appl Cryst 1999. —
6. P.S. Pao, M.A. Imam, L.A. Cooley, G.R. Yoder, Scr Met **23** (1989) 1455-1460.
7. S.R. Stock, Z.U. Rek, Y.H. Chung, P.C. Huang, B.M. Ditchek, J Appl Phys **73** (1993) 1737-1742.
8. K.T. Venkateswara Rao, R.O. Ritchie, Int Mater Rev 37#4 (1992) 153-185.
9. J.D. Haase, A. Guvenilir, J.R.Witt, S.R. Stock, Acta Mater **46** (1998) 4791-4799.
10. Jake D. Haase, Abbas Guvenilir, Jason R. Witt, Morten A. Langøy, Stuart R. Stock, in Mixed-Mode Crack Behavior, ASTM STP **1359** (1999) 160-173.
11. C.R. Patterson, II, "Synchrotron Polychromatic X-ray Diffraction Tomography of Aluminum Lithium 2090 T8E41," MS Thesis, Georgia Institute of Technology, 1999.

Synchrotron Radiation X-ray Microdiffraction Study of Cu Interconnects

X. Zhang, H. Solak, F. Cerrina
Electrical and Computer Engineering Department and Center for Nanotechnology, University of Wisconsin-Madison, Madison, Wisconsin 53706

B. Lai, Z. Cai, P. Ilinski, D. Legnini, W. Rodrigues
Advanced Photon Source, Argonne National Laboratory, Argonne, Illinois 60439

Abstract

We have used synchrotron radiation X-ray microdiffraction to study the microstructure and strain variation of copper interconnects. Different types of local microstructures have been found in different samples. Our data show that the Ti adhesion layer has a dramatic effect on Cu microstructure. On site electromigration test has been conducted and strain profile along the same interconnect line was measured before and after this electrical stressing. Cu fluorescence scan was used to find the mass variations along the line. Voids and hillocks can be clearly identified in this scan. X-ray micro-diffraction was used to measure the strain change around the interesting regions.

1.Introduction

Reliability is one of the most important issues in integrated circuit (IC) performance. Many failure mechanisms have been studied in the past and interconnect failure has been in the forefront of reliability issues for integrated circuits for the last two decades[1, 2]. In general, electromigration and mechanical stress related damage have been identified as the major causes for these interconnect failures in Al conductors[3, 4]. Copper is the candidate to replace Al alloys as the interconnect material for the advanced ICs [5, 6] because it has lower resistivity and higher resistance to electromigration failure. In order to understand the properties of Cu interconnects, it is very important to understand its material properties and behavior during electrical stressing.

Electromigration is the movement of atoms caused by flow of electrons. The phenomenon is attributed to the momentum transfer from electrons to atoms or *electron wind force*[3]. High current densities (10^5A/cm^2) in interconnects cause substantial mass transport, thus creating micro-voids that eventually may cause open circuit failure. It is also well known that electromigration is strongly influenced by the mechanical stress that exists in interconnects[4]. When the migration of atoms happens, material is accumulated at some places (usually close to the anode) and depleted at other places (close to the cathode). This movement of material induces a stress field in the lines that are confined by the substrate and passivation layer. In return, the stress field creates a force resisting the electromigration. Under certain conditions, these two forces can balance and electromigration will stop.

Many models have been developed to predict the evolution of stress in interconnects due to electromigration[7, 8, 9]. However, experimental verification is very difficult mainly because of the challenge of stress measurement in the extremely small volumes of lines with micron or submicron spatial resolution. The complex structure also requires a probe capable of penetrating several microns of dielectrics while maintaining good spatial resolution. X-ray diffraction is a direct technique for stress measurement and has been successfully used for

Mat. Res. Soc. Symp. Proc. Vol. 590 © 2000 Materials Research Society

detailed characterization of average stress in interconnect lines[10, 11]. However, the area sampled with traditional X-ray diffraction tools is too large to allow measurement of stress distribution with the required spatial resolution. The development of X-ray microdiffraction, based on extremely high brightness synchrotron radiation sources such as the Advanced Photon Source (APS) and the advances in X-ray optics such as the Fresnel zone plate (FZP) for hard X-ray radiation has opened new possibilities in this area[12, 13]. In these experiments, the photon density at the sample is increased to a level sufficient to acquire diffraction data from extremely small volume of the order of $0.3\mu m^3$. In the paper, we report the application of this state-of-art X-ray micro-beam diffraction technique to the study of electromigration and stress phenomena causing reliability problems in Cu interconnects.

2.Experiments

The Cu interconnect structures were fabricated using a lift-off process. After applying APEX-E photoresist to thermally oxidized Si wafers, a Leica electron beam lithography tool was used to directly write the patterns. Eight different linewidths were used in this experiment, varying from $5\mu m$ to $0.25\mu m$. The length of all lines were 1mm. After the photoresist development, an electron beam evaporator was used to deposit Cu(350nm) or Cu(350nm)/Ti(20nm) films on these wafers. An ultrasonic bath was then used to strip the resist and produce the interconnect patterns. SiN($0.2\mu m$) and SiO($0.4\mu m$) were deposited as a passivation layer using plasma-enhanced chemical vapor deposition (PECVD) at 350°C. Conventional optical lithography was used to open the contact hole, and Al was evaporated on the samples as the outer contact metal. Hence, the Cu interconnects were completely buried in the glass passivation layer. Finally, the wafers were annealed at 450°C for 30 minutes in a $N_2(90\%)/H_2(10\%)$ mixture. Figure1 shows a SEM cross section image of a $0.25\mu m$ passivated Cu line.

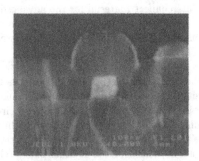

Figure 1: *A SEM cross section image of a passivated $0.25\mu m$ Cu line*

The X-ray microdiffraction experiments were conducted at Argonne National Laboratory. The phase-zone-plate-based Micro-Beam beamline 2-ID-D provides a spot of 1×5 μm^2 at a photon energy of 9.5keV. Our experimental setup is very similar to that described in a previous paper[13]. Figure2 shows the microdiffraction experiment setup on the 2-ID -D beamline. In the experiment described here, we used symmetric reflection geometry with the incident and diffracted beams making approximately the same angle with the surface normal. Therefore, the strain measurement in our experiment is sensitive only to the surface normal direction. In order to position the sample area of interest in the microbeam, a energy-dispersive Ge detector was used to detect the copper K_α(8.0keV) fluorescence. By scanning the sample in both horizontal and vertical directions and knowing the exact geometry of the pattern on the sample, we were able to accurately locate the microbeam and place the area of interest under it. An X-ray CCD camera on the 2θ arm of the goniometer was used to record the diffraction patterns from the sample. A typical example is shown in Figure3; the data files are in the format of 16-bit 1152×1242 gray-scale images. In our experiment, we used E=9.5keV photon energy to obtain the diffraction from a given family of lattice planes.

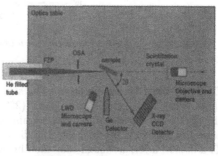

Figure 2: *Diffraction patterns from different interconnect samples; left: a photograph of the experiment station; right: a schematic highlight of the experiment station.*

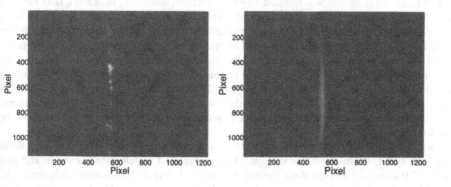

Figure 3: *Diffraction patterns from different interconnect samples; left: diffraction pattern pure Cu sample; right: diffraction from Cu/Ti sample.*

3.Results and analysis

Our experimental results show that the Ti adhesion layer has a dramatic effect on the microstructure of Cu film deposited on top of it. The X-ray diffraction patterns of the pure

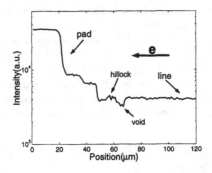

Figure 4: *Cu fluorescence linescan along a 2μm line after electromigration testing.*

Cu film and Cu/Ti bilayer are shown in Figure3. In the left panel, the as-deposited pure Cu film showed typical polycrystalline structure. In our experiment, we found diffraction spots from (111), (200), (220), (311), (222) plane families. According to our AFM data, the as-deposited pure Cu film has a grain size about 0.15μm before and about 0.4μm after annealing, which is very close to the dimension of its thickness. In the samples with the Ti adhesion layer, we found that the film has strong (111) texture and a grain size about 40nm; we did not observe any diffraction from other plane families from this Cu/Ti film. The sample has a 30 minutes 450°C annealing history.

The electromigration test was conducted on the Cu/Ti sample, the temperature setting was 250°C and the current density was about $7.8 \times 10^6 \text{A/cm}^2$. After 52 hours of electrical stressing, the total current was reduced to ~6% of its original value, indicating that electromigration had damaged most of the test patterns. Combining optical microscopy and Cu fluorescence, we could easily find the regions of interest. In Figure4, we show the fluorescence signal from Cu along one single 2μm line. The spatial resolution is 1μm. Since the intensity of this Cu fluorescence is proportional to the mass of the Cu under the X-ray beam spot, different types of features can be identified in this scan. The anode contact pad extending from 0 to about 25μm gives a very strong signal. The variation of intensity from 25μm to 70 μm shows the Cu depletion and Cu accumulation along the line, and a much more uniform intensity can be found after that region. The scan shows the existence of upstream voids and downstream hillocks.

An X-ray microdiffraction scan was then conducted around this area. The step sizes were 1μm and 4μm. In Figure we show the method we used to calculate the strain and the results of the strain variation along the same 2μm line before and after electromigration. It should be noted that our major concern in this experiment was to profile the strain distribution along the line rather than obtaining an absolute measurement. On the left, we show a composite of the typical diffraction pattern from our experiment, together with the best fit to the diffraction arc. The average strain of that spot $\Delta\epsilon$ was calculated by comparison of the center of this arc to a reference: $\Delta\epsilon = n \, s/(2r\tan\theta)$ where n is the difference between the two arcs in pixel number, s is the size of pixel which is 50μm in our case, r is the distance between the sample and the CCD camera which is 64.38cm, and θ is the Bragg angle.

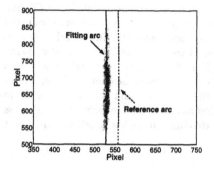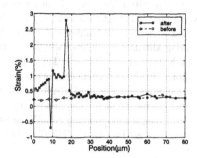

Figure 5: *Data analysis and strain calculations; left: typical diffraction pattern from Cu/Ti sample and the theoretical fitting; right: final results of strain variations along the same 2μm line before and after electromigration test.*

According to this setting, our strain resolution is

$$\Delta\epsilon = \frac{50}{2 \cdot 64.38 \times 10^4 \cdot \tan(18.25^o)} \sim 1.2 \times 10^{-4}$$

On the right, we show the final calculation results of the strain variation along the same 2μm line before and after electromigration. Comparing Figure4 and Figure, we conclude that the region we scanned in Figure is the region starting from about 40μm and beyond in Figure4. Significant strain variation around the void and hillock has been found. From our data, we found that there is a very large tensile-like deviation around the void which may not be explained as a tensile strain in Cu. We believe that the diffraction is actually from the TiCu alloy formed during annealing process. J.L. Liotard *et al.*[14] have done extensive studies on Cu/Ti bilayer using different methods and have shown that various CuTi alloys can be formed at the interface between Cu and Ti during the high temperature process. Our energy dispersive spectroscopy (EDS) data also show that there is a higher Ti peak at the void area, indicating that there is significant amount of CuTi left in the void. The strain becomes compressive around the hillock, which is close to the anode. Recently Wang *et al.*[15] have shown that compressive stress gradient builds up during the electrical stressing process and relaxes quickly after the stressing ended on the 10μm Al strip. In our experiment, the relaxation path was severed by the void and the compressive strain remained unchanged around the hillock long after the electrical stressing stopped. Figure also shows that the average strain has no significant difference beyond the void in the two different conditions, which is consistent with this stress relaxation mechanism.

4.Conclusions

In conclusion, in this paper we report the first microdiffraction data from Cu interconnects. The results show that X-ray micro-diffraction and micro-fluorescence can provide important insights in electromigration studies. In our experiment, we found that the 20nm Ti adhesion layer can significantly change the microstructure of Cu film deposited on top of it. The as-deposited Cu film did not show any special microstructures, but the Cu/Ti film showed strong (111) texture and fine grains. The microdiffraction experiment was conducted before and after the electromigration test. Significant mass flow and strain variation were found along the same line after electrical stressing. Large compressive strains were found around a hillock area, while the TiCu alloy complicated the stress state analysis around a void region. We did not observe any changes in the area far from the damage region, meaning that the stress relaxation in that region was quite rapid. It would be very interesting to extend our studies to include *in situ* microdiffraction measurements to investigate the stress evolution during electrical stressing. We are also extending our studies to other interconnect structures, such as Ta/Cu/Ta, TiN/Cu/TiN.

5. Acknowledgments

This project is supported by the Advanced Photon Source under Contract No. PO 062242-02.

References

[1] J.Curry, G.Fitzgibbon, Y.Guan, R.Muollo, G.Nelson, and A.Thomas, Proc. of the 20th Annl. Int. Reliability Symp., 6(1984)

[2] J.Klema, R.Pyle and E.Domangue, Proc. of the 20th Annl. Int. Reliability Symp., 1(1984)

[3] R.S. Sorbello, Mater. Res. Soc. Symp. Proc. 225, 3(1991)

[4] I.A. Blech, J. Appl. Phys. 47, 1203(1976)

[5] C.K. Hu, J.M.E. Harper, Int. Symp. on VLSI Tech. and Appl., 1997, 18-22

[6] D. Gupta, Materials Chemistry and Physics 41 (1995) 199-205

[7] M.A. Korhonen, P. Borgesen, K.N. Tu, and C. Li, J. Appl. Phys. 73, 3790(1993)

[8] C.L. Bauer and W.W. Mullins, Appl. Phys. Lett. 61, 2987(1992)

[9] Y.J. Park and C.V. Thompson, J. Appl. Phys. 82, 4277(1997)

[10] P.R. Besser and J.C. Bravman, AIP Conference Proceedings 305, 46(1993).

[11] P.C. Wang, G.S. Cargill, I.C. Noyan, E.G. Liniger, C.-K. Hu and K.Y. Lee, Materials Reliability in Microelectronics, Mater. Res. Soc. Proc. 427, 35(1996).

[12] B. Lai et al., Appl. Phys. Lett. 61, 1877(1992)

[13] H. H. Solak, Y. Vladimirsky, F. Cerrina, B. Lai, W. Yun, Z. Cai, P. Ilinski, D. Legnini, W. Rodrigues, J. Appl. Phys., 86(2), 884(1999)

[14] J.L. Liotard, D. Gupta, P.A. Psaras, P.S. Ho, J. Appl. Phys., 57(6), 1895(1985)

[15] P.-C. Wang, G.S. Cargill, I.C. Noyan, and C.-K. Hu, Appl. Phys. Lett. 72, 1296(1998)

ATTENUATION- AND PHASE-CONTRAST MICROTOMOGRAPHY USING SYNCHROTRON RADIATION FOR THE 3-DIM. INVESTIGATION OF SPECIMENS CONSISTING OF ELEMENTS WITH LOW AND MEDIUM ABSORPTION

F. Beckmann *, U. Bonse **
*Hamburger Synchrotronstrahlungslabor (HASYLAB), Deutsches Elektronen-Synchrotron (DESY), Notkestraße 85, 22607 Hamburg, Germany, felix.beckmann@desy.de
**Institute of Physics, University of Dortmund, 44221 Dortmund, Germany, bonse@physik.uni-dortmund.de

ABSTRACT

Attenuation- and phase-contrast microtomography using synchrotron radiation is applied to different samples demonstrating the advantages and limits of the two different contrast mechanism. Photon energies in the range of 8-25 keV and 60-100 keV are used. Scanning techniques employing a 2-dim. X-ray detector allow for investigation of larger specimens at high spatial resolution.

INTRODUCTION

Microtomography using synchrotron radiation has become a valuable tool for the 3-dim. investigation of samples in the fields of e.g. medicine, biology and material science [1]. Using the microtomography system at HASYLAB (DESY, Germany) attenuation-contrast (μCT) and, by adding an X-ray interferometer [2], phase-contrast microtomography (PμCT) [3,4] can be performed. At beamline BW2 photon energies in the range of 8 to 25 keV are used to reveal the 3-dim. structure of samples consisting of elements with low and medium absorption [5-7]. To investigate specimens consisting of material with higher absorption the apparatus is setup at beamline BW5 to perform μCT at photon energies in the range of 60 to 100 keV. First results using PμCT at a photon energy of 70 keV were recently presented [8,9].

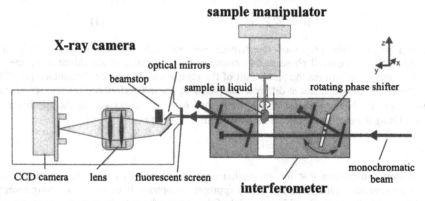

Figure 1: Experimental setup for X-ray microtomography using phase contrast (with interferometer) and attenuation contrast (without interferometer). Beamstop and optical mirrors are added to protect the CCD-chip from the incident X-rays for photon energies above 60 keV. At low photon energies (8 to 24 keV) they will be removed.

First we describe the experimental setup and present a short introduction to the different contrast mechanisms. Next we present first results of µCT obtained by scanning a 2-dim. X-ray detector at 24 keV. Furthermore, samples investigated at high energies are shown. Performing PµCT at 12 and 20 keV photon energy and µCT at 24, 60 and 70 keV the advantage of phase contrast against attenuation contrast is demonstrated.

EXPERIMENT

Setup

The experimental setup shown in Figure 1 consists of three main parts: the X-ray camera, the sample manipulator, and the interferometer. The 2-dim. X-ray detector uses a fluorescent screen (CdWO4 single crystal, 200µm thick) to convert X-rays to visible light. Using a standard objective a magnified optical image is then measured by a CCD-camera. The rotation and the x/z-position of the sample relative to the X-ray camera / interferometer can be altered by the sample manipulator. For phase-contrast microtomography the monochromatic beam is incident on the skew-symmetric Laue-case interferometer which contains the specimen in its upper interfering beam. The rotating phase shifter located immediately behind the beam splitter crystal serves to scan the overall phase of the interferometer. µCT is performed without the X-ray interferometer. A more detailed description of the used components can be found elsewhere [10].

Theory

Due to the high intensity and low divergence of synchrotron radiation projections can be measured by using parallel and highly monochromatic X-rays. Attenuation-contrast projections are obtained by measuring images with and without the sample at incremental sample rotations equally stepped between 0 and π. The tomographical reconstruction then yields the 3-dim. spatial description of the attenuation coefficient µ at position (x,y,z) of the specimen. Using the atomic number Z, mass density ρ and photon energy E it can be approximately expressed by:

$$\mu \propto Z^4 E^{-3} \rho \qquad (1)$$

For a single phase projection interference patterns with and without the sample are measured at different overall phase shifts introduced by the rotating phase shifter in Figure 1. Combining this set of images the phase shift of the sample can directly be determined [11,12]. By repeating this measurement at different sample rotations and performing the tomographical reconstruction the 3-dim. data set of the phase shift ϕ at position (x,y,z) of the specimen is obtained. Using the electron density σ the phase shift ϕ is given by:

$$\phi \propto \sigma E^{-1} \qquad (2)$$

Equation 1 illustrates that for attenuation contrast the photon energy has to match the sample characteristics quite closely to obtain appropriate contrast. Because of the strong energy dependence, only the most absorbing material of the specimen can be investigated in structural detail by µCT. However phase contrast, which decreases only linearly with increasing photon energy, can reveal the 3-dim. structure of differently absorbing material simultaneously. Furthermore, phase contrast is much in favor for samples consisting mainly of light elements. A more detailed comparison can be found elsewhere [9,10].

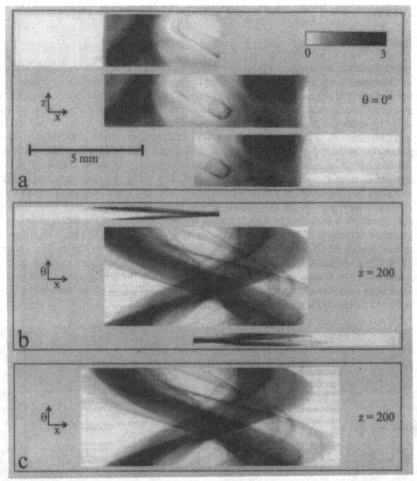

Figure 2: Investigation of large human bone sample consisting of a stapes with its surrounding skull bones. Three scans at different lateral position are performed using μCT at 24 keV photon energy (beamline BW2).
a: Projections of the scans at same angular increment are shown. The top radiogram shows the outer left, the bottom the outer right region of the sample.
b: Sinograms, 2-dim. representation of a line of radiograms at different sample rotation of the three scans are presented for the sample height 200 (1.2 mm). The outer scans consists of 90, the middle scan of 720 (hence a correspondingly larger height) projections.
c: The sinogram shown is obtained by combining the 3 sinograms presented in b.

RESULTS

Attenuation contrast

Several human bone samples were investigated in cooperation with U. Vogel, Middle Ear Lab., Department of Oto-Rhino-Laryngology, University Hospital "Carl Gustav Carus", Technical University Dresden (Germany). As part of the middle ear the 3-dim structure of the three ossicles (malleus, incus and stapes) were studied [13]. To reveal the 3-dim. structure of a stapes in its surrounding skull bones a scanning technique using the 2-dim. X-ray detector is applied. Three scans at different lateral position (x-direction), rough scans consisting of 90 projections in the outer regions and a fine scan consisting of 720 projection in the middle region, are performed. In a second step these scans are combined to build one huge scan. The tomographical reconstruction then yields the 3-dim. data set of the complete sample. Figure 2 demonstrates the measurement and combining procedure. In Figure 2a single projections at the same angular increment of the three scans are presented. For the upper projection the rotation axis of the sample was shifted to the left and for the bottom projection to the right. In Figure 2b the so called sinograms of the different scans for a special height are shown. The left and right sinograms consist of 90 projections equally stepped between 0 and π. The middle sinogram is based on 720 projections. The three sinograms of Figure 2b are combined to result in the final sinogram shown in Figure 2c which then is introduced into the tomographical reconstruction. Figure 3 shows a volume rendering of the complete 3-dim. data set of the sample consisting of 1945 x 1945 x 380 voxels each with an edge size of 5.9µm.

Figure 3: Volume rendering of a stapes part of the human middle ear using µCT at 24 keV at beamline BW2.

268

Fig. 4: 3-dim. rendering of a steel sample using μCT at 70 keV (BW5). Top surface is where the specimen fractured under tension.

Fig. 5: Volume rendering of a screw. The scan was performed using μCT at 60 keV (BW5).

High energy μCT was performed at beamline BW5 using photon energies in the range of 60 to 100 keV. Figure 4 presents the volume rendering of a steel sample investigated in cooperation with H.-A. Lauert, Department of Technical Mechanics, University of Bochum, Germany. The sample was fractured under tension. In the bulk, the accumulation of voids close to the surface are of interest. A tiny steel screw was investigated at 60 keV to demonstrate the high spatial resolution. The volume rendering given in Figure 5 shows the detailed surface of the sample.

Phase contrast

In cooperation with Prof. M. F. Rajewsky, Institute of Cell Biology [cancer research], University of Essen, Medical School, several rat trigeminal nerve samples were investigated by PμCT. The aim of the research was to reveal the 3-dim. structure of the early state of tumor creation [14-16]. In Figure 6 PμCT was performed at BW2 using 12 keV photon energy. The cracks shown are due to freeze-preserving of the specimen before embedding into wax. The volume rendering presented in Figure 7 shows the natural surface of another nerve. Three tomographical scans at different overall height of the sample were performed at BW2 using 20 keV photon energy. After reconstruction the three scans were combined to build one large data set.

CONCLUSIONS

The feasibility of different scanning techniques using a 2-dim. X-ray detector could be shown. By scanning in lateral direction, performing a precise scan in the middle (i.e. the most interesting volume) and rough scans in the outer section a region of interest method could be applied. Furthermore the setup was used at high photon energies to reveal the 3-dim. structure of specimens consisting of elements with higher absorption. The advantage of PμCT for low absorbing elements, e.g. biological samples, was demonstrated. Using 12 keV and 20 keV photon energies, the 3-dim. structure of rat trigeminal nerves were successfully studied. This

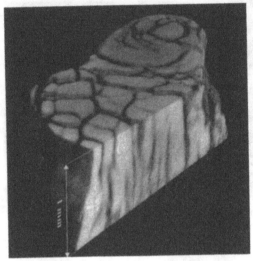

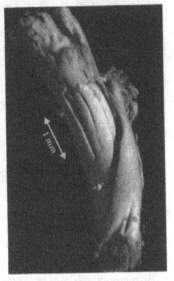

Fig. 6: 3-dim. rendering of a rat trigeminal nerve sample investigated by PμCT at beamline BW2 using 12 keV. The cracks are due to freeze-preserving the specimen before embedding into wax.

Fig. 7: Three tomographical scans at different height using PμCT at 20keV (beamline BW2) are combined. The volume rendering represent the natural surface of a rat trigeminal nerve sample.

demonstrates using PμCT at higher photon energies will allow for the simultaneous investigation of the 3-dim. structure of materials with greatly differing absorption. First results using PμCT at 70 keV are presented elsewhere [9,17].

ACKNOWLEDGEMENTS

For providing the bone sample of the human middle ear thanks are due to U. Vogel from the Middle Ear Lab., University Hospital "Carl Gustav Carus", Technical University Dresden, now at Fraunhofer-Institute for Microelectronic Circuits and Systems, Dresden. For the Fe-sample we thank H.-A. Lauert, Department of Technical Mechanics, University of Bochum. Special thanks are due to Prof M.F. Rajewsky, K. Heise and B. Kölsch, Institute of Cell Biology (Cancer Research), University of Essen Medical School, for kindly providing samples of rat trigeminal nerve.

We also thank W. Drube, H. Schulte-Schrepping of beamline BW2 of HASYLAB at DESY, Hamburg, T. Lippmann of beamline BW5, same institute, and M. Stampanoni, Institute of Biomedical Engineering, University of Zürich and ETHZ, for assistance during measurements.

Financial support by the Ministry of Science and Research of the Land NRW, Düsseldorf, is also gratefully acknowledged.

REFERENCES

1. U. Bonse and F. Busch, Prog. Biophys. molec. Biol. **65**, 1996, pp. 133-169.

2. U. Bonse and M. Hart, Appl. Phys. Lett. **6**, 1965, pp. 155-156.

3. A. Momose, Nucl. Inst. Meth. Phys. Res. A **352**, 1995, pp. 622-628

4. F. Beckmann, U. Bonse, F. Busch, and O. Günnewig, *Hasylab Jahresbericht II*, 1995, pp. 691-692.

5. F. Beckmann, U. Bonse and T. Biermann, *Hasylab Jahresbericht I*, 1998, pp. 885-886.

6. U. Bonse, F. Beckmann, M. Bartscher, T. Biermann, F. Busch, and O. Günnewig *Developments in X-Ray Tomography*, edited by U. Bonse (SPIE Proc. **3149**, San Diego, California, 1997) pp. 108-119.

7. P. Wyss, F. Beckmann, U. Bonse, B. Müller, J. Mayer, E. Wintermantel, W. Muster, DGZfP-Proceedings BB **67-CD**, 1999.

8. F. Beckmann, U. Bonse, T. Biermann, T. Lippmann, *Hasylab Jahresbericht I*, 1998, pp. 76-79.

9. F. Beckmann, U. Bonse, T. Biermann in *Developments in X-Ray Tomography II*, edited by U. Bonse (SPIE Proc. **3772**, Denver, Colorado, 1999) pp. 179-187.

10. F. Beckmann, *Ph.D. thesis*, University of Dortmund, Germany, 1998.

11. A. Momose, T. Takeda, and Y. Itai, Rev. Sci. Instrum. **66**, 1995, pp. 1434-1436.

12. F. Beckmann, U. Bonse, F. Busch, and O. Günnewig, J. Comput. Assist. Tomogr. **21**, 1997, pp. 539-553.

13. U. Vogel, *Ph.D. thesis*, Technical University of Dresden, Germany, 1999.

14. A. Yu Nikitin, L.A.P. Ballering, J. Lyons, and M.F. Rajewsky, *Proc. Natl. Acad. Sci. USA* **88**, 1991, pp. 9939-9943.

15. A. Yu Nikitin, J.-J. Jin, J. Papewalis, S.N. Prokopenko, K.M. Pozharisski, E. Winterhager, A. Flesken-Nikitin, and M.F. Rajewsky, Oncogene **12**, 1996, pp. 1309-1317.

16. F. Beckmann, K. Heise, B. Kölsch, U. Bonse, M. F. Rajewsky, M. Bartscher, and T. Biermann, Biophys. J. **76**, 1999, pp. 98-102

17. F. Beckmann, U. Bonse and T. Biermann, *Hasylab Jahresbericht I*, 1998, pp. 877-878.

X-ray phase contrast imaging study of activated carbon/carbon composite

Kenji Kobayashi, Koichi Izumi, Hidekazu Kimura, Shigeru Kimura, Tomoaki Ohira, Takashi Saito,[1] Yukari Kibi,[1] Takashi Ibuki,[2] Kengo Takai,[2] Yoshiyiki Tsusaka,[2] Yasushi Kagoshima,[2] and Junji Matsui[2]

Fundamental Research Laboratories, NEC Corporation, 34 Miyukigaoka, Tsukuba, Ibaraki 305-8501 Japan.

[1]*Functional Material Research Laboratories, NEC Corporation, 4-1-1 Miyazaki, Miyamae, Kawasaki, Kanagawa 216-8555 Japan.*

[2]*Faculty of Science, Himeji Institute of Technology, 3-2-1 Koto, Ako, Hyogo 678-1297 Japan.*

ABSTRACT

We have obtained the x-ray phase contrast images of the activated carbon/carbon (AC/C) composite used as electrodes in an electric double-layer capacitor (EDLC). To improve the spatial resolution, the beam size of x-ray transmitted from a sample were expanded by using asymmetric Bragg diffraction (asymmetric factor b) of analyzer crystal. Since the analyzer crystals were placed in (+, -) arrangements in both vertical and horizontal directions, the total magnification factor was $1/b^2$ for each direction. By using the 511 asymmetric Bragg diffraction of Si (100) analyzer with an asymmetric factor of 0.207, we could obtain the magnification factor of about 23.3. We applied this optical system to the AC/C composite. The phase contrast image corresponding to the carbon particles with the diameter of about 10 μm could be detected with the spatial resolution of 5 μm. Furthermore, we also obtained real-time images of bubble formation and movement during the overcharging of an EDLC with the time resolution of 30 frames number per second.

INTRODUCTION

X-ray phase contrast imaging has been attracting much attention as new imaging method [1-6]. Since this method is based on the real part of the refractive index $n=1- \delta +i \beta$, it is useful for the weak absorption materials. For example, the ratio between the refraction δ and the absorption β of x-ray is about 5000 for carbon at the x-ray energy of 15keV.

The availability of the intense, high-energy, and coherent x-rays produced in third generation synchrotron radiation (SR) facilities is expected to make this method suitable for industrial applications. In particular real-time imaging studies will become a valuable application for a characterization of electronic devices [4-6]. However, such real-time imaging studies have been restricted only to biological samples [4]. This is because the limitations of the available x-ray detectors for real-time observation. For example, the spatial resolution of the x-ray saticon camera, which is a popular detector for real-time imaging, is about 20 μm. Although

CCD cameras have high spatial resolution, they are not suitable for real-time measurements because of the long exposure time required. In order to use x-ray phase contrast imaging for real-time characterization of the electronic devices, we need a microscopy system with better spatial resolution.

In this paper we present a real-time x-ray phase contrast imaging system which can function as a microscopy. By using the asymmetric Bragg diffraction of analyzer crystals, the beam size of x-rays transmitted through the sample could be expanded. This system was used for imaging the activated carbon/carbon (AC/C) composite used as electrodes in an electronic double-layer capacitor (EDLC) [7]. We obtained an image showing contrast corresponding to carbon particles of 10 μm with the spatial resolution of 5 μm. We also succeeded in real-time imaging of the bubble formation and movement during the overcharging of an EDLC with the time resolution of 30 frames per second.

Experimental

The measurements were performed at Hyogo public beamline in SPring-8. The vertically polarized x-rays emitted from the 8-figure undulator were monochromatized by using a Si(111) double-crystal monochromator. The wavelength λ used was 0.826 (photon energy E = 15 keV). The beam size of x-ray was $1 \times 1 \text{mm}^2$ on the sample. Figure 1 shows the real-time phase contrast imaging system. This system is based on the method which is resolved the gradient of the phase shift by using the analyzer crystal. [3] In this system the beam size of x-ray transmitted from the sample is expanded by an analyzer crystal with an asymmetric factor b [8], which is defined by the following equation:

$$b = (\sin(\theta_B + \theta_\alpha))/(\sin(\theta_B - \theta_\alpha)) \qquad (1)$$

where θ_B is the Bragg angle, and θ_α is the angle between the crystal surface and Bragg diffraction plane. Since we used the Si (511) asymmetric Bragg diffraction of Si (100) crystal, the asymmetric factor b was 0.207. Moreover, the analyzer crystals were placed in (+, -) arrangements in both the vertical and horizontal directions, so that the total magnification factor $1/b^2$ was 23.3 for both directions. We obtained the phase contrast image by slightly changing the angle of the first and third analyzer crystals from the Bragg peak position. The magnified x-ray image was detected by x-ray camera (HX-250, Hitachi Denshi, Ltd.), which is a popular x-ray detector for real-time imaging. In this system the spatial resolution is limited by the energy resolution of the analyzer crystal, which is determined by the penetration depth of x-ray.

The sample used was the AC/C composite used as electrodes in EDLC with large electronic capacitance. This composite is suitable for evaluating this optical system because it consists mostly of carbon [7]. The sample size was $90 \times 48 \times 0.1$ mm3. A EDLC cell for real-time

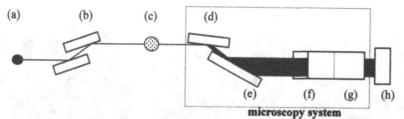

(e)   (f)   (g)   (h)

microscopy system

Fig. 1   Top of view of the optical system for x-ray phase contrast imaging: (a)x-ray source, (b)Si (111) double crystal monochromator, (c)sample, (d)-(g) Si (100) analyzer crystals, (h): detector. The asymmetric factor of the analyzer crystal is b=0.207 for λ =0.0826 nm. The transmitted x-ray beam was magnified to 23.3 times for vertical and horizontal directions by the 511 asymmetric Bragg diffraction of analyzer crystals (asymmetric factor b=0.207).

imaging was also prepared. The real-time imaging was performed while overcharging the EDLC at a constant voltage of 1.5 V.

RESULTS

We first obtained images of a Cu #2000 mesh in order to estimate the spatial resolution of this optical system. As shown in Fig. 2(a), the mesh patterns were observed clearly. Figure 2(b) shows the intensity profile along the line A-B in Fig. 2(a). The peak position was observed at regular intervals. The spatial resolution was estimated to be about 5 μ m, which corresponds to the penetration depth of x-ray for the analyzer crystals.

We then obtained x-ray phase contrast images of the AC/C composite. As shown in Fig. 3 (a), there was no detailed structure in the conventional absorption image. Figure 3(b) shows the image obtained without using the present microscopy system, simply by putting the detector far from the sample. Although the obtained image was different from the absorption image, it is difficult to identify the object in this image. On the other hand, in image obtained with the

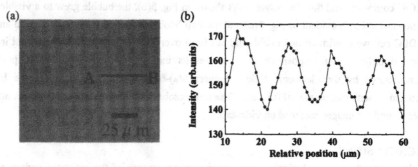

Fig. 2. (a): Magnified x-ray image of the Cu #2000 mesh.   (b): The intensity profile along the line A-B in (a).

(a)   (b)   (c)

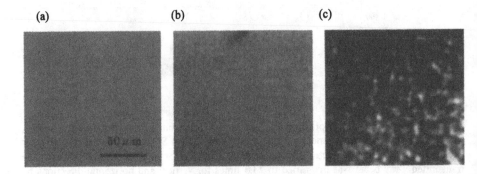

Fig. 3. X-ray image of the AC/C composite: (a) absorption image, (b) phase contrast image obtained using the conventional in-line method, (c) magnified phase contrast image observed using the present microscopy system.

present optical system the white-and-black contrast was observed clearly, as shown in Fig. 3(c). Scanning electron microscopy (SEM) revealed that the AC/C composite consists of the carbon particles about 10 μm in a diameter, amorphous carbon, and the void [7]. This contrast in Fig. 3(c) might be mainly caused by the difference of density between the carbon particles and the voids. The size of contrast regions in the image is consistent with the SEM results.

To improve the quality and the reliability of the EDLC, we need to know what the capacitor's properties are under severe conditions such as the overcharging. The overview of the EDLC cell prepared for real-time imaging is shown in Fig. 4. This cell has a unit structure composed from a pair of AC/C composite electrodes in the electrolyte (30 wt. % solution of sulfuric acid). Figures 5(a)-5(c) show the real-time image obtained during overcharging the capacitor cell at 1.5 V. After fifteen minutes a bubble formed at the point shown by the arrow in Fig.5(a). It is thought that this bubble is due to the electrochemical reaction between the AC/C composite and the electrolyte. As shown in Fig. 5(b), the bubble grew to a visible size. Another bubble is evident in Fig. 5(c) and the two bubbles eventually overlapped. In another EDLC cell we confirmed that bubble did not form over the whole of the electrode, but instead formed only at a few positions. These results indicate the existence of the positions concentrated by the electronic field. Figures 6(a)-6(d) shows the images of the bubble movement with the interval of 0.1 sec. The time resolution of this system is 30 frames number per second for images recorded on videotape.

CONCLUSIONS

In summary, we could obtain x-ray phase contrast images of the activated carbon/carbon composite by using the microscopy system composed of four analyzer crystals using

276

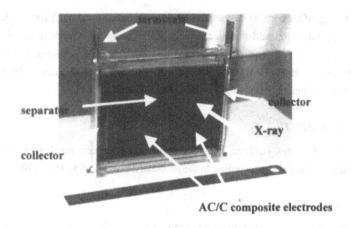

Fig. 4. Overview of the EDLC cell prepared for real-time imaging.

(a)   (b)   (c)

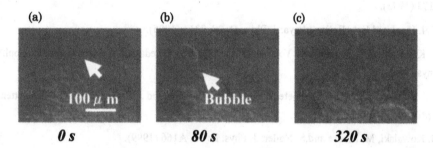

Fig. 5. Phase contrast images of bubble formation recorded while overcharging the EDLC.

(a)   (b)   (c)   (d)

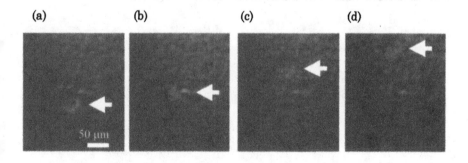

Fig. 6. Phase contrast images of bubble movements recorded with the time interval of 0.1 sec.

asymmetric Bragg diffraction. The spatial resolution obtained was about 5 μm. Furthermore, we observed the formation and movement of bubbles during the overcharging of electronic double layer capacitor cell with the time-resolution of 30 frames per second. Finally, the spatial resolution of this microscopy system will be improved drastically by using the asymmetric Bragg diffraction in the total reflection condition.

ACKOWLEDGEMENTS

We thank Drs. Y. Kubo, K. Sone, M. Yamagata, and M. Utsumi for their encouragement. This work has been carried out according to the proposal number C99A24XU-039N.

REFERENCES

1. A. Momose and J. Fukuda: Med. Phys. **22**, 375 (1995).

2. T.J. Davis, T.E. Gureyev, D. Gao, A.W. Stevenson and S.W. Wilkins: Phys. Rev. Lett. **74**, 3173 (1995).

3. V.N. Ingal and E.A. Beliaevskaya: J. Phys. D: **28**, 2314 (1995).

4. Y. Kagoshima, Y. Tsusaka, K. Yokoyama, K. Takai, S. Takeda and J. Matsui: Jpn. J. Appl. Phys. **38**, L470 (1999).

5. J-Y. Buffiere1, E. Maire, P. Cloetens, G. Peix, M. Salomé, and J. Baruchel: ESRF-Newsletter April No. **30**, 20 (1998).

6. G. Kowalski, M. Moore and S. Nailer: J. Phys. D: **32**, A166 (1999).

7. Y. Kibi , T. Saito, M. Kurata, J. Tabuchi and A. Ochi: J. Power Sources **60**, 219 (1996).

8. K. Kohra and M. Ando: Nuc. Instrum. & Methods **177**, 117(1980).

X-RAY TOPOGRAPHY STUDY OF SURFACE DAMAGE IN SINGLE-CRYSTAL SAPPHIRE

D. BLACK, R. POLVANI, K. MEDICUS AND H. BURDETTE
NIST, Gaithersburg, MD 20899, david.black@nist.gov

ABSTRACT

X-ray diffraction topography was used to investigate the relationship between sub-surface damage, near-surface microstructure, and fracture strength in a series of sapphire modulus of rupture (MOR) bars which had been fabricated to proof test fabrication processes. The strength of the bars was determined by failure in four point bending. The tensile surface of the bars was also examined using optical microscopy and non-contacting surface profilometry. Both show that the bars have good surface finish, with a typical RMS roughness of 0.7 nm. No correlation was found between RMS surface finish and fracture strength. Although the bars appeared to be indistinguishable, topographs taken prior to fracture testing revealed that they are of two distinct types. Type 1 has an oriented microstructure consisting of a pattern of linear features running the length of the bars. Type 2 was typical of well-polished sapphire, containing individual dislocations and occasional damage from handling. We attribute the Type 1 microstructure to fabrication damage that was not removed by subsequent processing and/or polishing. Fracture strength data showed that the Type 1 (damaged) bars had strengths less than 70 % of the bars without damage. Topography is sensitive to near-surface damage that can be correlated to fracture strength. Neither low magnification optical microscopy nor conventional surface finish statistics could be correlated to strength.

INTRODUCTION

High-strength at elevated temperatures makes single-crystal sapphire an ideal material for the windows or domes of IR seekers for ballistic missiles. However, the strength of sapphire can be affected by a variety of factors, including growth method, fabrication or finishing operations, and other processing practices such as annealing and doping. An important strength limiting factor is surface damage. Fabrication of sapphire components necessarily involves damaging the surface of the part to shape it. This damage can take the form of cracks, strains, crystallographic defects or compositional changes. To understand how subsurface damage affects strength, characterization techniques are needed which are sensitive to microstructural changes of the surface and near-surface in relation to fabrication and/or post-fabrication processes.

The Sapphire Statistical Characterization And Risk Reduction program (SSCARR) [1] was developed to provide a high-temperature fracture-strength data base for sapphire. The NIST contribution to this program has been to evaluate the utility of a variety of diagnostic methods, including x-ray diffraction topography, as characterization tools for near-surface damage. Results from the topographic examination are compared to "standard" optical characterization using Nomarski contrast and to surface finish measurements.

EXPERIMENT

Test pieces for this study were a set of thirty nine (39) MOR bars from the SSCARR program. The bars were nominally 3 mm x 4 mm x 45 mm with chamfered edges. The surface of the selected bars was to be of optical quality, representative of a specific fabrication process, and was parallel to the basal plane. Fracture testing was performed by the University of Dayton Research Institute under contract to the SSCARR program. The bars were fractured in four point bending as per ASTM standard C-1211-92 using fully articulated fixtures. Testing was performed at a temperature of 500 K in laboratory ambient air with a constant displacement of 0.5 mm/min. using an inner span of 10 mm and an outer span of 40 mm.

X-ray diffraction topographs were recorded from the surface of the bars prior to fracture testing. The topographs were taken at the NIST Materials Science Beam Line, X23A3, at the National Synchrotron Light Source at Brookhaven National Laboratory. An incident energy of 8 keV was used and the symmetric (0006) diffraction was recorded from the tensile surface of the bars. Low magnification optical microscopy was also used and surface finish was measured to compare to the topographs and to represent other inspection methods. The surface finish measurements were taken on a WYKO TOPO-3D [2] non-contact surface profiler with a Mirau interferometer. The system used phase-shifting interferometry (PSI) to measure the profile. In PSI, a known phase change is induced between the object and the reference beams and a solid-state detector is used to record fringes. From the intensity data, the phase of the wavefront and therefore the relative surface heights can be calculated directly [3]. The numerical data for the MOR bars was recorded using a 10x objective so that the RMS roughness values are area-averaged over 0.94 mm^2 of the surface. Three positions on each bar were measured and these values were averaged. In addition to purely numerical results, which are typical of standard practice, the area interferogram, from which the area-averaged value is calculated, was plotted. The optical microscopy was performed using a general-purpose laboratory microscope at 50X magnification and restricted to using Nomarski contrast for the imaging.

RESULTS

The topographs revealed that the 39 bars could be separated into two distinct groups, with 14 bars designated Type 1 and 25 bars designated Type 2. Figure 1 shows representative topographs of the two types. The Type 1 bars have a microstructure consisting of a series of linear features running the length of the bar with few other features. The microstructure in Type 2 bars shows a far less damaged surface. Individual dislocation images are visible, and subsequent damage from handling the samples can be detected.

The area-averaged RMS surface-finish values for 10 Type 1 and 4 Type 2 bars are shown in Figure 2. We found no correlation between surface finish and fracture strength. However, the interferograms from which the surface finish statistics were derived do show marked differences between the Type 1 and Type 2 surfaces. For example see Figure 3a and 3b. Low magnification Nomarski observations could not distinguish the two types.

The fracture data are shown in Table 1. If the data are taken as a whole, the arithmetic average strength is 594 MPa +/- 165 MPa (one standard deviation). Using the topographs to sort the data into the two Types, we see a large difference between their strengths. Type 1 bars

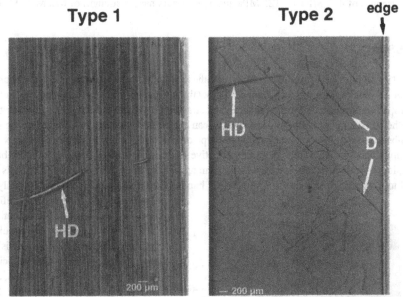

Figure 1. (0006) diffraction topographs of a representative Type 1 and Type 2 MOR bar. Handling damage (HD) is seen on both bars but individual dislocations (D) are seen only in the Type 2 bar.

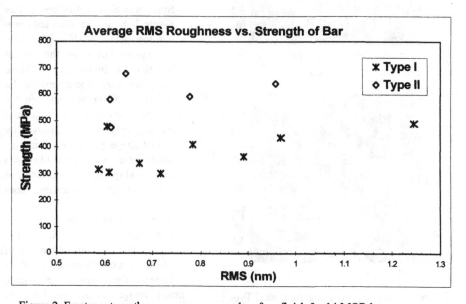

Figure 2. Fracture strength versus area averaged surface finish for 14 MOR bars.

have a strength of 453 MPa +/- 125 MPa and Type 2 bars have a strength of 672 MPa +/- 129 MPa.

DISCUSSION

Transmission electron microscopy has shown [4] that fabrication damage in sapphire consists of cracks, strains, and intense dislocations concentrations at and below ground surfaces; and of intense dislocation tangles, and strain at polishing marks. The damaged region below the surface of optically-polished crystals can extend 100 nm or more into the sample [5]. This subsurface damage contains exactly the type of crystallographic defects imaged by x-ray topography and so topography should be sensitive to any near-surface damage in MOR bars. In fact, it is usually just this type of polishing damage that must be eliminated from crystals for typical topographic examination. On the other hand, the Nomarski image is sensitive only to local changes in height. Therefore, local surface relief and/or reflections from internal surfaces are required to detect damage. On well finished surfaces there will not be sufficient surface relief for image contrast, and consequently, we found that low magnification optical microscopy could not distinguish between the two types of bars. Using the purely numerical values for surface finish was also ineffective, but as Figure 3 shows, that does not mean that the method itself is inherently inadequate, but rather the current practice is ineffective. More analysis is needed to determine if a numerical difference between the types of bar can be extracted from the interferograms. One possibility is that shape parameters, such as skewness and kurtosis could show a relationship that roughness, a surface weighted parameter, does not.

In contrast to optical microscopy and surface finish, x-ray topography has the advantage of producing a one-to-one image of the surface. Topographs combine large area coverage and high sensitivity to damage in one image. Higher

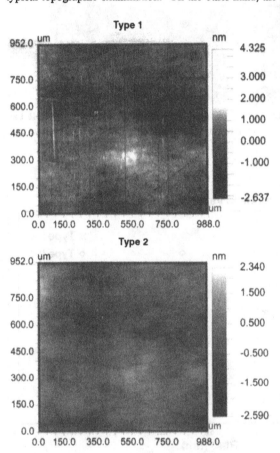

Figure 3. Area interferograms for typical Type 1 and Type 2 bars.

Sample	Strength (MPa)	Sample		Type 1	Strength (MPa)	Type 2	Strength (MPa)
		Table 1. Strength data for 39 MOR bars.					
1	301	21	642	1	301	1	688
2	490	22	601	2	490	2	698
3	338	23	669	3	338	3	688
4	412	24	601	4	412	4	1007
5	728	25	664	5	728	5	678
6	367	26	524	6	367	6	491
7	318	27	904	7	318	7	642
8	617	28	477	8	617	8	601
9	531	29	614	9	531	9	669
10	510	30	672	10	510	10	601
11	516	31	591	11	516	11	664
12	481	32	633	12	481	12	524
13	304	33	679	13	304	13	904
14	436	34	548	14	436	14	477
15	688	35	729	Average	453	15	614
16	698	36	581	σ	125	16	672
17	688	37	942			17	591
18	1007	38	676			18	633
19	678	39	811			19	679
20	491					20	548
		Average	530			21	729
		σ	180			22	581
						23	942
						24	676
						25	811
						Average	672
						σ	129

sensitivity surface finish measurements requires using higher magnification, thereby reducing the observed area. Observing a smaller area reduces the ability to image the distribution of the damage and can easily miss large features. Since topographs provide a large area view they can be used as maps to guide other techniques that rely on higher magnification.

The microstructure imaged in the topographs of Type 1 bars is consistent with residual damage from the fabrication process. Most likely this results from a surface grinding operation in which the sample is translated longitudinally under a grinding wheel. The damage created in this early processing step was not completely removed in the later steps. TEM has also shown that microcracks are associated with grinding damage. These cracks are too fine to be seen with the spatial resolution of topography but will certainly limit the fracture strength.

ACKNOWLEDGMENTS

This work is supported by the Ballistic Missile Defense Organization (BMDO) through the Air Force Metrology and Calibration Program, Heath, Ohio and by the Sapphire Statistical Characterization And Risk Reduction program, Huntsville, AL.

REFERENCES

1)"Final Report for the Sapphire Statistical Characterization and Risk Reduction (SSCARR) Program", in preparation, to be published by Nichols Research, Huntsville, AL.

2) Certain commercial equipment, instruments, or materials are identified in this paper in order to specify the experimental procedure adequately. Such identification is not intended to imply recommendation or endorsement by the National Institute of Standards and Technology, nor is it intended to imply that the materials or equipment identified are necessarily the best available for the purpose.

3) Creath, Katherine, "Phase-Measurement Interferometry Techniques" in Progress in Optics, Vol. XXVI, pp. 349-393, (1988).

4) B.J. Hockey, Proc. 20 of the British Ceramic Society, The British Ceramic Society, Stoke-on-Trent, G.B., pp. 95-115, (1972).

5) D. Black, R. Polvani, L. Braun, B. Hockey and G. White in *Window and Dome Technologies and Materials V*, edited by Randall W. Tustison (SPIE Proc. **3060**, Bellingham, WA, 1997) p. 102-114.

KEYS TO THE ENHANCED PERFORMANCE
OF MERCURIC IODIDE RADIATION DETECTORS
PROVIDED BY DIFFRACTION IMAGING

Bruce Steiner,* Lodewijk van den Berg,** and Uri Laor***

* NIST, Gaithersburg, MD 20899-8520
** Constellation Technology, Inc., Largo, FL 33777
*** NRCN, Be'er Sheva 84910, Israel

ABSTRACT

High resolution monochomatic diffraction imaging is playing a central role in the optimization of novel high energy radiation detectors for superior energy resolution at room temperature. In the early days of the space program, the electronic transport properties of mercuric iodide crystals grown in microgravity provided irrefutable evidence that substantial property improvement was possible. Through diffraction imaging, this superiority has been traced to the absence of inclusions. At the same time, other types of irregularity have been shown to be surprisingly less influential. As a result of the knowledge gained from these observations, the uniformity of terrestrial crystals has been modified, and their electronic properties have been enhanced. Progress toward property optimization through structural control is described.

INTRODUCTION

The point of departure for this study was NASA's interest in tracing the structural origins of the superior electronic performance of a first generation mercuric iodide crystal grown on Spacelab III, performance subsequently duplicated with second generation material on IML-1. The knowledge that proved key to understanding the nature of the space crystals, and then led to improved performance for terrestrial crystals, was derived from monochromatic diffraction imaging carried out on beamline X23A3 at the National Synchrotron Light Source at Brookhaven National Laboratory.

The performance of mercuric iodide for high energy radiation detectors is of particular interest because comparable energy resolution is not available at room temperature through other materials. Room temperature operation leads to instrumental simplicity and unattended operation, which enable a range of terrestrial as well as space applications.[1,2,3,4] Moreover, mercuric iodide detectors display exceptionally high resistance to lattice damage from the radiation that they are monitoring.[5] Optimization of the performance of these detectors will facilitate reliable monitoring of nuclear technology for nonproliferation of nuclear weapons, storage of sensitive nuclear materials, portal monitoring, nuclear medical treatment, and astrophysical observation.

EXPERIMENT

The diffraction imaging was carried out in Bragg geometry at 8 keV with beam divergence of less than an arc second. The images are correlated directly with the hole mobility lifetime product, $\mu_h \tau_h$. We have examined three generations of wafers of α mercuric iodide, differing from one another in purification and synthesis procedures. For the first two generations, the crystals grown both in microgravity and terrestrially from the same material were compared. The understanding achieved and the resulting model that incorporates it have enabled us progressively to enhance the performance of terrestrially grown mercuric iodide. The imaging provided the guidance needed for property optimization.

The necessary insight was derived in part from the structural information contained in the diffraction images themselves and in part from other observations that had not been synthesized into a coherent model prior to the unifying perspective provided by the diffraction imaging. From etching[6] and subsequent optical observation,[7] localized structural anomalies had been observed optically in wafers from terrestrial crystals grown both from

Table: PARAMETERS FOR CRYSTALS EXAMINED

	Terrestrial Growth	Space Growth
---	---	---
First Generation		(Spacelab III)
$\mu_h \cdot \tau_h$ (cm^2/V)	7×10^{-6}	42×10^{-6}
$\delta\,\theta(°)$	0.11	≤ 1.5
Second Generation:*		(IML-1)
$\mu_h \cdot \tau_h$ (cm^2/V)	21×10^{-6}	42×10^{-6}
$\delta\,\theta(°)$	0.7	0.06
Third Generation:**		
$\mu_h \cdot \tau_h$ (cm^2/V)	28×10^{-6}	
$\delta\,\theta(°)$	≤ 2.0	

* (Purification)
** (Purification & Stoichiometry)

mercury-rich and from stoichiometric material.[6] Two types of linear <110> arrays of small, round inclusions were found. Individual linear <100> anomalies also were observed. In addition, etching revealed a subgrain structure delineating abrupt changes in lattice orientation by 20' of arc. Gamma-ray diffraction of terrestrial wafers also displayed a family of relatively flat subgrains differing in orientation by 6' to 7' of arc.[2] For the second generation of material, calorimetry and resonance fluorescence spectroscopy of iodine-rich crystals provided strong evidence for iodine inclusions.[8]

In contrast to these observations of terrestrial wafers, the γ-ray rocking curves of wafers from the crystal grown on Spacelab III displayed only one grain, but one that was deformed, with substantial variation in lattice orientation. The connections among these various observations were not yet clear.

RESULTS AND DISCUSSION

Our own preceding work with NASA had noted that first generation terrestrial mercuric iodide crystals contained flat, low angle subgrains. This structure was found again in the present first generation terrestrial crystal, Figure 1 as well as in the γ-ray diffraction just noted. A comparable image of a wafer from the Spacelab III crystal, however, displayed only a single, highly irregular grain, permitting only a small fraction to be brought into diffraction at any one time, as in Figure 2.

Contrary to the expectations leading to the space experimentation, it was the far more irregular crystal with respect to lattice orientation that outperformed electronically the orientationally more uniform crystal. The hole mobility lifetime product for the space crystal was six times that for the terrestrial crystal, as indicated in the Table, along with the measured angular variation in lattice orientation, δθ.

The key to understanding these differences lay in closer examination of the diffraction images. The small features arrayed in rows in the terrestrial crystal are in fact always out of diffraction in the vicinity of mercuric iodide Bragg peaks and therefor could safely be ascribed to inclusions. Analysis of residues from terrestrial wafers strongly suggested that these inclusions were organo-metallic deposits;[9] features that could be identified with the earlier optical observations just noted. In contrast to this, the crystal grown in Spacelab III shown in Figure 2 is essentially devoid of inclusions. This led to the thesis that, contrary to expectations, the principal initial impediment to improved performance in terrestrial material was not irregularity in lattice orientation but rather the presence of inclusions. The evidence thus suggested that convective mixing of impurities was far more important to performance than was gravity loading during terrestrial crystal growth.

At the same time, these images suggest that the source of the characteristic subgrain structure in the first generation terrestrial crystals is the precipitation hardening derived from these inclusions. This turned our attention to the softness of the space crystal, requiring as a result the development of innovative handling procedures. Subsequent experimentation proved this thesis to be important.[10]

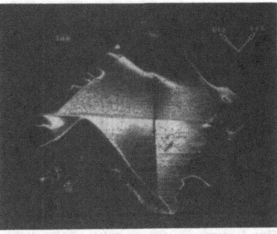

Figure 1. Asymmetric (1 1 12) image of first generation terrestrial wafer. The relatively large area in diffraction displays a high degree of lattice planarity with respect to rotation around a [1$\underline{1}$0] axis. The dark, vertical [110] median stripe delineates a sharp fold in the lattice by seven minutes of arc. Variation in the deposition of inclusions is visible as horizontal stripes.

These developments thus led to a second series of space experiments on IML-1, utilizing a second generation of material, synthesized and purified in house. The time remaining for preparation for this flight permitted effective attention to purification, but not to complete refinement of the stoichiometry. The resulting second generation material[11] indeed proved to be slightly iodine-rich.[8]

The performance of the second generation IML-1 space crystal duplicated that of the first generation Spacelab III crystal. At the same time, the second generation terrestrial crystal displayed marked improvement, though not yet to the level of the two space crystals.

The structural origins of these results also are visible in the diffraction images, in Figures 3 and 4. These are, respectively, images of the IML-1 wafer and a comparable terrestrial wafer. The uniformity of the terrestrial second generation crystal in these images clearly is influenced by a small but visible set of inclusions. In this case, however,

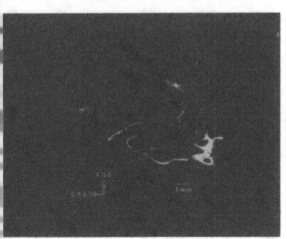

Figure 2. Asymmetric (1 0 10) image of wafer from first generation crystal grown on Spacelab III, comparable to the image of the terrestrial crystal in Figure 1. The relatively small area in diffraction displays variation in lattice orientation associated with the absence of the inclusions prominent in the terrestrial wafer.

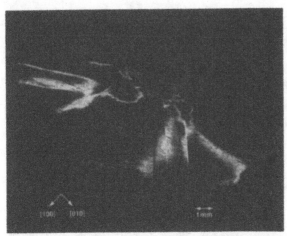

Figure 3. *Symmetric (008) diffraction image of wafer from the second generation crystal grown on IML-1. Repeated folding of the lattice around <110> directions is evident in the sharp vertical stripes on the right and the gentler low angle stripes on the left.*

the inclusions are iodine[8] rather than impurities.

This, too, has been confirmed, through the preparation of a third generation of material with close attention both to purity and to stoichiometry. The absence of inclusions in the image of the third generation terrestrial crystal, Figure 5, confirms that high degrees of purity and of stoichiometry have been achieved. At this point, the gap in performance between terrestrial and space crystals has been narrowed still further, as indicated in the lower part of the Table, although it has not yet been closed completely.[12]

The remaining gap in performance appears to represent a new, more subtle challenge. The electronically optimum chemical balance may not be complete stoichiometry, but something slightly different. This interesting thesis has led to refinement of the chemistry of the source material.

Two steps in this exploration are shown in Figures 6 and 7, which illustrate different levels of wafer polish and their consequences. In Figure 6, two types of defect distribution are evident. Of the four distinct sections in the wafer defined by the crystallographic directions that they took during growth, two contain inclusions, and two do not. Those parts of the crystal that grew in <111> directions display regular arrays of linear inclusions, while those that grew in other directions, both <001> and <101>, appear devoid of inclusions.

The linear inclusion arrays may reflect temporally periodic variation in the chemistry in the vicinity of the growing crystal. Alternatively, however, and perhaps more likely, this arrangement may reflect the important role played by even small changes in flow over a vicinal crystal surface steadily growing from solution, as described and analyzed by Chernov.[13-14] In the growth of a highly perfect crystal with few screw dislocations facilitating faceted growth, the spacing of steps on the growing surface appears to be

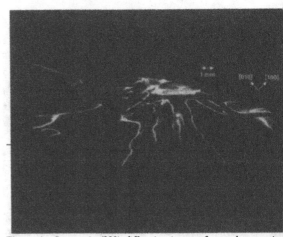

Figure 4. *Symmetric (008) diffraction image of second generation terrestrial wafer comparable to the IML wafer in the preceding image. The seed crystal is visible in the upper right hand corner. The restriction of diffraction to small areas in the region of new growth displays short range variation in lattice orientation.*

288

controlled sensitively by diffusion surrounding bunches of steps that develop spontaneously. In this instance, the deposition of inclusions, particularly those that inhibit crystal growth, will follow closely the local step density, which becomes spatially periodic, leading to the deposition of linear arrays of the inclusions,[15] precisely like those evident in Figure 6.

In Figure 7, a higher degree of polish of another wafer not only yields a smoother surface, but also highlights the striking difference in hardness between the sections with inclusions and those without.

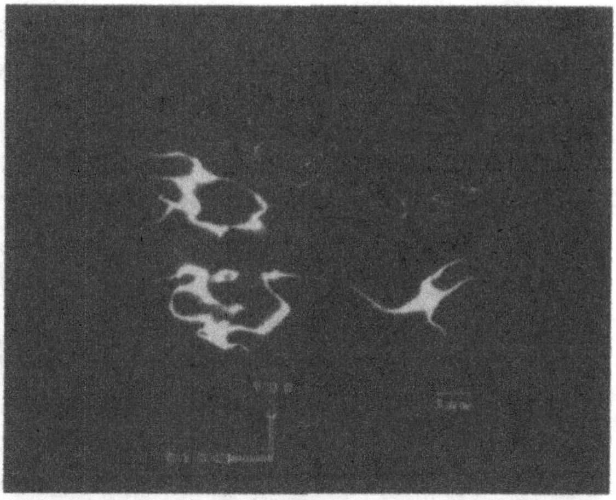

CONCLUSIONS

Figure 5. Asymmetric (1 0 10) diffraction image of third generation terrestrial wafer, displaying variation in lattice orientation associated with the absence of inclusions reflecting either impurities or imperfect stoichiometry.

The absence of convection in microgravity appears to be far more influential in determining the superior electronic properties of the space crystals at this stage than does the expected uniformity in lattice orientation permitted by the absence of *gravity loading* of the hot crystal during its growth. Through particular attention to purification and the achievement of stoichiometry, the effects of convection in terrestrial growth have now been mitigated. Although the performance of both generations of space crystals to date remains superior to the best of those grown on the ground, the difference has been reduced by more than 50 %.

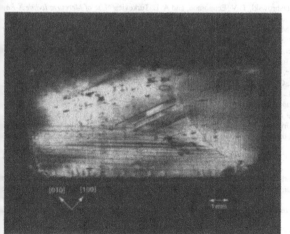

Figure 6. Symmetric (008) diffraction image of terrestrial wafer, displaying rows of inclusions in the regions of the boule that grew in <111> directions.

The progress to date leads to the expectation of further enhancement of electronic properties of terrestrial crystals. This progress suggests paths to surpass the properties of the superior space crystals grown to date. Advancement along these lines will be facilitated both by continued guidance from synchrotron science and by further space experimentation.

ACKNOWLEDGMENT

The Consortium for Commercial Crystal Growth at Clarkson University provided the financial and intellectual support necessary for determination of the state of the art of terrestrial α mercuric iodide

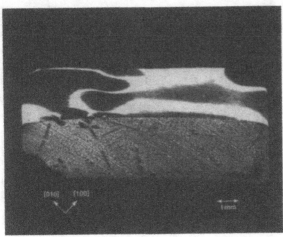

wafer uniformity at the time of initial growth in microgravity. This knowledge provided the stimulation of interest essential to the success of this work. Examination of the wafers fabricated from crystals grown subsequently in microgravity and associated crystals was made possible by partial support of the NASA Microgravity Sciences and Applications. Division.

Figure 7. Symmetric (008) diffraction image of more highly polished terrestrial wafer, displaying range of hardness associated with the presence of inclusions.

REFERENCES

1   J. S. Iwanczyk, N. Dorri, M. Wang, M Szawlowski, B. A. Patt, W. A. Warburton, B. Hedman, and K. O. Hodgson, *"The HgI₂ Energy Dispersive X-Ray Detectors and Miniaturized Processing Electronics Project"*, IEEE Trans. Nucl. Sci. 37, 198-202 (1990)

2   Jan S. Iwanczyk, *Advances in Mercuric Iodide -Ray Detectors and Low Noise Preamplification Systems"*, Nucl. Inst. Meth, Phys. Res. A 283, 208-214 (1989)

3   J. S. Iwanczyk, N. Dorri, M. Wand, M. Szawlowski, W. A. Warburton, B. Hedman, and K. O. Hodgson, *"Advances in mercuric iodide energy dispersive x-ray array detectors and associated miniaturized processing electronics"*, Rev. Sci. Inst. 60, 1561-1567 (1989)

4   J. S. Iwanczyk, Y. J. Wang, N. Dorri, A. J. Dabrowski, T. V. Economou, and A. L. Turkevich *"Use of Mercuric Iodide X-Ray Detectors with Alpha Backscattering Spectrometers for Space Applications*, IEEE Trans. Nucl. Sci 38, 574-579 (1991)

5   B. A. Patt, R. C.. Dolin, T. M. Devore, J. M, Markakis, J. S. Iwanczyk, N. Dorri, and J. Trombka, *"Radiation damage resistance in mercuric iodide X-ray detectors"*, Nucl. Inst. Meth. Phys Res. A 299, 176-181 (1990)

6   B Milstein, B. Farber, K. Kim, and W. F. Schnepple, *"Influence of Temperature upon Dislocation Mobility and Elastic Limit of single Crystal HgI₂"*, Nucl. Inst. Meth. 213, 65-76 (1983)

7   L. van den Berg, W. Schnepple, C. Ortale, and M. Schieber, *"Vapor Growth of Doped Mercuric Iodide Crystals by the Temperature Oscillation Method,"* J. Cryst. Growth 42, 160-165 (1977)

8   A. Burger, S. Morgan, C. He, E. Silberman, L. van den Berg, C. Ortale, L. Franks, and M. Schieber, *"A Study of Inhomogeneity and Deviations from Stoichiometry in Mercuric Iodide,"* J. Cryst. Growth 99, 988-993 (1990)

9   H. A. Lamonds, *"Review of Mercuric Iodide Development Program in Santa Barbara"*, Nucl. Inst. Meth. 213, 5-12, (1983)

10   Bruce Steiner, Lodewijk van den Berg, and Uri Laor, *"High resolution Diffraction Imaging of Mercuric Iodide: Demonstration of the Necessity for Alternate Crystal Processing Techniques for Highly Purified Material."* Mat. Res. Soc. Symp. 375, 259-264 (1995)

11   L. van den Berg, *Growth of Single Crystals of Mercuric Iodide on the Ground and in Space,"* Mater. Res. Soc. Proc. 302, 73-83 (1993)

12   Bruce Steiner; Lodewijk van den Berg, and Uri Laor, *"Enhancement of Mercuric Iodide Detector Performance through Increases in Wafer Uniformity by Purification and Crystal Growth in Microgravity,"* J. Appl. Phys. In press, October 15 issue (1999)

13   A. A. Chernov, *"How does the flow within the boundary layer influence morphological stability of a vicinal face?"*, J. Cryst. Growth 118, 333-347 (1992)

14   A. A. Chernov, *"Formation of crystals in solutions"*, Contemp. Phys. 30, 251-276 (1989)

15   A. A. Chernov, Yu. G. Kuznetsov, I.L.Smol'skii, and V.N. Rozhanskii, *"Hydrodynamic effects in growth of ADP crystals from aqueous solutions in the kinetic regime"*, Kristallografiya 31, 1193-1200 (1986); Sov. Phys. Crystallogr. 31, 705-709 (1986)

Dislocation motion around loaded notches in ice single crystals

D. CULLEN *, X. HU *, I. BAKER *, M. DUDLEY **
*Thayer School of Eng., Dartmouth College, Hanover, NH 03755,
**Dept. of Materials Science, SUNY at Stony Brook, Stony Brook, NY 11794

ABSTRACT

Synchrotron X-ray topography has been used to study dislocation behavior around a notch in single crystal ice during *in-situ* deformation at a constant strain-rate of 1 x 10^{-8}s^{-1} and a temperature of -8 °C. During deformation a dislocation depleted zone (DDZ) formed above the notch. Modeling the interaction between basal plane dislocation loops and the notch stress field suggested that this DDZ arose from dislocations gliding completely through the specimen.

INTRODUCTION

The interaction between dislocations and a crack tip is of great importance for understanding fracture. Dislocation free zones (DFZ), which have been observed ahead of crack-tips in numerous materials (e.g. [1] through [8]), can affect crack propagation and hence fracture [9]. However, these observations were either from thin foils in transmission electron microscopes or from surface observations, and are not necessarily indicative of bulk behavior. In this paper, x-ray topographic observations of a dislocation-depleted zone (DDZ) in bulk single-crystal ice are presented. A simple model, based on the interaction between basal plane dislocation loops and the notch stress field, is used to explain its formation.

EXPERIMENT

The experiments consisted of loading, under uniaxial tension, notched cuboidal ice specimens in their most brittle orientation (perpendicular to the basal plane) while topographic images where taken of the area adjacent to the notch. The ice used in these experiments was laboratory grown, single crystal ice with a chevron notch, between 2.5 mm and 3 mm long, that was cut into each specimen using a razor blade. Topographs of the notched specimens can be seen in Figure 1. The crystallographic orientation, coordinate system and dimensions of the specimens are shown in Figure 2. The specimen was strained in a loading jig based on one described previously by Hu et al. [10]. The temperature, -8°C, and displacement rate, 1 x 10^{-8}s^{-1}; were chosen to allow a series of topographs to be taken, using synchrotron x-radiation at Brookhaven National Labs, before the specimen fractured.

RESULTS AND DISCUSSION

Dislocation Depleted Zone Formation

The single crystal specimen was oriented so that the basal planes were viewed edge-on; the dark horizontal straight lines appearing on the images are projections of dislocations lying on these planes. In Figure 1(a), taken before deformation, a number of these dislocations are stacked in a vertical array (indicated by A). This array is a common feature and may have formed either during crystal growth or during subsequent specimen preparation. Figure 1(b) is after 3.8 hours of loading (1.05 MPa) with the load removed. Note that, in contrast to Wei et. al. [7], [8], there is no DFZ or DDZ visible ahead of the notch-tip, however, the loading conditions in this latter work were quite different.

The ice specimens can be difficult to orient exactly, and although prepared to have the basal plane perpendicular to the applied tensile stress, the c-axis is slightly off a true parallel to the loading direction. This small misalignment introduces shear stress onto the basal plane providing the impetus for the vertical dislocation array to glide to the right when specimen deformation begins. As the dislocations glide to the right they begin to interact with the notch-tip stress field and this retards their motion causing the vertical array to remain in the region adjacent to the notch-tip. The dislocation-depleted zone (DDZ) can be seen above the notch in Figure 1(b). To understand more clearly the physical mechanisms underlying the formation of this DDZ, a semi-quantitative analysis is presented below.

Dislocation Depleted Zone Model

Shearwood et al. [11], [12], Liu [13] and Jia [14] have all shown that dislocations on the basal plane, when exposed to a shear stress, move as partial or full hexagonal loops, comprised of 60° and screw dislocations. These straight dislocations arise because their energy is

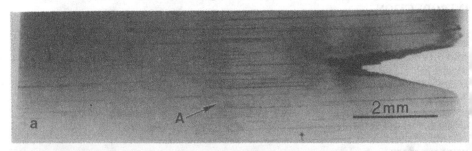

Figure 1. A pair of topographs before (a) and after (b) deformation. The diffraction vector is g = 10$\overline{1}$0. The vertical array is marked **A**, and the dislocation-depleted zone is marked DDZ.

minimized by lying in Peierls valleys [12]. The model uses this phenomenon, and the calculated notch-tip shear stress experienced by each segment of the hexagonal dislocation loop, to determine the formation and position of the DDZ. The DDZ is thought to form as dislocation loops (partial or full, Figure 3) expand on the basal plane and emerge from the free surface of the specimen.

Using the partial hexagonal loop configuration from Figure 3 (the full loop was also modeled, but the results did not agree well with the topographs in Figure 2), dislocation motion was modeled on three basal planes, 1.0 mm, 1.5 mm, and 2.0 mm (BP-1, BP-1.5 and BP-2, respectively) above the notch-tip. In the following discussion, BP-1.5 is specifically considered, but the general discussion and equations used to describe the stress state and dislocation motion are applicable to all three planes with only slight geometrical modifications.

Consider BP-1.5 (Figure 2) under an applied stress. The two segments, AB and BC, of a hexagonal dislocation loop are lying on this plane, possibly as part of the vertical dislocation array. The projection of this partial loop onto the front surface of the specimen, which is the edge-on view of the basal plane dislocation loop, is represented by the segment AC. There are three possibilities for the Burgers' vector b of dislocations in the basal plane: $[1\bar{2}10]$, $[11\bar{2}0]$ and $[2\bar{1}\bar{1}0]$. In this case, $b = [1\bar{2}10]$ is not possible since the diffraction vector $g = 10\bar{1}0$ and the dislocations would not be in contrast. Thus b can only be $[11\bar{2}0]$ or $[2\bar{1}\bar{1}0]$. If b is $[11\bar{2}0]$, segments AB and BC are in screw and 60° orientations respectively, while if b is $[2\bar{1}\bar{1}0]$, AB and BC are in 60° and screw orientations respectively. Since the stress field about the notch-tip is asymmetrical, both cases were modeled and it was determined that the orientation found in case b more accurately reproduced the features found in topograph Figure 1(b).

The applied normal stress (σ_o), its orientation relative to the notch-tip, and the orthogonal axes, 1, 2 and 3 used for the analysis, are indicated in Figure 2. The notch length $a \ll H$ (specimen height), and the thickness $t \ll (W$ (specimen width), H) hence the notch is loaded in plane-stress mode I and the shear stress at some point (r, θ) around the notch-tip can be approximately represented using Eq. 1 from linear elastic fracture mechanics.

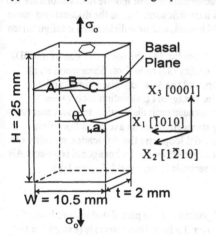

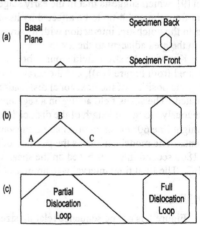

Figure 2. The geometry, dimensions, coordinate system, and crystallographic orientation of the specimen. ABC is a partial hexagonal dislocation loop.

Figure 3. The basis of the DDZ model. (a) time = T1, (b) T2 > T1, and (c) T3 > T2. The dislocation loops expand and pass through the free surface of the specimen creating a DDZ.

293

$$\sigma_{31} = \frac{2.55\sigma_0}{\sqrt{2\pi}}\left(\frac{a}{r}\right)^{1/2} f(\theta) \qquad r < a \qquad (1)$$

where

$$f(\theta) = \sin\left(\frac{\theta}{2}\right)\cos\left(\frac{\theta}{2}\right)\cos\left(\frac{3\theta}{2}\right) \qquad (2)$$

r is the distance from the notch-tip to a point on the basal plane of interest and θ is the angle between the plane of the notch and this point.

On BP-1.5, r can be written as a function of θ:

$$r = \frac{1.5}{\sin\theta} \qquad r < a \qquad (3)$$

Normalizing Eq. (1) with respect to the applied stress and substituting r from Eq. 3 yields

$$\sigma_{31}/\sigma_0 = \frac{2.55}{\sqrt{3\pi}}(a\sin\theta)^{1/2} f(\theta) \qquad (4)$$

which is applicable when $r < a$. The normalized shear stress, σ_{31}/σ_0, for BP-1, BP-1.5 and BP-2 are plotted as a function of θ in Figure 4. From this figure it can be seen that:

(1) The shear stress, σ_{31}, changes sign in front of the notch-tip.

(2) The shear stress reaches its maximum value in a region that shifts to the right as the distance from the notch-tip increases.

(3) The shear stresses are equal in magnitude, but opposite in sign above and below the notch.

As the dislocation loops move closer to the notch-tip the stress field experienced by these dislocations, which was initially due only to the far-field stress, now has a large component supplied by the notch-tip and the stress can be expected to change the positions and shapes of the dislocation loops. This behavior can be viewed to some extent using the Ashby and Embury model [9], which suggests that in the early stage of deformation most mobile dislocations are too distant from the notch-tip to be affected by its stress intensification; but as the dislocations move closer to the notch-tip, interaction with its stress field begins, and new dislocation configurations arise in the area adjacent to the notch.

Knowing the stress field around the notch-tip from Eq. (4), the observed size of the DDZ (measured from Figure 1(b)), and the screw to 60° mobility (M) ratio form Shearwood et. al. (1991), the position of the hexagonal dislocation loop, and the M at any given time can be calculated. The stress field acting on a segment of the hexagonal dislocation loop varies continuously along the length of the dislocation, but because each dislocation segment maintains its straight line 60° or screw orientation, the varying stress field can be averaged from finite elements. The model calculates the position of the partial loop, and the dislocation mobility (M) over 1800 second intervals based on the shear stress experienced by the hexagonal segments AB and BC. The final time interval was limited to 1080 seconds to agree with Figure 1(b).

Model Results

Figure 5 is a time sequence relating size and position of the partial dislocation loops for BP-1, BP-1.5, and BP-2 as calculated by the model over 0.5 hour time intervals (except the last interval, which is 0.3 hours to agree with the end of the test). The M calculated for the three basal planes was approximately 1.45 $\mu m\ s^{-1}\ MPa^{-1}$ for the 60° dislocations and 0.82 $\mu m\ s^{-1}\ MPa^{-1}$ for the screw dislocations.

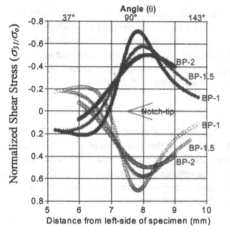

Figure 4. Normalized shear stress (σ_{31}/σ_o) as a function of θ and distance from left-side of specimen. σ_{31}/σ_o is displayed over its respective valid range for the basal plane of interest. Note: the notch-tip is meant to provide a reference for horizontal, not vertical, distance.

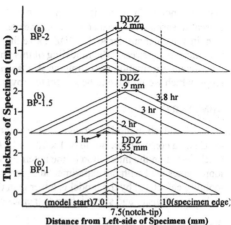

Figure 5. A top view time sequence of basal plane hexagonal dislocation loop expansion displaying the location and formation of the DDZ as calculated by the model for (a) BP-2, (b) BP-1.5 and (c) BP-1. A thickness of 0 mm represents the specimen front and that of 2 mm the back of the specimen. The indicated time is the same for all three basal planes.

When these results are compared to those of Shearwood et. al. [12], the M calculated by the model is approximately 5 times lower. Two possible reasons for this are: (1) the uncertainty in basal plane dislocation density due to the edge-on orientation of the topographs. Any increase in this density relative to that found in the Shearwood et. al. experiment would depress dislocation mobility. (2) The possibility that localized stress inhomogeneities [15] could have skewed the Shearwood et. al. results. In addition to the work of Shearwood et. al. many others ([16], [17], [18]) have previously determined dislocation mobilities, but with the conventional x-ray sources used it is likely that dislocation recovery (during the long exposure time) would have affected the results.

In addition to calculating dislocation mobility, the model also determined the position of the DDZ. Referring again to Figure 5, an obvious shift of the partial hexagonal dislocation loop to the right as distance from the notch-tip increased, can be seen. This rightward shift can also be distinguished in topograph Figure 1(b) and, as indicated by the shear stress curves in Figure 4, the partial hexagonal loop follows the path of maximum shear stress. In summary, the model yields physically reasonable values for dislocation M, and DDZ position, suggesting that the basis for the model and the assumptions made in it are sound.

The Vertical Dislocation Array Below the Notch-tip

In contrast to the DDZ observed above the notch, a vertical array of dislocations can be seen below the notch (Figure 1(b)). A likely explanation for the formation of this array would be a reversal of the mechanism proposed to explain the DDZ above the notch-tip. The shear stress below the notch-tip, which is opposite in sign to that above, would cause any dislocation loops present to shrink. As the distance from the basal plane increases, the vertical array shifts to the right in a manner similar to the DDZ above the notch and this shift tends to follow the region of maximum shear stress. The modeling of this phenomenon was not attempted because of insufficient information regarding the size, shape, position and number of hexagonal dislocation loops, which would have to be known to provide a starting point for the model.

SUMMARY

Dislocation motion around a notch in ice single crystals was modeled at a temperature of -8 °C and a constant strain-rate of 1 x 10^{-8} s^{-1}. An analysis of the formation of a DDZ above the notch, based on the interaction of the notch-tip stress field with a partial hexagonal dislocation loop, agrees well with the experimental observations in regards to size and position. The mobility calculated by the model is lower than that determined by others, but differences in dislocation density and stress inhomogeneities could account for this.

ACKNOWLEDGMENTS

Grant OPP-9526454 from the National Science Foundation and Grant DAA-H04-96-1-0041 from the Army Research Office supported this research.

REFERENCES

1. S. Kobayashi and S.M. Ohr, Scripta Metall. **15**, 343 (1981).
2. S. Kobayashi and S.M. Ohr, J. Mater. Sci. **19**, 2273 (1984).
3. T. Luoh and C.P. Chang, Acta. Mater. **44**, 2683 (1996).
4. J.A. Horton and S.M. Ohr, J. Mater. Sci. **17**, 3140 (1982).
5. S.M Ohr, Mater. Sci. Eng. **72**, 1501 (1985).
6. S.M. Ohr, H. Saka, Y. Zhu, and T. Imura, Phil. Mag. A **57**, 677 (1988).
7. Y. Wei, and J.P. Dempsey, *Mechanics of Creep Brittle Materials, vol. 2*, Cocks, A. C. F. and Ponter, A. R. S. eds. (Barking Essex: Elsevier Applied Science, 1991) p. 62.
8. Y. Wei, and J.P. Dempsey, *Scripta Metall.* **32**, 949 (1995).
9. M.F. Ashby and J.D Embury, *Script. Met.* **19**, 557 (1985).
10. X. Hu, K. Jia, F. Liu, I. Baker, and D. Black, , *Applications of Synchrotron Radiation Techniques to Material Science II*, ed. by L. J. Terminello, N. D. Shinn, G. E. Ice, K. L. D'amico, and D. L. Perry, *Proc. Mater. Res. Soc.* (Pennsylvania: Materials Research Society, 1995) **375**, p. 287.
11. C. Shearwood, M. Ohtomo, and R.W. Whitworth, Nature **319**. 659 (1986).
12. C. Shearwood, and R.W. Whitworth, *Phil. Mag.* A **64**, 289 (1991).
13. F. Liu, Ph.D. Thesis, Dartmouth College, Hanover, NH (1992).
14. K. Jia, 1994, M.S. Thesis, Dartmouth College, Hanover, NH (1994).
15. D.M. Cole, private communication, U.S. Army, CRREL, Hanover, NH (1999).
16. A. Fukuda and A. Higashi, Crystal Lattice Defects 4, 203 (1973).
17. C. Mai, *Comptes Rendus Hebdomadaies des Seauces de l'Academic des Sciences* (Paris), Ser. B, Tom. **283** (22), 515 (1976).
18. J. Perez, C. Mai, and R. Vassoille, J. Glaciology 21 (85), 361 (1978).

GROWTH TWINS IN Yb$_x$Y$_{1-x}$Al$_3$(BO$_3$)$_4$ AND Nd$_x$Gd$_{1-x}$Al$_3$(BO$_3$)$_4$ CRYSTALS OBSERVED BY WHITE-BEAM SYNCHROTRON RADIATION TOPOGRAPHY

X.B.Hu, J.Y.Wang, M.Guo, S.R.Zhao, B.Gong, J.Q.Wei and Y.G.Liu
State Key Laboratory of Crystal Materials, Shandong University, Jinan 250100, P.R.China,
E-mail: xbhu@icm.sdu.edu.cn

ABSTRACT

Growth twins in Yb$_x$Y$_{1-x}$Al$_3$(BO$_3$)$_4$ (YbYAB) and Nd$_x$Gd$_{1-x}$Al$_3$(BO$_3$)$_4$ (NGAB) crystals were observed by white-beam synchrotron radiation topography combined with chemical etch. It was found that growth twins in YbYAB crystals are of inversion types in which two twinned pairs have the central inversion relationship. This kind of twinning was visible in X-ray topography not by 'domain contrast' but by 'boundary contrast' stemming from the kinematical X-ray diffraction at the boundary. Growth twins in NGAB crystals are of 180° rotation types in which the twofold symmetric operation axis is parallel to the [0001] axis. This kind of twinning often shows black-and-white contrast in X-ray topography which originates from the different structure factors between twinned pairs. In addition, the formation mechanisms of growth twins are discussed.

INTRODUCTION

In the past ten years, many researchers have focused on the self-frequency-doubling (SFD) crystal Nd$_x$Y$_{1-x}$Al$_3$(BO$_3$)$_4$ (NYAB). As we know, the NYAB crystal is characterized by its large emission cross section and rather high non-linear optical coefficient [1,2]. In addition, it has excellent chemical and physical properties, which are essential to laser devices [3,4]. However, it is very difficult to grow optical grade NYAB crystals because of the complexity of the structure and the high viscosity of the borate systems. Although NYAB crystals have been successfully grown by the flux method [5,6,7,8], the existence of various defects in flux grown NYAB and its insufficient size have limited its applications. Thus, searching for new numbers and growing high perfection crystals of the NYAB family are still important tasks for developing self-frequency-doubling laser crystals and realizing their commercial application. For this purpose, YbYAB and NGAB crystals have been grown successfully by flux method in our laboratory.

Table 1 lists the ion radii of several rare-earth elements. From the table, it can be seen that YbYAB and NGAB crystals have a smaller difference between cation and doped ion radii than does NYAB crystal. Thus, it is believed that YbYAB and NGAB crystals probably have higher perfection than does NYAB. In this paper, we examine the growth twins in these crystals by chemical etch method and transmission synchrotron topography.

Table 1. The ion radii of several rare-earth elements

Elements	Gd	Y	Yb	Nd
Ion radii (nm)	0.938	0.893	0.858	0.995

Mat. Res. Soc. Symp. Proc. Vol. 590 © 2000 Materials Research Society

EXPERIMENTAL

YbYAB and NGAB crystals belong to the same space group R32 with hexagonal lattice constants of a=0.9293nm, c=0.7245nm and a=0.9398, c=0.7284nm, respectively. Both of them melt incongruently and could be grown only by the flux method. The detailed growth process has been described[9].

A few wafers, approximately 1mm in thickness with their surface parallel to $(10\bar{1}1)$ plane, were prepared using standard mechanical cutting and polishing techniques. Then these wafers were etched in hot concentrated H_3PO_4 at 200°C for about 20min and observed in an optical microscope. The wafers, used for synchrotron topographic observation, are about 0.2mm in thickness. Synchrotron topographic studies were performed using transmission synchrotron topography at Beijing Synchrotron Radiation Laboratory.

RESULTS AND DISCUSSION

By chemical etch, twins were detected in the $(10\bar{1}1)$ slice of YbYAB crystal. Fig.1(a) shows the etch pattern of $(10\bar{1}1)$ slice. The twin can be easily identified because the morphology of the etch pits in twin region (denoted by TW) is different from that in the host crystal region. The etch pits in both twin and host crystal regions are triangles but assume different orientations. Thus the twin boundaries (denoted by TB) can be clearly seen. Due to the fact that the twin has penetrated through the full thickness of the slice, from the traces of the twin boundaries on both surfaces, we can find that the twin boundaries are parallel to the $(0\bar{1}11)$ lattice plane.

Fig.1(b) shows the etch pattern of the $(10\bar{1}1)$ slice in another region. The polysynthetic twins are clearly observed. They appear as the host crystal and twin arrangement alternatively.

In order to determine the twin structure, we examined the same slice using the transmission synchrotron topography after it was thinned and polished once again. Fig.2(a) shows the transmission synchrotron topograph of the slice in $1\bar{3}21$ reflection. The arrows in the topograph indicate the twin region corresponding to Fig.1(a). The fringes in the lower

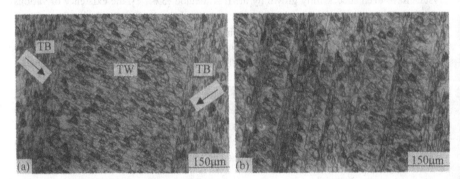

Fig.1 The growth twins in $(10\bar{1}1)$ slice of YbYAB crystal observed by chemical etch. (a) The etch pattern of band-shaped twin. The arrows indicate the twin boundaries. (b) The etch pattern of polysynthetic twins. It can be seen that the host crystals and the twins appear alternately in the slice.

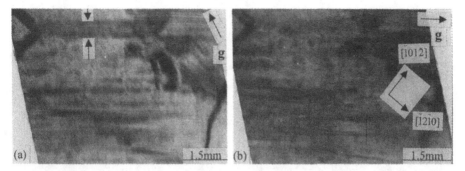

Fig.2 The transmission synchrotron topograph of the ($10\bar{1}1$) slice in YbYAB crystals. (a) $1\bar{3}21$ reflection. The twins are visible in form of the X-ray kinematical diffraction contrast stemming from the twin boundaries. (b) $1\bar{4}32$ reflection.

part of the topograph correspond to the region of polysynthetic twins in Fig.1(b).

Fig.2(b) presents another transmission synchrotron topograph of the ($10\bar{1}1$) slice in $1\bar{4}32$ reflection. Compared with Fig.2(a), the twins in YbYAB crystal are invisible in this topograph. It should be noticed that the diffraction vector **g** is parallel to the twin boundaries in this topograph.

Based on the diffraction features of the twins in X-ray topography, we know that the twins in YbYAB crystal are visible in the topograph not by 'domain contrast' but by 'boundary contrast', i.e. the twins appear in the topograph in the form of X-ray kinematical diffraction contrast due to the lattice strain stemming perhaps from the impurity incorporation in the boundary. In this case, the X-ray diffraction topographic contrast normal to the twin boundaries is similar to that of growth striations. When the diffraction vector **g** is parallel to the boundaries, the twin boundaries are invisible since the lattice strains resulted from the impurity incorporation or the inhomogeneity distribution of doped atoms are always perpendicular to the boundaries and occur exactly in the diffracting planes in such case.

As a result, we determine the twins in YbYAB crystal to be inversion types. In general, inversion twins do not exhibit 'domain contrast' in X-ray topography so long as the anomalous scattering is not so strong. For twins in YbYAB crystals, due to the impurity or doped solute atoms precipitating on the boundaries, the twin boundaries turn into the X-ray kinematical contrast.

Because the occurrence of the growth twins in YbYAB crystals was accompanied by the serious strain in their boundaries, we deduce that the growth twins were induced by growth bands.

Fig.3 shows transmission synchrotron topographs recorded from the ($11\bar{2}0$) wafer of NGAB crystal in different reflections. The orientation contrasts resulting from twinning were clearly observed. Obviously, the whole morphology of the wafer was present in Fig.3(a) corresponding to $\bar{4}156$ reflection; whereas only a part of the wafer appeared in Fig.3(b) and 3(c) with $5\bar{3}\bar{2}1$ and $\bar{3}5\bar{2}2$ reflection respectively.

In order to understand the possible twinning structure of NGAB, a rhombohedral structure of space group R32 should be considered. The twinning under consideration can be described by the two kind of twin elements. One is the vertical mirror plane normal to a two-

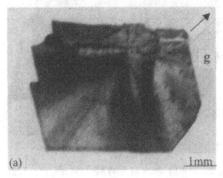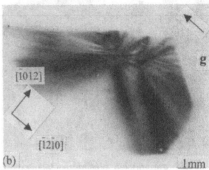

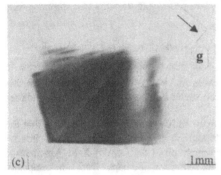

(a) 1mm (b) 1mm

(c) 1mm

Fig.3 The transmission synchrotron topograph of the $(10\bar{1}1)$ slice in NGAB crystals. (a) $\bar{4}156$ reflection; (b) $5\bar{3}\bar{2}1$ reflection; (c) $\bar{3}5\bar{2}2$ reflection.

fold axis of structural symmetry. The other is the twofold rotation axis parallel to the threefold axis of structural symmetry. For the former case, due to the twofold axis in R32 space group, the twinning is equivalent to an inversion twin. According to Friedel's rule, inversion twins do not induce contrast so long as anomalous scattering is not strong. Twinning is visible only by dynamical fringe contrast of its boundaries due to the phase shift in the structure factor. This is not the case for the twinning in the NGAB crystal. Thus this twinning is ascribed to 180° rotation twinning in which the twofold axis is along the threefold axis of structural symmetry. In this case, the 180° rotation operation leads to the interchange of the obverse and reverse rhombohedral lattice. Furthermore it leads to maximum contrast of the two twin domains in X-ray topography.

For the two kinds of rhombohedral lattices in twinned orientation, the twin can be described by means of two different centered hexagonal cells. The 'obverse' with centering points in (2/3,1/3,1/3) and (1/3,2/3,2/3) and the 'reverse' with centering points in (2/3,1/3,2/3) and (1/3,2/3,1/3). The two modes of 'centering', however, are interchanged by the twinning, and so are the 'reflection condition', which is expressed as follows:

$$-h+k+l = 3N \quad \text{(for 'obverse' domain)} \qquad (1a)$$

$$h-k+l = 3M \quad \text{(for 'reverse' domain)} \qquad (1b)$$

From the reflection condition, we can find that for the $\bar{4}156$ reflection, both of conditions are fulfilled simultaneously in the obverse and reverse domains. No domain contrast appears in Fig.3(a). But for $5\bar{3}\bar{2}1$ reflection, the reflection conditions are not

300

simultaneously fulfilled, only the obverse domain is imaged, the reverse is not. This leads to black-and-white domain contrast in Fig.3(b). In $3\bar{5}22$ reflection, only the reverse domain is imaged in Fig.3(c).

Fig.4(a) shows the transmission synchrotron topograph of the $(11\bar{2}0)$ wafer in $12\bar{1}6$ reflection. From the topograph, the obverse and reverse domains obey the reflection condition simultaneously and twin contrast is not present. However, for the $0\bar{1}15$ reflection, the reflection condition is not simultaneously fulfilled for two domains. In this topograph, the reflection condition is fulfilled only by the 'reverse' domain, which shows strong reflection contrast. The $0\bar{1}15$ reflection is not present in the 'obverse' domain and it seems that the 'obverse' domain should not diffract. But the result of the observation is not the case and the 'obverse' domain exhibits still weak diffraction in the topograph. It is believed that the weak diffraction of 'obverse' domain originates from the contribution of its third harmonic reflection $0\bar{3}315$.

Besides the growth twinning, the growth bands are the main defects in NGAB, and their interfaces are parallel to the $(0\bar{1}11)$ plane as shown in Fig.4(a). It should be noted that the growth bands exhibit the strong diffraction in the 'obverse' domain and relatively weak diffraction in 'reverse' domain. It is well known that the growth bands arise from fluctuations in lattice parameter during crystal growth. The lattice is distorted normal to the growth front by inclusion of impurities or inhomogeneous distribution of solute atoms. For example, a rapid fluctuation in temperature will cause a large fluctuation in impurity or solute atom content. Furthermore it will lead to fluctuation in the lattice parameter. In X-ray topographs, the contrast of the growth bands reflects the strain level. Obviously, the strain in the 'obverse' domain is larger than that in 'reverse' domain. Therefore, the twinning in NGAB crystal is thought to be caused by the formation of growth bands.

Although the growth twin in YbYAB crystals is structurally different from that in NGAB crystals, both of them were formed by same mechanism. In the flux mixture of YbYAB or NGAB, the B_2O_3 is often doped as an additive in order to increase the dissolving ability of the solvent with respect to solute. But the high viscosity of B_2O_3 will easily result in the inhomogeneous distribution of solute atoms, which is responsible for the formation of growth bands and twins.

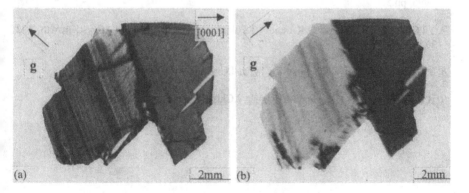

Fig.4 The transmission synchrotron topograph of the $(11\bar{2}0)$ wafer in NGAB crystals. (a) $12\bar{1}6$ reflection; (b) $0\bar{1}15$ reflection.

CONCLUSION

The growth twins in YbYAB and NGAB crystals have been studied by white-beam synchrotron radiation topography and chemical etch method. The results are summarized as follows:

1. The growth twins in YbYAB crystals are of inversion types and are visible in X-ray topography by the kinematical diffraction on their boundaries

2. The growth twins in NGAB crystals are of 180° rotation twins in which the two-fold symmetric operation axis is parallel to the structural three-fold symmetric axis.

3. The growth twins in YbYAB and NGAB crystals are considered to be induced by the growth bands.

ACKNOWLEDGEMENTS

The work was supported by National Natural Science Foundation (contract No. 68890235) and a grant for State Key Program of China.

REFERENCES

1. Z.D.Luo, A.D.Jiang, Y.C.Huang and M.W.Qiu, Chin. Phys. Lett. **6**, 440(1989).

2. L.M.Dorozhkin, I.I.Kuratev, N.I.Leonyuk, T.I.Timchenko and A.V.Shestakov, Soviet Tech. Phys. Lett. **7**, 1297(1981).

3. T.Omastsu, Y,Kato, M.Shimosegawa, Optics Communications **118**, 302(1995).

4. B.S.Lu, J.Wang, H.F.Pan and M.H.Jiang, J.Appl.Phys. **66**, 6052(1989).

5. M.Iwai, Y.Mori, T.Sasaki, S.Nakai and N.Sarukura, Jpn.J.Appl.Phys. **34**, 2338(1995).

6. J.Y.Wang, H.F.Pan, J.Q.Wei, Y.G.Liu, L.L.Tian, The 7th China-Japan Symposium on Science and Technology of Crystal Growth and Materials, Shanghai, China, November 1, 1996, pp.5.

7. V.I.Chani, K.Shimamura, K.Inoue, T.Fukuda and K.Sugiyama, J.Cryst.Growth **132**, 173(1993).

8. S.T.Jung, D.Y.Choi, J.K.Kang, S.J.Chung, J.Cryst.Growth **148**, 207(1995).

9. B.S.Lu, J.Wang, H.F.Pan and M.H.Jiang, J.Chin.Synth.Cryst. **16**, 195(1987).

AUTHOR INDEX

304

SUBJECT INDEX

absorption edge, 253
actinide, 27, 39
adsorption, 189
alumina, see aluminum oxide
aluminum(-), 227, 241, 247, 253
  iron, 91
  nitride, 195
  oxide, 9, 195, 279
APS, 27, 33, 39, 91, 131, 151, 157, 247, 259
ARPES, 57
attenuation contrast, 265

bacteria, 27, 33
ball milling, 71
band structure, 57
battery, 17
beam monitoring, 125
Beijing Synchrotron Radiation Laboratory, 297
binary alloys, 71
biotic redox chemistry, 27
boundary contrast, 297
bulk diffraction, 157

calcium sulfonate, 63
CAMD, 3, 45, 71
capillary waves, 165
carbon, 273
catalysis, 77
cavitation, 131
charge
  order, 103
  transfer, 55
CHESS, 151
chevron structure, 201
clathrates, 51, 145
clay, 39
composite, 157, 273
copper, 71, 157, 227, 259
creep, 131
crystal uniformity, 285
crystallization, 137
CVD (diamond), 125

damage, 157
deformation, 131, 227, 247, 253, 259, 291
density of states, 91
desulfouibrio desulfuricans, 27, 33
detector performance optimization, 285
diamond, 125
diesel particulates, 63
diffraction imaging, see imaging

dislocation(s), 291
  free zones, 291
domain contrast, 297

electric properties, 285
electrolyte, 45
electromigration, 259
electron density, 145
electronic, 57
ESRF, 125, 219, 227, 241
EXAFS, see XAFS
exhaust (engine), 3
extracellular precipitates, 33

Fe_3Al, 91

geological materials, 9
GIXD, 177, 207, 213
GIXS, 201
glass, 151
grazing incidence x-ray
  diffraction, see GIXD
  scattering, see GIXS
growth twin, 297

HASYLAB, 177, 207, 213, 265
heterostructures, 219
hexane, 165
Hg-Au phase diagram, 183
high energy synchrotron radiation, 241

ice, 291
imaging, 285
  real time, 273
  x-ray, 265, 273, 297
inclusions (see precipitates)
indium gallium arsenide, 151
inelastic scattering, 91
interconnect, 259
interface, 165
  internal-, 213, 219
  liquid-liquid, 165
interfacial behavior, 201
interferometer, 265
internal
  load transfer, 157
  stress, 157
iron(-), 71
  aluminum, 91
  sulfur chemistry, 33

structural modification, 213
structure, 119
 factor, 297
superlattices, 219
surface
 complexation, 9
 damage, 279
 finish, 279
 structure, 183
surfactant, 177
synchrotron (see storage ring)

texture, 227, 253
thermoelectrics, 51, 57, 145
thermomechanical processing, 227
thin film, 195
three-dimensional (3-D), 227, 241, 247, 253, 265
time resolved, 113
titanium, 213
 oxide(-), 219
  silicon oxide, 77, 83, 119
tungsten sulfide, 151
twin, 297
twist grains boundary structure, 201
2-D ordering, 177

ultrafast, 125
uranium, 9

vibrational entropy, 91

water, 165
 -alkane, 165

waveguide, 213

XAFS, 9, 17, 33, 39, 45, 51, 77
XANES, 3, 9, 17, 27, 39, 51, 71, 77, 83, 113, 119, 189
XAS, 39, 51, 63, 113, 119
xerogel, 77, 83
x-ray
 absorption, 63, 253, 285
  near edge structure, see XANES
  spectroscopy, see XAS
 detection, 125
 diffraction, 113, 119, 145, 151, 157, 189, 195, 207, 213, 219, 227, 241, 247, 253, 259, 279, 285, 291, 297
 fluorescence analyses, 3
 imaging, see imaging
 microdiffraction, 253, 259
 microprobe, 3
 microscope, 227, 273
 reflectivity, 219
 scattering, 195
  resonant, 91, 103, 183
  small angle, see SAXS
  surface, 165
 spectra-microscopy, 3
 topography, 279, 285, 291, 297

zeolite, 39
zirconium oxide-silicon oxide, 83

Printed in the United States
By Bookmasters